THE QUANTIFYING SPIRIT IN THE 18TH CENTURY

THE QUANTIFYING SPIRIT IN THE 18TH CENTURY

Edited by

Tore Frängsmyr, J.L. Heilbron, and Robin E. Rider

UNIVERSITY OF CALIFORNIA PRESS

Berkeley Los Angeles Oxford

University of California Press
Berkeley and Los Angeles, California

University of California Press
Oxford, England

Library of Congress Cataloging Publication Data

The Quantifying spirit in the 18th century / edited by Tore Frängsmyr, J.L. Heilbron, and Robin E. Rider.
p. cm.
Includes bibliographical references (p.). ISBN 0-520-07022-4 (alk. paper)
1. Science—History—18th century. 2. Technology—History—18th century. I. Frängsmyr, Tore, 1938- . II. Heilbron, J.L. III. Rider, Robin E.
Q125.Q34 1990
509'.033—dc20
90-10893
CIP

This volume is number 7 in the series *Uppsala studies in history of science.*

Printed in the United States of America
1 2 3 4 5 6 7 8 9

The paper used in this publication meets the minimum requirements of American National Standard for Information Sciences—Permanence of Paper for Printed Library Materials, ANSI Z39.48-1984. ∞

Contents

Preface vii

Introductory Essay 1
J.L. Heilbron

Part 1. Rationalization of Knowledge 25

1. The Mathematical Philosophy 27
Tore Frängsmyr

2. The Broken Circle 45
Gunnar Broberg

3. Systematics and the Geometrical Spirit 73
John E. Lesch

4. Measure of Ideas, Rule of Language:
Mathematics and Language in the 18th Century 113
Robin E. Rider

Part 2. More Exact Sciences 141

5. Late Enlightenment Meteorology 143
Theodore S. Feldman

6. Accuracy, Rhetoric, and Technology:
The Paris-Greenwich Triangulation, 1784–88 179
Sven Widmalm

7. The Measure of Enlightenment 207
J.L. Heilbron

Part 3. Wider Applications 243

8. The Changing Role of Numbers in 18th-Century Chemistry 245
Anders Lundgren

9. The Most Confused Knot in the Doctrine of Reproduction 267
James Larson

10. Labs in the Woods: The Quantification of Technology During the Late Enlightenment 291
Svante Lindqvist

11. The Calculating Forester: Quantification, Cameral Science, and the Emergence of Scientific Forestry Management in Germany 315
Henry E. Lowood

12. Society in Numbers: The Debate over Quantification in 18th-Century Political Economy 343
Karin Johannisson

13. The Laplacean View of Calculation 363
Roger Hahn

Bibliographical Afterword 381
Robin E. Rider

Index 397

Illustrations follow page 242

Unless otherwise specified, illustrations appear by courtesy of the Uppsala University Library or The Bancroft Library, University of California, Berkeley.

Preface

This book is a result of a collaborative research project between historians of science in Berkeley and Uppsala. We began with two objectives. One was to study quantification and mathematization in the 18th century in order to cast light on the character of its science. The other was to give special attention to Sweden as an exemplar of the close connection between science, utility, and economics that marked the activities of the quantifying spirit. We believe that we have shown that the development in leading European countries during the 18th century can be complemented usefully with examples from a small country like Sweden.

The project has progressed through continuous contacts between the two groups of scholars, led by John Heilbron and Tore Frängsmyr. We have discussed the different papers in joint seminars, in Berkeley and in Uppsala, in correspondence between the groups, and within each group. The geographical distance between the two groups sometimes made for problems of communication; it takes time to send papers and reflect upon criticism, especially when ordinary duties intervene. But the stimulating experience of so unusual a transatlantic project more than compensated for the difficulties.

One constant source of stimulation and instruction was the challenge of combining the different views of the two groups. In Sweden, history of science has a humanistic background, against which science is studied as a cultural and social factor. The Berkeley group, in keeping with the emphasis in the United States, is inclined to concentrate on concrete applications or illustrations of quantification in different disciplines. In our efforts to combine these two perspectives, we have had many lengthy and fruitful discussions. We have learned something from each other. We knew, of course, that history of science must bring together a general cultural outlook and a collection of detailed episodes and illustrations. We have learned from each other ways of accomplishing this difficult task.

The editing turned out to be too much work for two. Robin Rider has been good enough to act as co-editor as well as to contribute a chapter and a bibliographical essay.

It is a pleasure to acknowledge the skill and dedication of the staff of the Office for History of Science and Technology at Berkeley, and especially Diana Wear, who produced the camera-ready copy for this book.

Financial support for the project came from the United States National Science Foundation and the Humanistisk-samhällsvetenskapliga forskningsrådet and Riksbankens Jubileumsfond in Sweden.

Tore Frängsmyr
J.L. Heilbron
Robin E. Rider

Introductory Essay

By J.L. Heilbron

In a famous essay on the usefulness of mathematics, the almost perpetual secretary of the Paris Academy of Sciences, Bernard le Bovier de Fontenelle, ascribed the improvement of literature in his time to the spread of the geometrical spirit. To whatever it touched it brought order, clarity, and precision. "A work on ethics, politics, criticism, and, perhaps, even rhetoric will be better, other things being equal, if done by a geometer."[1] When these words were written in 1699, they had little purchase outside their own rhetoric. Sixty years later, according to the *Encyclopédie* of Diderot and d'Alembert, the idea had caught on: people had come to realize that the mathematical method, the method of exact definition and proof of propositions, was not restricted to mathematics. The big change had come with the writings of Leibniz's intellectual heir Christian Wolff. "Wolff made people see in theory, and especially in practice, and in the composition of all his books, that the mathematical method belongs to all the sciences, is natural to the human mind, and leads to discoveries of truths of all kinds."[2]

From about 1760 on, the spirit Fontenelle and Wolff conjured up did indeed come to invest not only the sciences but also their applications to fields as distant as artificial languages and forest management. Even in the exact sciences, such as astronomy, optics, and

I thank my fellow editors for discussions about the structure of this introduction.

1. Fontenelle, *Histoire du renouvellement de l'Académie royale des sciences en mdcxcix et les éloges historiques*, 2 vols. (Amsterdam: Pierre du Coup, 1719–20), *1*, 14.

2. *Encyclopédie, or dictionnaire raisonné des sciences, des arts et des métiers*, 17 vols. (Paris and Neuchâtel: Briasson, Faulche, et al., 1751–65), s.v. "Méthode (logiques)," *10*, 446 (1765).

mechanics, we find a newly effective emphasis on precision. We believe that our thesis, that the later 18th century saw a rapid increase in the range and intensity of application of mathematical methods, is an important one. It amounts to specifying the time and surveying the routes by which what may be the quintessential form of modern thought first spread widely through society.

We translate l'esprit géometrique as "the quantifying spirit," which we understand to include the passion to order and systematize as well as to measure and calculate. We recognize that much posturing accompanied the exercise of this passion, that "mathematical" or "geometrical" did not mean the same things to all parties, and that numbers have their rhetoric, a truth we instantiate where appropriate. A novel feature of our work, which we consider to be a strength, is the inclusion of botanists and encyclopedists among the quantifiers. They qualify for inclusion in two ways: by the sheer quantity of the material they handled and by the unimpeachable warrant of the *Encyclopédie*. There we read that the spirit of a geometer is one of "method and justice." "It is a spirit of computation and of slow and careful arrangement (*combinaison*), which examines all parts of an object one after another and compares them among themselves, taking care to omit none." The encyclopedist gives the example of biological classification, with recommendations about the ratio of the numbers of genera to those of species that the naturalist should try to establish.[3]

All systematized or quantified knowledge and information current in the 18th century qualify for our consideration. To cover the topic we would have had to compile an encyclopedia. We have preferred to compose case histories and to concentrate on the second half of the century. In order to point up what we take to be characteristic of that period, however, most essays include material about earlier relevant developments and some extend coverage to the end of the Napoleonic era. In the high and wide realm of philosophical and programmatic discourse, rhetorical continuity runs through the

3. Art. "Géometrie," *Encyclopédie, 7*, 628 (1757), and "Méthode," ibid., *10*, 458, 460 (1765).

century, from Fontenelle through d'Alembert and Wolff to Condillac. But we perceive even here an important inflection around 1760. As T. Frängsmyr shows in the first chapter of our book, Wolff's notion of mathematical method, which derived from Greek geometry, proved unequal to his ambitions and fell victim to the principled criticisms of Kant. Thereafter, as R. Rider explains in chapter 4, analysis and algebra, which, in contrast to geometry, had an instrumentalist bias, became the exemplar of the mathematical method.

This instrumentalism was a key ingredient of the quantifying spirit after 1760. Everywhere we see an increased emphasis on the practical uses of number and system. In sciences already quantified, like astronomy and surveying, new instruments and organizations arising from governmental and military needs brought greater precision and larger operations; S. Widmalm examines the case of English geodesy in chapter 6. Likewise, interest in promoting public health and agricultural yields pushed the standardization of meteorological instruments and the study of climatology, as explained by T. Feldman in chapter 5. The bureaucracies of the centralizing nations and principalities developed an enormous appetite for statistics (K. Johannisson, chapter 12) and a compulsion toward rational exploitation of natural resources (H. Lowood, chapter 11). The explosion of information inspired the invention and development of artificial techniques of control from natural history (J. Lesch, chapter 3) to lexicography (G. Broberg, chapter 2). Further examples of instrumentalist approaches are found in chemistry (A. Lundgren, chapter 8) and technology assessment (S. Lindqvist, chapter 10).

It may be useful to bring together some of the results relating to practical operations of the geometrical spirit after 1760 and to extend them with a few further examples. The survey proceeds from instruments and instrumentation through applications to statistics, land use, and trades to the information explosion. It concludes with observations on the social support that underlay these developments.

INSTRUMENTS AND INSTRUMENTATION

The late Enlightenment rediscovered and fruitfully exploited the complementary character of truth and quantification first made manifest in the epicycles of the ancient astronomers. An example from physics will illustrate both the timing and the content of this turn toward instrumentalism. In the 1740s Alexis Claude Clairaut worked out a theory of capillarity based on a force between particles of water and of glass that acted over distances as great as the radius of capillary tubes. He could not think of a way to discover the precise form of this law, and he considered his failure a blemish. If the unique correct form were inaccessible, how could the theory support inquiry into the fundamental powers of nature? Fifty years later, Laplace broke his head over the subject. Unable to do better than Clairaut, he made a virtue of necessity, redid the physics to require a force that acted only at insensible distances, and rearranged his calculations so that they did not depend on knowing the precise form of the force law. This trick, or insight, which would have been a fundamental flaw in a realist theory, became the basis of Laplace's molecular physics, which became part of what might be called the Standard Model of the physics of the late 18th and early 19th centuries.[4]

The Standard Model invoked imponderable special fluids to carry the various "forces" apparently differentiated in nature. Electrical theory, with its one (or two) repellent and attractive fluids, served as exemplar. The theory was proposed by several people around 1770, demonstrated by Charles-Augustin de Coulomb in 1785, and generally accepted by the end of the century. The matter of heat, caloric, came into existence in the 1770s and 1780s almost in parallel with the electric fluids; the discoveries of latent and specific heats then not only reinforced the old idea of a matter of heat but also provided opportunities for extensive and intricate quantitative experiments.[5]

4. Pierre Simon de Laplace, *Œuvres complètes*, 14 vols. (Paris: Gauthier-Villars, 1878–1912), *4*, 349–51; *14*, 217–8 (texts of 1805–6); cf. J.J. Bikerman, "Theories of capillary attraction," *Centaurus, 19* (1975), 182–206, on 184–5. The theories of Clairaut and Laplace are set forth in unnerving detail in J.L. Heilbron, *Some activities of the quantitative spirit around the end of the 18th century* (Berkeley: Office for History of Science and Technology, University of California, 1990), chap. 1.

5. Douglas McKie and Niels H. de V. Heathcote, *The discovery of specific and la-*

Two fluids for magnetism and one for light completed the list of fundamental imponderables, to which might be added phlogiston (for as long as it existed), the ethers supposed to mediate gravitation and the interactions of light and matter, radiant heat, and a gravimetric fluid. As the cases of electricity, magnetism, light, and heat evidenced, the invocation of the fluids made possible computations of a sort familiar from the theory of gravitation. "This exchange of *lumières*," wrote J.B. Biot in 1803, referring to the then recent increase of the use of mathematics in physics, "is the certain proof of the perfection of the sciences."[6]

Most of the leading proponents of the Standard Model, including Biot, made clear that they understood it in an instrumentalist sense. They conceded that even in the exemplary case, electricity, they could not determine how many, if any, fluids operated; and that every explanation of heat phenomena in terms of caloric had an equivalent account in terms of matter in motion.[7] In return for the convenience of thought and the purchase for calculation afforded by the Standard Model, quantifying physicists or chemists surrendered their claims to truth. In this they found themselves in agreement with the epistemologies of Hume and Kant, and perhaps also with Condillac's teaching that clear and simple language, not intuitions of truth, conduces to the advancement of science.

Calculation involves not only the software of theory but also the hardware with which numbers are garnered from nature. A dramatic increase in the precision of the instruments of physical science occurred during the 18th century: in the apparatus of the oldest exact

---

tent heats (London: Arnold, 1935), 54–121; Robert Fox, *The caloric theory of gases from Lavoisier to Regnault* (Oxford: Oxford University Press, 1971), 6–11; Johann Carl Fischer, *Geschichte der Physik seit der Wiederherstellung der Künste und Wissenschaften bis auf die neuesten Zeiten*, 8 vols. (Göttingen: Röwer, 1801–8), 7, 524, 529; Georg Wilhelm Muncke, "Wärme," in J.S.T. Gehler, *Physikalisches Wörterbuch, 10* (Leipzig: Schwickert, 1841), 52–1178, on 56f.

6. Jean-Baptiste Biot, *Essai sur l'histoire générale des sciences pendant la révolution française* (Paris: Duprat and Fuchs, 1803), 22.

7. J.L. Heilbron, *Electricity in the 17th and 18th centuries. A study in early modern physics* (Berkeley: University of California Press, 1979), 446–8; Fox, *Caloric theory*, 23–4, 28–37; A. Seguin, "Observations générales sur le calorique," *Annales de chimie, 3* (1789), 182–4, 189, 207, 221–4.

sciences, astronomy and geodesy; in the meteorological instruments invented in the 17th century, like the barometer and thermometer; and in devices created more recently, or applied to new purposes, like the calorimeter, electrometer, and chemical balance. This class of instruments might be called "measurers." A second class, which included "explorers" like the air pump and the electrical machine, produced artificial phenomena for demonstration, investigation, and measurement. A third class, "finders," were measurers and explorers with direct application to practical affairs: the telescopes, chronometers, theodolites, and so on that made the increasing trade in the instruments of navigation, surveying, and mensuration. The improvement in accuracy of the measurers and the increase in power of the explorers depended on the growing market for better finders as well as on demand for instruments for instruction and entertainment. The instrument business of the 18th century was driven by raison d'état, that is, by a need for finders of position on land and sea, for the guidance of the army, the navy, and the tax collector, and by raison tout court, that is, by the wish to see and perhaps to study the novelties brought to light by enlightened natural philosophers.

In the case of measurers, increase in accuracy can itself be measured. During the century from Tycho Brahe to John Flamsteed, the fineness of graduation improved by a factor of 3, from 1 minute to 20 seconds of arc. (These were of course the best instruments; the ordinary sectors of 1700 were divided to 10 minutes of arc.) During the 18th century graduation improved by a factor of 200, from 20 seconds to a tenth of a second.[8] This transformation built on improvements in metalworking and, above all, in mechanical means of subdividing angles, the so-called ruling engines, among which Jesse Ramsden's model of 1773 perhaps represented the largest advance. The drama of angular division may have been dramatic only to

8. Allan Chapman, "The accuracy of angular measuring instruments used in astronomy between 1500 and 1850," *Journal for the history of astronomy, 14* (1983), 133–7; cf. W.D. Hackmann, "Instrumentation in the theory and practice of science: Scientific instruments as evidence and as an aid to discovery," Istituto e museo di storia della scienza, Florence, *Annali, 10:1* (1985), 87–115, on 104–5, and figs. 2 and 3.

instrument-makers and their more precise clients; but certain consequences of better measurers could not fail to arouse the interest of a wider public. Astronomers brought to light the aberration of stars (which has a maximum value of 20 seconds of arc), the nutation of the earth (which has a maximum value of 9.2 seconds and a period of 19 years), and the secular change in the inclination of the earth's axis to the plane of its motion (which amounted to around 45 seconds of arc *per century*).[9]

A similar story can be told about clocks. Huygens' pendulum clock, which embodied a great leap in chronological performance, was accurate to perhaps 10 seconds a day. During the 18th century, improved escapements and temperature compensation increased accuracy by almost two orders of magnitude. Chronometers available in 1800 could keep time to better than a fifth of a second per day. Among the practical consequences of this accuracy were the determination of longitude at sea to within 2 minutes of arc, or a mile or two in most latitudes, as compared with discrepancies of as much as a hundred miles in 1750; and the specification of the length of a seconds pendulum (which many people proposed as a basis of weights and measures) to within one part in a hundred thousand.[10]

Barometers and thermometers improved in parallel with clocks and sectors. Around 1730 it was deemed unnecessary to correct barometer readings for temperature or to affix anything sturdier than paper scales to thermometers, which in any case were seldom calibrated between fixed points. By 1780, the best barometers could be read to a few thousandths of an inch, an improvement of a factor of 10 in 50 years; and, after compensation for temperature, capillarity, the curve of the miniscus, and so on, these readings gave the true value of the pressure within the accuracy of observation. The best

9. James Bradley, *Miscellaneous works and correspondence* (Oxford: Oxford University Press, 1832), 1–41, on 5, 11, 14, 21–2, 29–31; S.P. Rigaud, "Memoirs," ibid., xii–xiii, lxii–lxx; J.B.J. Delambre, *Histoire de l'astronomie au dix-huitième siècle* (Paris: Bachelier, 1827), 403–6, and s.v. "obliquité."

10. David W. Waters, "Nautical astronomy and the problem of longitude," in John G. Burke, ed., *The uses of science in the age of Newton* (Berkeley: University of California Press, 1983), 143–69, on 146, 160–1; C.J.E. Wolf, ed., *Mémoires sur la pendule*, 2 vols. (Paris: Gauthier-Villars, 1889–91), *1*, 17–64.

thermometers of the 1780s were literally incomparably better than those in use in the first third of the century. Careful and uniform procedures for finding the fixed points and meticulous division between them resulted in instruments that could be read to a hundredth of a degree and that, moreover, gave the same readings when immersed in the same temperature bath.[11] Among the consequences of these improvements of wider interest was the perfection of barometric hypsometry—the determination of heights by barometer readings compensated for temperature. By 1790 anyone interested and proficient could obtain the height of a mountain to within 0.5 percent by carrying good meteorological instruments to its top.[12]

The electrometer offers another example of the fact and the consequences of the acceleration in accuracy of instruments during the second half of the 18th century. Around 1750 electrometers came into existence, without standards or standardization, and without much agreement on the part of their makers about what they were measuring. Then the need to standardize measurement—of the settings of machines for trade, of the shocks given in medical treatments, of the degree of atmospheric electricity—produced a strong demand for reliable and sensitive instruments. The demand was met in the 1780s, notably by Alessandro Volta, whose most sensitive straw electrometer registered about 40 volts per degree. The contact of silver and zinc develops about 0.78 volt. Using mechanisms invented during the 1780s for multiplying small charges and driven by galvanism (or by the urge to disprove it), Volta managed to amplify the effect of a single zinc-silver junction until it stimulated his electrometer.[13] His compulsion for the quantitative enabled him to make a

11. W.E.K. Middleton, *The history of the barometer* (Baltimore: Johns Hopkins University Press, 1964), 178–9, 188, 197, 243–5; *A history of the thermometer and its use in meteorology* (Baltimore: Johns Hopkins University Press, 1966), 80, 119, 127–8; Maurice Daumas, *Les instruments scientifiques aux Xviie et Xviiie siècles* (Paris: PUF, 1953), 273, 280–1.

12. T.S. Feldman, "Applied mathematics and the quantification of experimental physics: The example of barometric hypsometry," *Historical studies in the physical sciences, 15:2* (1985), 127–97, on 164–81.

13. Heilbron, *Electricity,* 82, 493.

discovery that has made a qualitative difference in the history of the world.

The quantitative information secured by the instruments itself grew in quantity. It is very easy to record the readings of thermometers and barometers and to compute averages to crowds of decimals, but difficult to arrange them usefully. One method increasingly applied during the last third of the century was the tabular display. Johann Carl Fischer's monumental *Geschichte der Physik*, the longest work on the subject ever published (eight volumes, 7,500 pages, all published between 1801 and 1808), gives even those who only turn its pages a vivid impression of the growing use of numbers and tables in physics during the Enlightenment. Fischer divides the modern period into two parts, from the time of Newton until the discovery of the various types of gases, and from the discovery until the time of his writing. Take the case of heat.[14] There are four tables of data from the first period, sixteen—and many untabulated numbers—from the second. Where the tables cover the same subject in both periods, the detail is much greater for the later: for example, temperatures to a tenth of a degree and expansion coefficients to three or four figures in the first period, temperatures to a hundredth of a degree and expansions to six figures in the second. There are also many new subjects for tabulation in the second period: specific heats, warming by radiant heat, heat of exothermic reactions, expansions of fluids, pressure of superheated steam, and so on. In a word, or rather a number: Fischer needed four times as many tables to set out work done on heat in the last third of the 18th century as he needed to display the measurements of the first two-thirds.

Meteorology offers a luxurious example of rampaging numbers. Before the middle of the 18th century, people who liked to measure the weather tended to do so desultorily, with imperfect instruments, and with little regard for the ways and means used by others. Efforts to establish more regular observing with standard equipment always failed. The Royal Society of London collected reports from all over

14. Fischer, *Geschichte der Physik* (Göttingen: J.F. Röwer, 1801, 5, 10, 21–2, 43, 61; 7, 284, 289, 292–3, 295, 306, 311–20, 329–30, 334, 340, 393–5, 416, 419, 441–2, 500, 508, 512, 567–8.

Britain and from parts of the continent in the early 18th century; the miscellaneous and mismatched returns allowed only qualitative generalizations, such as "Pisa's prodigious rains make it the 'Piss-pot of Italy'." In a second try, begun in 1723, the Society distributed thermometers, barometers, and rain boxes, all by the same maker, and detailed instructions for their use. It received in return ill-digested, capricious, inaccurate, and worthless reports and registers from people who had soon felt the inconvenience of reading several instruments several times a day, rain or shine.[15] Much changed after mid-century. Tabular displays improved in quality and increased in number; climatologists obtained intercomparable instruments; the Royal Society's earlier objective, surveying the weather by network rather than by occasional reliable informant, was realized on the continent, where the resources of centralized and enlightened government had been brought to inspire observers to stay close to their barometers. (These matters are considered in the detail they inspire in the essay by T. Feldman.) The Societas meteorologica palatina, which took the lead on the continent, also distinguished itself by publishing some of the results of its observers in graphical form.[16]

THE PLAY OF THE SPIRIT

In 1778 there appeared a pseudonymous work entitled *Recherches et considérations sur la population de la France*. The author rhapsodized as follows: "Experiment, research, calculation are the probe of the sciences. What problems could not be so treated in administration! What sublime questions could not be submitted to the law of calculation!"[17] The probe or key to intelligent administration, the

15. J.L. Heilbron, *Physics at the Royal Society during Newton's presidency* (Los Angeles: William Andrews Clark Memorial Library, 1983), 104–9.

16. David Cassidy, "Meteorology in Mannheim: The Palatine Meteorological Society," *Sudhoffs Archiv, 69* (1985), 8–25, on 16–22.

17. Quoted in Josephine Hecht, "L'idée de dénombrement jusqu'à la Révolution," in Institut national de la statistique et des études économiques, *Pour une histoire de la statistique, 1* (Paris: Imprimerie nationale, [c. 1976]), 21–81, on 52; Fernand Faure, "France," in John Koren, ed., *History of statistics. Their development and progress in many countries* (New York: Macmillan, 1918), 217–329, on 257.

most important number in the Kingdom—a number customarily kept secret by governments up to the middle of the 18th century—was the number of the subjects of the King. This strategic number provided an index to the strength of the state. The population could be considered an instrument, and its reading an indicator of health or decline. Lavoisier and the intendant des Pommelles likened it to a thermometer, "the thermometer of public prosperity," a pleasant image since, in French usage of the time, "température" was to the air what "tempérament" was to the bodily humors, that is, an indicator of condition or temper, and climate was a recognized factor in public health.[18] Other protostatisticians saw an analogy to the barometer. The Paris Academy of Sciences recognized a connection with science, and opened its pages during the last years of the Ancien Régime to tables of births, deaths, and marriages recorded for various districts marked out on the great map of France drawn up under the direction of its members, the Cassinis.[19]

There is a further analogy between population statistics and meteorology. Coordinating a national census presented many of the same problems as organizing an international survey of the weather. The first effort at a thorough census in France was the instruction sent in 1697 to the thirty-two intendants of the administration of Louis XIV to report the number of towns, villages, hamlets, and inhabitants within their jurisdictions. The intendants had neither the means nor the inclination for steady observation of the demographic barometer. They returned a medley of bad numbers taken by hearth rather than head counts, or from old tax rolls; nine of the intendants did not bother to estimate total populations; one omitted all towns,

18. *Encyclopédie*, s.v. "Température," *16*, 59 (1765). "Température" as a degree of heat gained currency in the latter 18th century; see M.J. Brisson, *Dictionnaire de physique*, 2d ed., 6 vols. (Paris: Librairie économique, an VIII), *6*, 165.

19. Hecht, "L'idée," 57, on analogies to meteorological instruments; A.P. Dionis de Séjour, Marquis de Condorcet, and P.S. de Laplace, in Académie des sciences, Paris, *Mémoires*, 1783, 703–5; Laplace, *Œuvres*, *11*, 35 (text of 1783); A.M. de La Michodière, "Essai pour connaître la population du royaume," Académie des sciences, Paris, *Mémoires*, 1783, 705–18; 1784, 577–92; 1785, 661–89; 1786, 703–17; 1787, 601–10; 1788, 755–67.

another all servants; and the whole suffered from errors in multiplication and transcription.[20]

The physical difficulties with which census takers of 1700 had to contend, like poor communications, ameliorated during the course of the century. Not so the psychological. The people did not see the beauty in acting as a barometer of public prosperity; they saw only the certainty that their taxes would increase when their prosperity was better known. Des Pommelles, writing in 1789, concluded that the opposition of the people made a head count impossible.[21] For this reason, and because of the cost, the central administration did not push for a general census. Still they wanted a demographic index. They hit upon a clever method of approximation. In 1772, just before Turgot came to power as controller-general of finance, his predecessor ordered that the intendants forward every year, retrospective to 1770, the numbers of births, deaths, and marriages that took place in their jurisdictions. Turgot then demanded that they make a head count in the leading towns and a few neighboring country parishes. Division of this head count by the average number of births in the same regions over the preceding decade or quinquennium would give a coefficient—let us call it k—by which all the data about births for all France could be converted into a figure that might be defined as the French population.[22]

Values of k founded on various samples congregated around 26, the number used in the estimates published by the Academy of Sciences. The question naturally arose whether the method had any merit. Turgot's mathematical advisor, the marquis de Condorcet, sought an answer, but without much success. Laplace then took it up, and managed to solve it by reducing it to operations with the standard apparatus of the game theorist, that is, an urn filled with very large numbers of black balls and white balls. Let the black balls stand for the existing population, the white balls for the average number of births. The records returned by the intendants gave the

20. Faure, "France," 250–7; Jacques Dupâquier and Eric Vilquin, "Le pouvoir royal et la statistique démographique," in *Pour une histoire,* 83–104, on 85, 87.

21. Dupâquier and Vilquin, "Le pouvoir," 91–2, 101.

22. Faure, "France," 259–64.

head counts for a sample s, supposed representative, of the total population t, and the number of births, b_s and b_t, for the sample and for the total. The numbers s and b_s can be likened to the number of black balls and of white balls, respectively, taken from the urn in $s + b_s$ separate draws. Imagine now a new play at the urn, giving b_t white balls and an unknown number t of black balls. What is the probability that t lies within any arbitrary number of $(b_t / b_s)s = ks$? The answer depends on the size of the multiplier k. To be on the safe side, Laplace recommended taking samples of a million or more. When this was done just after the turn of the century, k came out to be 28.3.[23]

The introduction of annual reporting of births and deaths to the central administration around 1770 marked an epoch in French statistics—the advance from occasional, imperfect, one-time, static surveys to continuous data collection and useful time series.[24] The later 18th century was in fact a watershed in the gathering of all sorts of quantitative statistics, industrial (particularly in the low countries) as well as demographic. The first national statistical bureau, appropriately called the "department of tables" (*Tabellverket*), came into existence in Sweden in 1749; it owed its precocity to the availability of a corps of inexpensive and experienced workers, the parish pastors, who had been keeping registers of their parishioners for fifty years or more. (K. Johannisson sets forth in chapter 12 the Swedish case and its connection with English political arithmetic.) Norway and Denmark took their first national censuses in 1769 and set up their own bureaus of tables in 1797. The central German states, Mecklenburg, Hesse-Darmstadt, and Bavaria, and also Austria, counted themselves in 1776 and 1777. In 1791, a colonel of militia mobilized the clergy of Scotland by threatening to quarter his troops

23. Laplace, *Œuvres, 11*, 42–43 (text of 1783); La Michodière, "Essai," 1783, 705–7; Dupâquier and Vilquin, "Le pouvoir," 91; Marcel Reinhard, "La population de la France et sa mesure, de l'ancien régime au consulat," in Commission d'histoire économique et sociale de la révolution française, *Mémoires et documents, 18* (1968), 259–74.

24. Pierre Chanu, "Les sources statistiques de l'histoire de France (Basse Normandie). Problème de méthode," *Annales de Normandie, 15* (1965), 1–14, on 11–2.

on them if they did not send him inventories of their parishes. "The Ministers have it in their choice [said he], either to write to the Colonel, or to treat his soldiers." This maneuver brought the soldier-statistician the wherewithal to compile twenty-one volumes in eight years.[25]

Back in France, the spirit of the Revolution puffed new vigor into the statisticians, who had it decreed, in 1791, that every administrative unit should furnish each year, during November and December, the name, age, birthplace, residence, profession, and other means of subsistence of all citizens living in its territory. Three out of 36,000 communes replied. No one succeeded in enforcing the decree until 1801, when a number was obtained about a million less than Laplace had calculated. Another count, in 1806, came out almost 800,000 larger than he had allowed. Napoleon liked these readings of the thermometer of public prosperity, but did not care to try again, lest the results reveal a decline in population and an argument for critics of his stewardship. The census of 1806 was followed by that of 1821. Meanwhile the British faithfully fulfilled the requirement of a decennial census, which they took upon themselves for the first time in 1801.[26]

The land is easier to measure than the people. Our quantifying spirit came to ground in the late 18th century not only in exact cartography and geography (as in the cases of the Ordnance and metric surveys) but also in the exploitation of old forests and virgin lands. Two examples are worth our attention, one from the Old World, where the quantifying landscaper faced constraints imposed by centuries of unregulated growth, and one from the New World, where the United States felt itself free to cut up the unsettled western territories as it pleased.

25. Hecht, "L'idée," 63–70; articles by R. Meyer, Armin Julin, Adolf Jensen, A.N. Kiaer, and Edvard Arosenius in Koren, *History*, 85–6, 125–6, 201–2, 447–8, 538–48, respectively; Philippe Moureaux, *Les préoccupations statistiques du gouvernement des pays bas autrichiens, et le dénombrement des industries dressé en 1764* (Brussels: Université de Bruxelles, 1971), 27, 270–95, 511–3; R.L. Plackett, "The old statistical account," Royal Statistical Society, *Journal, A149* (1986), 245–51, on 249.

26. Faure, "France," 277–87; Athelstane Baines, "Great Britain and Ireland," in Koren, *History*, 365–89, on 367.

The European mathematical landscapers made their greatest conquests in the princely forests of Germany. The products of these forests provided an essential element in the budgets of German states and principalities and an equally important item in the well-being of its inhabitants. By the end of the Seven Years' War in 1763, enlightened princelings had recognized the need for prudent management of the remaining forest lands. (The following is derived from H. Lowood's account in chapter 11.) *Forstwissenschaft* required an inventory of the number and sorts of trees in the prince's domain, a plan for harvesting and reforestation so as to maximize yield while guaranteeing constant productivity, and a method for estimating the value of the cut timber. Foresters developed methods of calculating wood mass from estimates of tree height; the more daring among them had recourse to the integral calculus to handle the irregular shapes of standing trees; others idealized the shapes to truncated cones; and all recognized the utility of the fiction of a *Normalbaum,* or standard tree, to which the natural specimens could be assimilated.

Two aspects of this management are relevant to the quantifying spirit. One is the amount of quantitative dog work *Forstwissenschaft* demanded. The manuals set out exemplary tables—paradigms in the true sense—to be filled in with details about the location, type, and yield of every commercial species grown in the forest. A representative example concerns the value of pine stands of a particular domain in the Jägerthal. It has space for 400 numbers; and it is only one of many tables in the manual. The second aspect of interest is the compulsion to remake the forest to place the different species where they grew best so as to maximize not only the harvestable wood mass but also its quality and hence its value. Nature and custom prevented the full realization of this plan, but its design suggests what the mathematical landscaper might do when given a free hand.

That happened in the newly liberated British colonies in America. The several states had to honor the grants of land they had made to the officers and men of the Revolutionary army and the debts they owed to foreign governments. By 1785 most of the states had ceded most of their claims to western territories to the federal government. That year Congress passed its first ordinance directing the method for surveying and cutting up its new lands for disposal to soldiers, creditors, and settlers. The ordinance represented a great victory for the quantifiers, led by Thomas Jefferson, who insisted that the lands be surveyed into equal spans before being offered, and subdivided into parts affordable by small farmers; their opponents, primarily southern aristocrats and plantation owners, favored large grants to companies or wealthy individuals who would undertake to divide it up into such shapes and by such boundaries as suited their interests.[27] A similar compact between geometers and democrats developed during the metric reform.

The Ordinance of 1785 stipulated that a Geographer's Line be run due west beginning at the intersection of the Ohio River with the western boundary of Pennsylvania, and that every 6 miles lines be run due south until they met the Ohio. These north-south strips, known as ranges, were to be cut into squares 6 miles on a side by parallels to the Geographer's Line. The squares, known as townships, might be further divided into 36 lots of one square mile each, later called sections. In each such township the federal government reserved for parks or other purposes four lots symmetrically placed, and, for a school to be run by the settlers, an additional lot at the center.[28]

It is not possible to make the boundaries of the townships parallels of latitude and meridians of longitude and also to have all townships 6 miles square irrespective of their distance north or south of any east-west baseline. Surveyors adjusted to this impertinence of the earth's curvature by incorporating correction lines parallel to the baseline at 24-mile intervals. The middle townships are slightly larger, the northerly ones slightly smaller. At the correction line, the shorter-than-average sides terminating the townships below were expanded into the longer-than-average sides terminating the townships above.

27. C. Albert White, *A history of the rectangular survey system* (Washington, D.C.: G.P.O., n.d.), 10–1; Hildegard Binder Johnson, *Order upon the land: The U.S. rectangular land survey and the upper Mississippi country* (New York: Oxford University Press, 1976), 41–3.

28. For this and the next paragraph, White, *History,* 11–30, including the text of the Ordinance of 1785 (on 11–4).

The earth still exacts a penalty for the presumption of those who would cut it into equal pieces. Motorists in the Great Plains, who can speed from east to west along roads that go straight as an arrow forever, must stop every 24 miles where north-south boundary roads jog suddenly at the correction lines.[29]

Jefferson's original proposal would have divided the western territories into what he called "hundreds"—squares with sides of 10 nautical miles—rather than the townships of 36 square statute miles Congress eventually authorized. Each hundred was to contain 1,000 Jeffersonian acres and so on, for Jefferson, who invented the American system of pennies, dimes, and dollars, championed decimal division almost as strongly as democracy. He and the professor of mathematics who helped him work out his system of rigid squares recommended it for its order and clarity, and as an obstacle to cheating. They argued that irregular lots inspired fraud, and could point to the experience of Massachusetts, which discovered that holdings in the country typically held 10 percent more land, and often 100 percent more land, than had been granted. Friendly surveyors set the boundaries where their clients wished. The square grid made the practice much more difficult and allowed purchasers to get more or less what they paid for.[30]

Industry did not escape attempts at rationalization induced by the quantifying spirit. Its ingredients of inventorying and systematization expressed themselves in the *Dictionnaire des arts et métiers* of the Paris Academy and in the articles and plates of the *Encyclopédie*. The first volumes of both sets came into print in the early 1760s, obedient to our periodization and consequent to a rush to secure priority.[31] The inventory was to be the first step toward a rationalization that would be accomplished by artisans and philosophers in unlikely collaborations. The grandest vision in this direction was possessed by a

29. White, *History*, 119; Johnson, *Order upon the land*, 56–8. The offset amounts to about 160 feet at a correction line at latitude 40°.

30. Johnson, *Order upon the land*, 42–3; William David Pattison, *Beginnings of the American rectangular land survey system, 1784–1800* (Chicago: University of Chicago, Department of Geography, 1957), Research Paper 50.

31. Bertrand Gille, "L'Encyclopédie, dictionnaire technique," *Revue d'histoire des sciences pures et appliquées*, 5 (1952), 26–53.

French general, Jean-Baptiste de Gribeauval, who beginning around 1765 plumped for standardized armaments with standardized parts. His notion of uniformity or interchangeability came to the United States via its champion of numeracy, Thomas Jefferson.[32] To realize interchangeability required great advances in precision in machine tools and their application. The enlightened French general saw truly if distantly: the exactness that his quantifying and military spirit knew to be achievable was indeed accomplished by force of arms, by gun-makers to the U.S. Army.

Even in the rudimentary, gross industry of the 18th century a program to rationalize through experiment and measurement may be discerned. The interrelations of physics, chemistry, mathematics, self-help, nonconforming education, and invention in industrializing Britain, especially the Midlands, during the later part of the 18th century have been examined at length. Clock-makers, the very model of the exact machinist, made the textile industry tick; instrument-makers created the precision tools that made possible the realization of Watt's engines.[33] Here the quantifying spirit expressed itself in the balance sheet: not only in the counting house, but also in the engineering shop, where, as in the practice of John Smeaton, designs and processes were optimized by systematic variation of pertinent parameters that changed output in measurable ways. This approach contrasted with that of French engineers, like Coulomb or Charles Borda, who liked to work from the principles of analytic mechanics. Watt perhaps occupies an intermediate position.[34] From our point of view, the old dispute over the quantity of science that figured in the industrial revolution derives its smoke as well as its fire from a false

32. David A. Hounshell, *From the American system to mass production* (Baltimore: Johns Hopkins University Press, 1984), 25–6.

33. A.E. Musson and Eric Robinson, *Science and technology in the industrial revolution* (Manchester: Manchester University Press, 1969), 28, 115–8, 145–6, 429–30, 435–40; A.R. Hall, "Engineering and the scientific revolution," *Technology and culture, 2* (1961), 333–41, on 337–8.

34. D.S.L. Cardwell, "Science, technology and industry," in G.S. Rousseau and Roy Porter, eds., *The ferment of knowledge. Studies in the historiography of eighteenth-century science* (Cambridge: Cambridge University Press, 1980), 449–83, on 468–72, and *From Watt to Clausius* (Ithaca: Cornell University Press, 1971), 46, 72–84.

and poorly defined dichotomy. One did not need to proceed from the axioms of the *Principia* to a new machine, or wish to do so, to share a primary trait of the men of science of the late 18th century: an instrumentalist use of mathematics.

In chapter 10, S. Lindqvist schematizes the advance of the quantifying-instrumentalist approach to technological improvement by examining episodes in Swedish experience. To dramatize the difference between academic science and engineering imperative, the episodes take place in "labs in the woods"—or, anyway, on location, where facilities for large-scale experiment existed. In the case of water power, the very considerable advance in effective quantification between the early 18th and the early 19th centuries is made clear by a comparison between the 25,000 disjointed, inaccurate, useless experiments undertaken by Christopher Polhem on his own initiative and the sustained, precise, theoretically motivated and mathematically analyzed measurements made under the direction of Pehr Lagerhjelm on a commission from the Swedish Ironmasters' Association. Studies of charcoal-burning show the same pattern: some rough, qualitative assessments around 1750 and a thorough, careful, quantitative investigation by the Ironmasters around 1810. The Ironmasters were by no means the first in Sweden to explore productivity carefully and quantitatively. Lindqvist gives as the earliest example an effort to apply brain to brawn: a detailed study of the efficiency of human muscle power, carried out in the naval yards at Karlskrona in the early 1770s.

We touch upon quantification in technology in two additional contexts. A. Lundgren (chapter 8) points to the increased use of the balance in mineralogy and pharmacy after the middle of the 18th century and its more frequent presence in chemical laboratories. Here an instrument applied primarily to improve technique entered an environment where it was to effect—or to help effect—a revolution in physical principles. R. Rider's example (chapter 4) goes the other way. She moves from the elevated reaches of the theory of universal languages to the very practical problem of devising an efficient, long-range optical telegraph. The signaling codes, which found employment during the Napoleonic wars, disclose evidence of the operation of the quantifying spirit.

FURTHERMOST AFIELD

The *Encyclopédie* required 28 folio volumes, 71,818 articles, and 2,885 plates to do justice (or, as many thought, injustice) to the knowledge of the time.[35] That is no doubt impressive; but to say the truth, Diderot and d'Alembert were but indolent encyclopedists. Johann Heinrich Zedler's *Grosses vollständiges Universal-Lexikon*, completed in 64 volumes just as the *Encyclopédie* began to appear, could not stop with completion, and began to issue "necessary" supplements, which reached the letter "C" in four volumes, and died there. And, as G. Broberg demonstrates beyond peradventure in chapter 2, these *summae* were made possible only by suppressing most of what was known. Take the notorious case of insects. John Ray, writing in the 1690s, guessed that there might be as many as 20,000 species of them. Fifty years later, Pieter van Musschenbrock made it 130,000; another forty years, and it had reached 875,000. The total number of species of plants, animals, polyps, and microscopical creatures evidently exceeded the number of insects. How survey it all?

Encyclopedists and lexicographers had an easy and obvious method: they could list their information alphabetically and leave synthesis to the reader. The chemists, perplexed with some hundreds of mixed salts and the prospect of many more, found their salvation in the new gases, the composition of water, and the linguistics of Condillac. Their nomenclature of 1787 and its purely instrumentalist definition of "element" were so well constructed that they survived the discovery that their most controversial ingredient, oxygen, is not, as its carefully chosen name implied, the acidic principle.[36] Natural

35. Robert Darnton, *The business of Enlightenment. A publishing history of the Encyclopédie (1775–1800)* (Cambridge: Harvard University Press, 1979), 7.

36. F.L. Holmes, *Eighteenth-century chemistry as an investigative enterprise*

historians handled their multitudes by arithmetic: classification by numbers of flowers, teeth, nipples, and toes—anything superficial and denumerable. The system of Linnæus was a godsend to naturalists at sea in the quantity of their own discoveries. His way was not only arithmetical but also geometrical. The placement of species identified by arithmetical criteria under genera and higher orders permitted a two-dimensional layout of God's plan. It fit perfectly the instrumentalist character of the quantifying spirit. The physicist Biot made it the first great achievement of 18th-century science, "a universal systematic method," capable of arraying a vast amount of information "down to the smallest detail."[37] As J. Lesch shows with a wealth of examples in chapter 3, the Linnæan method became the instrument of classifiers from the medical through the mechanical sciences. The binomial approach to knowledge was as characteristic of the Age of Reason as silk stockings.

Natural species did not exhaust the species of nature. With human help, plants and animals disinclined to breed in the wild could be made to father and mother hybrids. In chapter 9, in our closest approach to sex, J. Larson describes J.G. Koelreuter's crossing of two species of tobacco plant and analyses of the resulting bastards. More interesting for our purposes than Koelreuter's deductions about the transformability of plants was his quantitative method. His compulsion to measure equaled that of his contemporary Ezra Stiles, president of Yale University, who took the temperature of the weather twice a day for thirty years, weighed his children regularly before breakfast, and counted 888 houses, 439 warehouses, 16 stills, 77 oxen, 35 cows, and 1,601 sheep on supernumerary walks in Rhode Island one autumn.[38] Koelreuter followed the progress of hybridization by minute measurement of flower parts.

(Berkeley: Office for History of Science and Technology, 1989); Maurice P. Crosland, *Historical studies in the language of chemistry* (London: Heinemann, 1962), 168–92, 221–2.

37. Biot, *Essai*, 9.

38. Patricia Cline Cohen, *A calculating people. The spread of numeracy in early America* (Chicago: University of Chicago Press, 1982), 110–2.

The rapid progress of the quantifying spirit during the last three decades of the Ancien Régime indicates wide social support. It is scarcely a challenge to discover in the cultural and political history of the 18th century developments that created or strengthened this support. The more obvious include the rise of the benevolent despots of central Europe, with their cameralist bureaucracies and programs of economic rationalization; the multiplication in France of enlightened philosophers and their fellow travelers, intent on transforming the arts, sciences, education, and government; the acceleration of industrial innovation in Britain owing to a happy conjunction of capital, skilled labor, natural resources, and expanding markets; and, throughout the Atlantic countries, that insistence on guiding society by the lamp of human reason, however dim it might be, rather than by the light of revelation or the radiance of sun kings, that marked the "age of democratic revolutions."

The tempo of these developments picked up around the middle of the 18th century. The exemplary despots of Prussia and Austria then established their positions. In 1751 the *philosophes* published the first volume of their mightiest weapon, the *Encyclopédie*. The precise origins of great social revolutions do not lend themselves to exact dating. Nonetheless, the quantitative instrument of the historian, the time scale, demands the effort. The balance of contemporary learned opinion seems to incline toward the same date for the quickening of the industrial revolution in Britain that Arnold Toynbee proposed a century ago: 1760.[39] That is over precisely the identical date that the historian of democratic revolutions settled on for the beginning of his saga,[40] and that we have identified as the watershed in the activities of our quantifying spirit.

Links between these wider movements and ours may be made at several levels. The need of the increasingly bureaucratic state to organize itself and control its resources gave an impulse to the

39. R.M. Hartwell, "The causes of the industrial revolution: An essay in methodology," in R.M. Hartwell, ed., *The causes of the industrial revolution in Britain* (London: Methnen, 1967), 53–79, on 59, 79.

40. R.R. Palmer, *The age of the democratic revolution. A political history of Europe and America, 1760-1800* (Princeton: Princeton University Press, 1959), 22–4.

collection and analysis of vital and other statistics; to forestry and rational agriculture; to surveying and exact cartography; and to public hygiene and climatology. Industrial innovation encouraged quantitative experiment. The political philosophy of the Enlightenment and its application in revolution spread notions of reason and nature compatible with the rationalization realizable by numbers. The interconnection of the themes of revolutionary rhetoric, numeracy, bureaucratic imperative, and precision measurement is evident in the course and nature of the reform of weights and measures effected during the French Revolution, which is the subject of chapter 7.

The institutions of benevolent despotism and enlightened bureaucracy provided many niches for mathematicians. One set of niches harbored salaried academicians. Their opportunities increased rapidly shortly after 1750, when the rate of founding of academies doubled over what it had been in the previous half-century. This enlargement continued until the French Revolution and affected the quality as well as the quantity of academicians. The insistence upon expertise in a science as a condition for admission, although not always imposed, did create an environment more favorable than earlier more lenient times to the advancement of quantifying spirits.[41] Other important niches opened in armies, navies, and state industries.[42]

The upward mobility of the mathematically talented is nowhere better illustrated than in the career of Laplace, who has already appeared several times in these few pages, as a contributor to the Standard Model, to the theory of population statistics and (tacitly) to the metric reform. We end our book with R. Hahn on Laplace's career as we began with T. Frängsmyr on Wolff's. We thus frame our accounts of the quantifying spirit with scientific biographies that, between them, encapsulate the transformation documented in the balance of the book.

41. James E. McClellan III, *Science reorganized. Scientific societies in the 18th century* (New York: Columbia University Press, 1985), 243–50, and "The Académie royale des sciences, 1699–1793: A statistical portrait," *Isis, 72* (1981), 541–67, on 552–3.

42. Roger Hahn, "Scientific research as an occupation in 18th century Paris," *Minerva, 13* (1975), 501–13, and "Scientific careers in 18th-century France," in Maurice P. Crosland, ed., *The emergence of science in Western Europe* (New York: Science History Publications, 1976), 127–38.

PART 1

RATIONALIZATION OF KNOWLEDGE

1 The Mathematical Philosophy

By Tore Frängsmyr

There are two distinct opinions regarding the status of science in the 18th century. One dismisses the period as an uninteresting interval between the breakthroughs of the 17th century and the expanding industrialism of the 19th. The opposing view holds that the 18th century was the time when all the important work of the previous century bore fruit.

The dismissers argue that nothing really new emerged in the 18th century. "By many historians the century has been deplored for allegedly producing science that was boring, unoriginal, lacking in rigour and overspeculative." It has been castigated as "comparatively undistinguished in its science"; if its science is included in its reputation, the *siècle des lumières* cannot be freed from "an element of dullness."[1]

In direct opposition to these nay-sayers, other historians have identified much of the intellectual life of the 18th century with the fostering of its heritage from the scientific revolution. The Enlightenment thus climaxes the development of early modern science.[2] The most recent general assessment of the role of science in the Enlightenment gives particular weight to the century's efforts to consolidate and organize what it had received. "The creation of the new scientific disciplines was probably the most important contribution of the Enlightenment to the modernization of science, and one that we might easily overlook. It marks the Enlightenment as a period of

1. G.S. Rousseau and Roy Porter, eds. *The ferment of knowledge: Studies in the historiography of eighteenth-century science* (Cambridge: Cambridge University Press, 1980), introduction.

2. Preserved Smith, *A history of modern culture*, 2 vols. (New York: H. Holt and Co., 1930–4), 33–5.

transition between the old and the new."[3] That careful and informed students of the 18th century have reached diametrically opposed evaluations of its science is largely a consequence of their differing ideas about what counts as science. In this volume we sidestep the difficulty by restricting our attention to the working out of a particular method, or, as it might better be called in many instances, of an appeal to a slogan. That slogan, as Galileo formulated it, holds that "the book of nature is written in the language of mathematics."[4]

In the 18th century, this proposition inspired the work of the most productive mathematicians and underlay the program of the most ambitious philosophers, who proposed to extend it beyond the book of nature to the books of man—that is, to all exercises of human reason.[5] The languages of reason and nature were thereby to agree. We see two phases in the 18th-century elaboration of this proposition. Until about 1760, "mathematics" was applied to the general exercise of reason (as opposed to mathematics itself and the quantified parts of astronomy and physics) more as a symbol and slogan than as a useful tool. In this first phase the philosophy of Leibniz's epigone, Christian Wolff, is exemplary. In the last third of the century, however, the equation of reason with mathematics inspired instrumentalist and quantitative approaches to a wide range of problems in various sciences, arts, and technologies. We take the activities of this quantifying spirit to have been progressive, in the sense that under its inspiration the achievements of the scientific revolution were extended and reinterpreted in a way that made possible rapid advance in mathematics, physical sciences, and statistics (conceived as the sciences of the state) in the early 19th century.

In what follows, Wolff's method is seen as representative of the process by which reason was "naturalized" during the first half of the 18th century. The purpose is to provide both an introduction and a contrast to the applications of the quantifying spirit later in the century that form the subject matter of the other essays in this volume.

3. Thomas L. Hankins, *Science and the Enlightenment* (Cambridge: Cambridge University Press, 1985), 5.

4. Smith, *A history of modern culture.*

5. Ibid., 118.

UNIVERSAL MATHEMATICS

Leibniz wanted not only to see the world as a machine with a "static structure of matter in geometrical and mechanical relation"; he also wanted to give exact mathematical expression to all the processes and changes that took place in the world. Through the infinitesimal calculus he found a constant law that remained valid for an infinite number of processes and changes. The elements constituting the universe, nature, and human life were not particles of matter, but forces and motions. Each element or individual substance—Leibniz called them "monads"—was complete in itself, a microworld or a mirror of the universe, and yet intimately related to all other individuals in a system of pre-established harmony. The mathematical system came by its harmony naturally.

Given this structure of the world it was natural for Leibniz to take another step: to organize human knowledge in a mathematical pattern. Drawing both on the Lullian tradition that sought a universal language and on his own mathematical work, Leibniz wanted to create a *mathesis universalis* or *ars characteristica*, which would represent, combine, and produce all possible knowledge. This language would serve not only as a philosophical method but also as the key to a true encyclopedia. In the *mathesis universalis*, each human concept or idea corresponded to a different symbol or combination of symbols. By combining these symbols according to mathematically defined rules, Wolff's method could produce long lists of new truths and confirm old ones. The new language became for Leibniz both a means of communication and a new kind of logic, a way to formulate statements and an instrument of reason.[6]

The most influential philosopher in the Leibnizian tradition was Christian Wolff. Wolff was not an original thinker. Instead, he took the foundation of his system from his teacher Leibniz, systematized it, and put it within the grasp of any educated person. Because of his

6. Louis Couturat and L. Leau, *Histoire de la langue universelle* (Paris: Librairie Hachette, 1903); Louis Couturat, *La logique de Leibniz d'après des documents inédits* (Paris: F. Alcan, 1901), esp. chaps. 2–4; cf. C.I. Lewis and C.H. Langford, *Symbolic logic* (New York: The Century Company, 1932), 5ff.

firm insistence on the rationalism that was characteristic of his "mathematical method," Wolff came to be regarded as a philosopher of the Enlightenment. Mathematics held symbolic value as the driving force behind the progress of science, and also the progress of reason and free thought. As Newton had given the universe a mathematical structure, so Wolff aimed to subsume human and society within a vast mathematical system. He pushed his mathematical method into every possible area: philosophy and science, theology and ethics, politics and public finance.

Wolff's "Rational Thoughts"

Many historians of philosophy have described Wolff as one of the most influential philosophers of his time, without, however, paying much attention to his ideas. He has been called "the master of the Enlightenment in Germany," and even outside Germany; Diderot sang Wolff's praises in his *Encyclopédie.* Paul Hazard made Wolff "the intellectual leader of Germany" and a general-purpose sage admired by all nations. Lewis White Beck portrays Wolff as a rationalist in the Enlightenment spirit: "Wolff is the best German representative of a general movement of thought towards deism, utilitarianism and free thought that was sweeping over Europe as a whole."[7]

These characterizations of Wolff and his influence claim both too much and too little. Too much, because terms like "rationalism" and "mathematics" did not necessarily imply Enlightenment, nor did Wolff's philosophy. It was the *form* of mathematics that interested him, its way of pursuing an argument to its proof. The characterizations claim too little because the association of Wolff's brand of rationalization with the Enlightenment neglects the vast effort at systematization of all knowledge inherited from Leibniz and the 17th century.

7. Paul Hazard, *European thought in the eighteenth century* (London: Hollis, 1954), 40; Lewis White Beck, *Early German philosophy: Kant and his predecessors* (Cambridge, Mass.: Belknap Press, 1969), 256–76. Cf. Tore Frängsmyr, "Christian Wolff's mathematical method and its impact on the eighteenth century," *Journal of the history of ideas, 36* (1975), 863–8.

Wolff turned to the connection between mathematics and philosophy early in his career. He obtained his doctorate in 1703 with a dissertation on ethics based on mathematical foundations. He soon came into contact with Leibniz, who nominated him for election to the Academy of Sciences in Berlin. In 1707 Wolff became professor of mathematics at Halle and later assumed the chair of philosophy. In Wolff's work the one subject cannot be distinguished from the other. As a mathematician he hardly made an original contribution, but his diligence yielded bulky volumes summarizing the contemporary state of knowledge—for instance, his *Anfangsgründe aller mathematischen Wissenschaften* (6 vols., 1710) and *Elementa matheseos universae* (5 vols., 1713–5).

Wolff announced his own particular specialty—the application of mathematics to philosophy—in his *Ratio praelectionum Wolfianarum in mathesin et philosophiam universam* (1718). This line of work culminated in a book in German with the monumental title *Vernünfftige Gedancken von Gott, der Welt und der Seele des Menschen, auch allen Dingen überhaupt* (1720). The book, usually known as his "German metaphysics," sets out Wolff's philosophy virtually in its entirety; it brought him wider renown and prefigured a whole literary genre during the 18th century. Wolff himself published a series of *Vernünfftige Gedancken* on various topics; elsewhere philosophers followed his lead and published "rational thoughts" on one subject after another.[8]

Wolff's stature as a liberal theologian has also contributed to his reputation as a rationalist philosopher of the Enlightenment. In Halle, the dominant theology was the conservative strain of Pietism developed by August Hermann Francke, which sought renewal within the Church. Competition came from influential orthodox

8. Hans Werner Arndt, "Rationalismus und Empirismus in der Erkenntnislehre," in Werner Schneider, ed., *Christian Wolff, 1679–1754: Interpretationen zur seiner Philosophie und deren Wirkung* (Hamburg: F. Meiner, 1983), 31–47. The same collection contains six articles about Wolff's influence, among them Notker Hammerstein, "Christian Wolff und die Universitäten: Zur Wirkungsgeschichte des Wolffinanismus im 18. Jahrhundert," 266–77. See also Beck's chapter "A generation of epigoni," in *Early German philosophy*, 276–305, and Tore Frängsmyr, *Wolffianismens genombrott i Uppsala* (Universitet; Stockholm: Almqvist & Wiksell, 1972), with an English summary: "The emergence of Wolffianism at Uppsala University, 1720–1760."

pastors and from a group of "transitional theologians," who attempted the feat of mixing orthodoxy, Pietism, and rationalism.

Wolff found himself at odds with this eclectic combination. In 1721 he delivered a formal address on the teaching of Confucius, *Oratio de sinarum philosophia practica*, which emphasized its strict morality.[9] By studying human nature, but without invoking divine revelation, Confucius had attained a lofty moral position. From this Wolff argued that morality could be comprehended by reason alone. Here he linked up with the deism that had been spreading steadily since the mid-17th century and that often saw in Confucianism the confirmation of its ideas. Wolff's address contained nothing new or sensational, but under the circumstances it was taken as a challenge. The Pietists retaliated by accusing Wolff of atheism; condemnatory sermons echoed in the churches and protests ascended to higher levels; the highly regarded theologian Johann Franz Buddeus, orthodox but favorably disposed to the Pietists, wrote an indictment of Wolff's views. The campaign finally succeeded in 1723, when Frederik Wilhelm I banished Wolff from Halle. Wolff retreated to Marburg, a little university town in the landgraviate of Hessen, an environment more conducive to his reputation. Within a few years he was being hailed as the greatest mathematician and philosopher of his age; students on their grand tours included a stop in Marburg to attend his lectures; scientific academies and universities throughout Europe wooed him. He began to write in Latin, which permitted access to a wider international audience. The 1730s saw the appearance of the great classical works, *Philosophia prima sive ontologia* (1730), *Cosmologia generalis* (1731), *Psychologia empirica* (1732), *Psychologia rationalis* (1734), *Theologia naturalis* (2 vols., 1736–7), and *Philosophia practica universalis* (2 vols., 1738–9). Again, they contained little new, but developed and refined Wolff's ideas in different philosophical disciplines and generated popularity, renown, and followers for their author.[10]

9. Wolff's *Oratio de sinarum philosophia* is now published in a Latin-German parallel edition, *Rede über die praktische Philosophie der Chinesen*, ed. Michael Albrecht (Hamburg: Meiner, 1985), with a 100-page introduction. See also Beck, *Early German philosophy*, 258ff.

10. Anton Bissinger, *Die Struktur der Gotteserkenntnis* (Bonn: H. Bouvier, 1970).

Wolff's reputation as a rationalist and freethinker derived more from his banishment than from his writings. His position was enhanced when Frederick the Great ascended the throne of Prussia and decreed that salvation was everyone's responsibility. One of Frederick's first acts was to recall Wolff to Halle and to shower him with honors. Wolff assumed positions as professor of law, vice-chancellor of the university, privy councillor, and, in three years, chancellor. In 1745 he was ennobled as baron von Wolff. The summons from Frederick the Great, prince of Enlightenment, confirmed Wolff's status as an Enlightenment philosopher.

The Mathematical Method

The characteristic aspect of Wolff's philosophical teaching was its strictly logical construction. He had a manic capacity for arranging and systematizing everything, step by step, clearly and paradigmatically, accurately and in detail. He worked from the principles of contradiction (*principium contradictionis*) and of sufficient reason (*principium rationis sufficientis*). He regarded the principle of contradiction as the first law of philosophy from which all other propositions could be derived. With the principle one could decide whether something was possible; that is, everything that might be held to have sufficient reason. Philosophy was the science of the possible, with the task of showing how and why things are possible.[11]

In working out his science, the philosopher had to observe certain rules. No principles could be employed unless adequately proved, and no new ones allowed unless derived from proven principles. No departure could be made from the meaning that words had generally acquired, and if new words or concepts were required, they had to be accurately defined. Intrinsically different objects and phenomena had to be given different names. Starting with axioms, clear definitions, and distinctions, deductive methods would link truths with one another and thus reach irrefutable conclusions. In this way results

11. *Vernünfftige Gedancken von den Kräfften des menschlichen Verstandes* (Halle: Renger, 1723; 7th ed., 1733), 1. Cf. *Cogitationes rationales* (Frankfurt and Leipzig: Renger, 1730), 1.

obtained in philosophy would be as reliable as those in mathematics. In none of this did Wolff go beyond Leibniz, or, for that matter, Aristotle, Thomas Aquinas (whose influence he acknowledged), Descartes, Spinoza, or Tschirnhausen, whose *Medicina mentis* he had studied in detail. The mathematical method nevertheless became known as Wolffian, because Wolff systematized it as never before. In his hands it became an all-purpose means for establishing truth in any sphere.[12]

Wolff's epistemology is quite simple. Human knowledge outside Christian revelation can be acquired in three ways: by experience (historical knowledge), by reason (mathematical knowledge), or by a combination of the two (philosophical knowledge). The last of these three methods is preferable; the other two have value only to the extent that they can be of use to philosophy. Our senses awaken our most general concepts, which are innate in our consciousness, and provide many fresh ideas; reason clarifies them and puts them into context. Thus every philosophical discipline has both empirical and rational components. For Wolff, philosophy came to mean the same thing as science, as the German term for philosophy, "Welt-Weissheit," suggests.

Mathematics is not a part of Wolff's philosophy. Instead, it serves as an instrument of knowledge in its own right—a method, a means rather than an end. Mathematics, in both form and inherent logic, provides a model for all areas of human knowledge. From this point of view, the mathematical method was identified with the philosophical one, and assumed a central role in the scheme of all Wolffian philosophers, even if they did not all see mathematics in the same subtle light as Wolff did. For the imitators, in fact, mathematics at times meant little more than computation.[13]

Wolff held dear the idea of a system of knowledge. Knowledge was not merely a sudden insight or an idea, but arose from step-by-step deduction leading to reliable conclusions. The complex, reliable system therefore became an emblem of Wolff's entire philosophy—

12. *Philosophia rationalis sive Logica*, 2d ed. (Frankfurt and Leipzig: Renger, 1732), 53–71; the "Discursus preliminaris" is the best explanation of Wolff's goal.

13. Ibid., 28–52, "De partibus philosophiae."

both theoretical (logic, metaphysics, psychology, natural theology, physics, and technology) and practical (law, ethics, economics, and politics).[14]

For Wolff, mathematics was the *method* and philosophy the *content.* This distinction is essential to an appreciation of his philosophy and its place in 18th-century thought.

For and Against Enlightenment

Wolff's working life falls into two periods. In the first he was active at Halle, wrote in his native German, formulated a more popular philosophy, and spoke so freely on religious matters that he was accused of atheism and banished. In the second period he lived in Marburg, wrote books in Latin, put forward a more theoretical philosophy explicitly addressed to the world of learning, and appeared in religious matters as an orthodox and unrelenting apologist. The watershed came somewhere around 1730. In the earlier phase he followed the precepts of the Enlightenment; in the latter he opposed them.

Wolff's German books aimed at a larger public are characteristic of his enlightened period. They have a pronounced practical and utilitarian tone and allot philosophy a "liberating" role, fully comparable to later manifestos of the French Enlightenment. Good examples are plentiful in his *Vernünfftige Gedancken von der Menschen Thun und Lassen, zu Beförderung ihrer Glückseligkeit* (1720). A critical indication of his popular, liberal position was his vacillation between physicotheology and teleology, two doctrines not easily distinguished and therefore often confused. Physicotheology rested on the ingenious organization of the natural world in general, but it often singled out particular features in nature as evidence of a creator behind the order. This search for an all-ordering God behind the exquisite subtlety of nature could justify, and even motivate, all scientific research. Where *order* served as a fundamental principle in physi-

14. Carl Günther Ludovici, *Kurtzer Entwurff einer vollständiger Historie der Wolffischen Philosophie* (Leipzig: J.G. Löwe, 1736), 113ff. Cf. Bissinger, *Die Struktur,* 49f.

cotheology, so *intention* emerged as essential in teleology. If everything fulfills a definite purpose, there must be at least an implicit ordering principle—a God who intended the order. Many physicotheologists came to confuse proofs based on general order and those based on intent, and to subsume teleology under physicotheology. In either case, a "popular" proof of the existence of a divine creator was secured.

To see Wolff as a great representative of physicotheology, however, is to overplay his hand. Wolff attacked both the physicotheological and teleological approaches.[15] He recognized the legitimacy of demonstrating divine intent in nature, but demanded a prior proof of God's existence and of the creation of the world as a free divine act. Wolff's method required that the theoretical foundation be established first. Yet a few years later, Wolff appeared as a full-fledged physicotheologist in his great work *Vernünfftige Gedancken von der Absichten der natürlichen Dinge* (1723). Had his views suddenly changed? No, he had distinguished between two sets of readers—academic and popular. Wolff's major works in Latin do not feature physicotheology. In a foreword to a translation of Nieuwentyt's well-known book on physicotheology (1731), he referred to his own work in German and underlined that it was intended for readers who did not speak the languages of the erudite. Whatever he said, in practice he inferred God's perfection from the usual inventory of the natural scientist. He described the cosmos and all its bodies, the sun and the fixed stars, the earth and the planets; he dealt with air and winds, different kinds of precipitation, rainbows, lightning, and thunder. The quantitative philosopher will want to know that his evidence includes "the number of heavenly bodies."[16]

In sum, although Wolff demanded strict standards of logical proof in his mathematical method, he allowed himself to break the rules when addressing an unlearned audience. The end justified the means,

15. *Ratio praelectionum Wolfianarum in mathesin et philosophiam universam* (Halle: Renger, 1718), 153–8.

16. Wolff's preface reprinted in *Gesammelte kleine philosophische Schrifften* (Halle: Renger, 1736), 515f. Cf. Bissinger, *Die Struktur*, 232f.

for the religious objective was all-important. He made an even more significant adjustment in his major work *Theologia naturalis*, an extended exercise in apologetics, rapidly becoming his specialty. He devoted more than 350 pages to demonstrating the false steps and absurd notions of atheism.[17] He did not confine his attack to unbelievers but also took up cudgels against every conceivable kind of philosophical freethinker: atheists, fatalists, deists, naturalists, anthropomorphists, materialists, paganists, Manichæans, Spinozists, and Epicureans. Just as he had previously used the mathematical method to lend positive support to philosophy or theology, he now used it to expose and exclude those who were not orthodox. Fourteen years earlier he had been banished and accused of atheism; now Wolff attacked those who adopted the viewpoint he had then espoused. Yet he held fast throughout these sea-changes to his mathematical method and philosophical rationalism.

WEIGHING IN THE BALANCE

Which side of Wolff attracted his contemporaries—the mathematical-rationalist or the teleological-apologetic? The answer must be that many accepted both sides, viewing them as two parts of the same entity. The mathematical method tied them together, and became the guiding principle for Wolff's pupils, even for those most inclined to apologetics.[18]

One of the first of these pupils was Ludwig Philipp Thümmig, appointed professor in Kassel on the master's recommendation. His extensive *Institutiones philosophiae Wolfianae* (2 vols., 1725–6) differed little from Wolff's own account, and functioned as an authoritative source for later Wolffians. Another pupil, regarded as even more significant, was Georg Bernhard Bilfinger, professor of philosophy first at Tübingen, then at St. Petersburg, and again at Tübingen. His

17. "De atheismo aliisque agnatis erroribus," *Theologia naturalis II* (Frankfurt and Leipzig: Renger, 1737), 369–726.

18. Max Wundt, *Die deutsche Schulphilosophie im Zeitalter der Aufklärung* (1945; new ed., Hildesheim: G. Olms, 1964), 199–230, and Beck, *Early German philosophy*, 276–305.

Dilucidationes philosophiae de Deo, anima humana, mundo et generalibus rerum affectionibus (1725) followed up, in title and contents, Wolff's *Vernünfftige Gedancken von Gott, der Welt und der Seele des Menschen.* Bilfinger emphasized the twin foundation stones of reason and experience, and stood ready to point out correspondences between natural science and Christian faith. Alexander Gottlieb Baumgarten, first in Halle, then at Frankfurt, wished above all to apply Wolff's philosophy to aesthetics. His well-balanced *Metaphysics* (1739), extremely faithful to Wolffian systematics, amounted to a reference book or compendium of Wolffian doctrine. In Leipzig, Johann Christian Gottsched and Carl Günther Ludovici wrote historical works on Wolff but were not themselves leading philosophers. Israel Theophil Canz at Tübingen did much to inject the doctrine into theology, as did Jacob Carpov at Jena, who sought to prove divine revelation using Wolff's teleological method. Important expositions of the mathematical method were also offered by Friedrich Christian Baumeister at Gölitz and Georg Heinrich Ribov at Göttingen.

Outside Germany Wolffianism found adherents in England, France, the Netherlands, Switzerland, and Scandinavia. In Denmark, Wolff's doctrines exerted influence through a number of mathematicians and philosophers: Christian Hee, who stayed for a time with Wolff at Marburg; Friedrich Christian Eilschow; and, above all, Jens Kraft, eminent as both a mathematician and a philosopher, who published a series of textbooks in imitation of Wolff. In Sweden, too, mathematicians introduced Wolff's ideas, which spread to the Swedish university at Åbo (Turku) in Finland.[19]

In Sweden

The tendency of Wolff's disciples to tread confidently in their master's footsteps, to accept both his philosophical and his theological views, and to tie them together via the mathematical method, was muted in Sweden by resistance to the casual combination of philosophy and theology. Sweden's true introduction to Wolff came through

19. Frängsmyr, *Wolffianismens genombrott*, 11–44.

the efforts of the mathematicians Anders Celsius and Samuel Klingenstierna, whose reputation at Uppsala University recommended their opinions to many students. In a manual of arithmetic (1727), Celsius spoke warmly of the existence of the mathematical method and also of its importance to other disciplines.[20] By the following year, in a *pro gradu* dissertation on the existence of the soul or the intelligence, Celsius completely embraced the philosophy of Wolff. He praised not only Wolff's method but also followed the master in setting out his dissertation in traditional Euclidean form, with short propositions under such headings as *axioma, theorema, definitio, observatio, demonstratio*, and *scholion*. The dissertation argues that the soul can be proved to exist by means of Wolff's philosophical laws, and Wolff himself is called "the greatest philosopher of our time," *philosophus nostra aetate summus*.[21]

Several theses reminiscent of Celsius' approach appeared over the next few years, all of them loud in praise of the mathematical method and bearing the imprint of the mathematical form favored by Wolff. Early in June 1729, the brothers Erik, Nils, and Johan Gottschalk Wallerius and their close friend Olof Hammaræus performed as respondents to four Wolffian dissertations. Presented first *pro exercitio*, two years later all four dissertations were defended *pro gradu*, under Celsius and Klingenstierna as tutors. Several of the dissertations describe the mathematical method without employing it; one stresses that certain philosophical knowledge is impossible without it. Jacob Friedrich Müller, professor of philosophy at Giessen, was quoted in support of this statement, although, unknown to the author, Müller had just defected to the camp of Wolff's denigrators.[22]

Other dissertations under Celsius follow the same pattern. One deals with the subject of "incomprehensible books," which provided

20. Celsius, *Arithmetica eller räkne-konst* (Uppsala, 1727), preface.

21. *De existentia mentis,* pres. J.J. Hermansson, resp. A. Celsius (Uppsala, 1728), 7.

22. *De natura et utilitate scientificae*, pres. S. Klingenstierna, resp. J.G. Wallerius (Stockholm, 1731), 4, 7, 10. Cf. *De demonstratione*, pres. E. Alstrin, resp. E. Wallerius (Stockholm, 1729). All these dissertations appear in Frängsmyr, *Wolffianismens genombrott*, chap. 3, 64–87.

the opportunity to laud Wolff as an example of lucidity and intelligibility.[23] Evidence suggests that Celsius was behind the most detailed presentation of Wolff's philosophy, a dissertation in two parts (1731–2) on the subject of "how to attain worldly happiness through philosophy." The subject may appear novel, but the theme is familiar. The philosophy that can promote profit and happiness is contained in logic and mathematics. Hence the mathematical method is superior to earlier instruments and Wolff is the greatest of philosophers, outshining lesser lights like Plato, Aristotle, and Descartes. With Wolff's method, the author "brushes aside the weapons of atheism and defends the truths of the Christian religion." *Ad pleniorem scientiam*, the dissertation demonstrates that mathematics and physics are of fundamental importance to all other sciences, technology and mechanics, medicine, law, economics, and military subjects such as fortification and pyrotechnics.[24]

Samuel Klingenstierna had already earned a name for himself as a mathematical genius when he set off to study abroad in 1727. He went first to Marburg to hear the much admired Wolff. It is recorded that the pupil much impressed the teacher. When the chair of mathematics at Uppsala became vacant, Klingenstierna applied for it. Because Marburg was within the native state of King Fredrik I of Sweden, Wolff had access to the king and recommended Klingenstierna in the strongest terms. Appointed to the post in 1728, Klingenstierna did not take up his professorship until 1731, when he returned to Sweden with newly purchased books and fresh reports on the fashionable new philosophy. His were popular lectures, and students sought him as tutor for their doctoral theses. These theses are filled with quite simple Wolffian propositions about the excellence of the mathematical method or the role of contingency in the creation of the world, and frequently make reference to Wolff and Bilfinger. Mathematics thus linked up with philosophy, notably in Klingenstierna's seminars *in philosophiam naturalem*.[25]

23. *De obscuritate librorum*, pres. A. Celsius, resp. O.J. Petræus (Uppsala, 1731), 11.

24. *De felicitate civili per philosophiam comparanda*, pres. A. Celsius, resp. J. Råfelt (Uppsala, 1732), passim. Quotations from and references to Wolff abound.

25. Uppsala University, *Catalogus lectionum publicarum* (1731).

But enthusiasm for Wolff and the use of mathematics in philosophy went too far. Alert theologians, worried by the excessive spread of rationalism, castigated Wolff as "heathen and atheistic." In 1732 the chancellor of Uppsala, Gustaf Cronhielm, warned professors against dealing too casually with the new philosophy inspired by Wolff. The professors responded with promises to be careful and to protect their young students. Two years later the tone of admonition grew sharper. The chancellor decreed that professors should not preside at theses outside their own disciplines, a rebuke directed particularly at the professors of mathematics, who had been so ready to interpret the Wolffian philosophy. A year later, students were required to give a declaration of faith when enrolling at the university, to guard the purity of the doctrine. Such censorship and official criticism encouraged attacks on Wolff. His opponents asserted that mathematics and philosophy did not belong among the fundamental sciences; they saw all questions as ultimately teleological in nature. Philosophy, they insisted, should stick to its time-honored role as the handmaiden of faith.[26]

The situation soon changed radically. Cronhielm was succeeded as chancellor by Gustaf Bonde, known for his deep interest in mysticism, alchemy, and Platonic mathematics. An admirer of physicotheology and of Wolff, he later published three volumes of "reflections on the wonders of God in nature." On a visit to Uppsala University in 1738, Bonde stressed the desirability of teaching the younger generation *theologia naturalis* as a timely defense against "atheists and indifferentists." The new chancellor thus recommended what the old one had forbidden. Bonde went even further when he engineered an offer to Wolff of the most highly regarded professorship in the university (which Wolff declined). Bonde's permissiveness sanctioned latent interests among the faculty. Petrus Ullén, professor of philosophy, became the first important figure in this new phase. By the time of his death in 1747, he had presided over a hundred theses, a third of which were Wolffian through and through. Ullén was no original

26. The university chancellor's letter reprinted in Claes Annerstedt, *Upsala Universitets historia*, app. 3 (Uppsala: Almqvist & Wiksell, 1910), 198.

thinker. He praised mathematics for its ability to clarify and present problems in easily grasped diagrams and figures, and he insisted on the importance of the Wolffian philosophy to theology. He vehemently attacked all tendencies toward deism or "indifferentism"; true to the later Wolff, he used rationalism as a defense of orthodoxy and against the ideas of the Enlightenment.[27]

An even louder champion of Wolffianism was Nils Wallerius, who started out as a mathematician and physicist, continued as a philosopher, and ended up as a theologian. He succeeded Ullén as professor of philosophy in 1746. Within the compass of philosophy he included logic, metaphysics, psychology, and natural theology, all slavishly arranged in accordance with Wolff's system. The mathematical method was fundamental in all philosophy, but mathematics must yield to theology, lest it lead to materialism and atheism. Wallerius shared the concern of both philosophers and theologians over *theologia polemica*—the struggle against Enlightenment philosophy. In 1755 Wallerius received a new chair in theology devoted to uncovering and combatting heretics. The new professor was to repudiate all freethinkers, "such as atheists, naturalists, deists, anti-scripturalists, indifferentists and other unbelievers."[28] Wallerius warmed to the task: Moravian Brethren, Socinians, pantheists, and mystics, too—here he named Jakob Böhme, Paracelsus, Robert Fludd, and Johann Conrad Dippel—fell under his flail. In an essay on the repulsiveness and wickedness of materialism, Wallerius ranted at the ancient atomists Democritus, Epicurus, and Lucretius and their modern successors Thomas Hobbes and Pierre Bayle. "*O stupida ingenia, sive mente ac ratione*," you are so blind that you cannot imagine anything beyond the bounds of the material.[29] In 1756, Julien Offray de La Mettrie, the leading contemporary materialist, came in for particularly severe criticism.[30] Wallerius decried the deists from Locke to Hume as

27. Frängsmyr, *Wolffianismens genombrott*, 110–29.

28. Wallerius's program printed in Annerstedt, *Upsala Universitets historia*, app. 3, 417f.

29. *Diss. mala et incommoda materialismi sistens*, pres. N. Wallerius, resp. C.P. Eurenius (Uppsala, 1754), 4, 7, 11.

30. *Diss. sistens particulam primam vindiciarum religionis contra D:num De La Mettrie*, pres. N. Wallerius, resp. G. Nordén (Uppsala, 1756–65), vi, 76.

"naturalists"—Wallerius' favorite epithet for his opponents—and Voltaire as "the greatest fraud of the day."[31]

Wallerius' guiding ambition was to reconcile mathematics, philosophy, and theology. At hand were all the necessary tools: a professional grasp of mathematics, philosophy, and theology; a passion for system; and Wolff's method. His eminent elogist in the Royal Swedish Academy of Sciences, Torbern Bergman, showed restrained appreciation of Wallerius' contribution, and a like opinion was expressed in another contemporary biography: "Had he lived fifty years earlier and in a more scholastic era, his memory would have been even more illustrious."[32]

By Kant

Wolff's use of mathematics usually made the best impression on those who knew the least mathematics. During the late 1740s, in a celebrated dispute between the supporters of Newton and those of Wolff, Pierre Louis Moreau de Maupertuis was the main adversary of the Wolffians. The most telling attack on Wolff came from Kant, however, who as usual went right to the core and challenged the mathematical method itself.

The Berlin Academy had posed the question of whether metaphysical truth could be equated with mathematical truth and, if it could not, what sort of truth it then was. Kant replied with the treatise, *Untersuchung über die Deutlichkeit der Gründsätze der natürlichen Theologie und der Moral* (1762, published in 1764), which struck a decisive blow to Wolff's philosophical teachings.[33]

Kant had been influenced by Wolff's philosophy, but had never

31. Nils Wallerius, *Prænotiones theologicae*, 6 vols. (Uppsala, 1756), e.g., 13f.

32. *Upsala Ärkestifts Herdaminne, 1* (Uppsala), 426–7; Torbern Bergman, *Åminnelse-tal öfver Nils Wallerius* (Stockholm: L. Salvius, 1765), 24.

33. Reprinted in Kant's *Gesammelte Schriften* (Berlin: G. Reimer, 1902–), *2*. Cf. G.G. Fülleborn, *Zur Geschichte der mathematischen Methode in der deutschen Philosophie*, in *Beyträge zur Geschichte der Philosophie, 5* (Züllichau and Freystadt: F. Fromann, 1795), 108–30; Giorgio Tonelli, "Der Streit über die mathematischen Methode in der Philosophie in der ersten Hälfte des 18. Jahrhunderts und die Entstehung von Kants Schrift über die 'Deutlichkeit'," *Archiv für Philosophie, 9* (1959), 37–66; Hans-Jürgen Engfer, "Zur Bedeutung Wolffs für die Methodendiskussion der deutschen Aufklärungsphilosophie: Analytische und syntetische Methode bei Wolff und beim vorkritischen Kant," in W. Schneider, ed., *Christian Wolff*, 48–65.

followed it slavishly. The treatise of 1762 marked his rejection of the mathematical ideal in philosophy. In it Kant draws his well-known distinction between analytic and synthetic propositions, and concludes that metaphysics is not synthetic like mathematics but rather analytic. The synthetic structures of mathematics cannot be transferred to philosophy and its "mathematical method," for philosophy, unlike mathematics, does not have at its disposal definitions and axioms from which to proceed methodically. Application of the mathematical method to the field of philosophy had thus given rise to errors and mistakes: no dogma in philosophy could be likened to the definitions and axioms of mathematics.

Kant's criticism swept away the very foundations of Wolff's influence. Wolff's mathematical method depended on the possibility of applying procedures of mathematical proof to all philosophical and theological questions. Now Kant argued that the basic structures of mathematics and philosophy were different. Wolff's system crumbled.[34]

Several conclusions emerge. Wolffian thinking became a fashionable philosophy in the first half of the 18th century, and its exponents and detractors both saw the mathematical method as the essential element of his doctrine. It is important to distinguish between form and content in our analysis of the Wolffian mathematical method, and to recognize that Wolff's philosophy fulfilled different functions in different situations. While "rationalism" and "mathematical method" were popular rhetorical flourishes in the 18th century, they also stood for a well-defined means of attaining certain intellectual objectives. Both friends and enemies of the Enlightenment appropriated that method, since mathematics as a method of proof promised results in all areas of human knowledge. That it collapsed so readily under Kant's attack reflected that Wolff's mathematical method had already proved inadequate: it failed to provide tools for revamping critical philosophy, or for creating instrumentalist science, or for solving practical problems facing the bureaucratic states of the late 18th century. The form and content of mathematics itself, however, would continue to speak to these needs.

34. Cf. Raffaele Ciafardone, "Von der Kritik an Wolff zum vorkritischen Kant: Wolff-Kritik bei Rüdiger und Crusius," and Norbert Hinske, "Wolffs Stellung in der deutschen Aufklärung," in ibid., 289–305, 306–20, resp.

2 The Broken Circle

By Gunnar Broberg

The word "quantitative" applies to natural history during the second half of the 18th century in two distinct but related ways: as characteristic of the object and of the method of study. As for object, the sheer number of known and estimated forms forced new approaches to the storage and retrieval of information; as for method, these new approaches were instrumentalist and, in the dominant system of Linnæus, mathematical. These features—the overwhelming flow of information and the determination to inventory and survey it for useful purposes—characterize much of the learned activity of the late Enlightenment. In what follows, I first bring Ernst Cassirer's opposition between *l'esprit de système* and *l'esprit systématique* to bear, and indicate the scope of the problem of order faced by the encyclopedists of the Enlightenment. I then turn to natural history, reporting the inflation of information and some of the problems of keeping it together. Finally, I look at the paradoxes of natural history at the turn of the century.

ENCYCLOPEDISM AND ORDER

The classificatory project of 18th-century natural history runs parallel to and was in many ways inspired by the development of the encyclopedic enterprise of the same period. Natural historians and encyclopedists shared similar ambitions for totality, coherence, and order. Both camps had to ask the same questions: How much is there in the world to know? Is it proper to measure all the riches of creation? If so, by what "method" should they be described? Is there a correspondence between what we can know and the fundamental nature of things? Parallel to the rapid increase of information came the epistemological shake-up of the early Enlightenment. Natural history as well as general encyclopedism moved from the finite world of the baroque to the infinite world of modern times.[1] If "encyclopedia" means "the circle of knowledge," the circle was broken during the latter part of the 18th century. Natural history opened out similarly.

In following this development we shall find it useful to adopt a distinction d'Alembert employed in his introduction to *La grande encyclopédie*. By *esprit de système* he meant the old a priori physics connected with scholasticism, rationalism, and deduction, the sort of thing that Christian von Wolff did in the first half of the 18th century.[2] *Esprit systématique* was connected with experiments, empiricist ideals, and a readiness to go down the toilsome path from the Lockean world; travelers on this journey had to accept a demarcation between what we can know and what we cannot know, between physics and metaphysics, and to try not to mingle both sides. But that is what often happened. As Cassirer wrote in *The philosophy of Enlightenment*: "This difference in mode of thinking [between *esprit de système* and *esprit systématique*] does not mean a radical transformation, it amounts merely to a shifting of emphasis. This emphasis is constantly moving from the general to the particular, from principles to phenomena. But the basic assumptions remain..., the self-confidence of reason is nowhere shaken. The rationalistic postulate of uniformity dominates the mind of this age."[3]

The 18th century manifested its rationalism in encyclopedias. Natural history helped to determine their structure. The Linnæan encyclopedia had one of its roots in the religious idea of Creation as the Book of Nature opened to the naturalist-reader. The same ambition to take possession of what had been discovered, indeed conquered, characterized both the *Systema naturae* and the great French

1. Marjorie Nicolson, *The breaking of the circle: Studies in the effect of the "new science" upon seventeenth-century poetry*, rev. ed. (New York: Columbia University Press, 1960).

2. See Tore Frängsmyr, chap. 1 in this volume.

3. Ernst Cassirer, *The philosophy of the Enlightenment* (Princeton: Princeton University Press, 1951), 22f.

Encyclopédie. It was not mere coincidence that d'Alembert's and Diderot's great work began at the same time that Georges Louis Buffon's *Histoire naturelle* and the geodetic survey of France started. In his *Discours préliminaire*, d'Alembert alluded—and most appropriately—to the new success of natural history.[4] Each side could profit from the other. Linnæan natural history, with its clear structure, universal language, and simplicity, made a good model for encyclopedias. Its rationality and practicality gave it entrée everywhere.

According to Zedler's magnificent *Grosses vollständiges Universal-Lexicon aller Wissenschaften und Künste* (1732–54), an encyclopedia is "a total idea of all sciences, which the ancients brought together so as to show the relationship they have to one another."[5] Zedler settled for something less. Both he and Ephraim Chambers, the author of the *Cyclopedia* (1728), arranged their works not systematically but alphabetically, though, to be sure, not without long discussions about the systematic aspect: "It may be even said, that if the system be an improvement upon the Dictionary, the Dictionary is some advantage to the System; and that this is perhaps the only way wherein the whole circle, or body of knowledge, with all its parts and dependencies, can well be delivered." Chambers, who was influenced by Locke and Newton, went on to explain the perfection that characterizes a classical work in contrast to a modern one. The classic had so little to compare itself to that it could devote all its attention to order and its own perfection, whereas the latter—"since experiments are endless"—will never reach so far. "To have philosophy complete, we should have the order, precision and distinctness of the old; and the matter and *copia* of the new."[6] Chambers opposed order to abundance of material.

The opposition Chambers developed was connected with the decline of rationalistic systems of sciences in the Leibniz-Wolff

4. Jean Le Rond d'Alembert, *Inledning till Encyclopedien* (Stockholm: Carmina, 1981), 84.

5. Johann Heinrich Zedler, *Grosses vollständiges Universal-Lexicon aller Wissenschaften und Künste*, 64 vols. and 4 supplements (Halle and Leipzig: J.H. Zedler, 1732–54), *8* (1734), 1138.

6. Ephraim Chambers, *Cyclopedia*, ed. Rees (London, 1778; original ed., London: J. & J. Knapton, 1728), "Preface," xxii.

tradition, which were especially important in Germany from 1720 to 1765.[7] These often extremely elaborate maps of knowledge lost much of their appeal when examined in the light of English empiricist and French sensualistic epistemology, more precisely the French *Encyclopédie* and later Kant's criticism. The *Encyclopédie* has often been taken as the starting point for a new concept of knowledge because of its antimetaphysics, its updated information, its optimism, its democratic call for action, and many more things, including its empiricism.

The *Encyclopédie* was a joint adventure with many collaborators and two principal editors, Diderot and d'Alembert, who did not agree about everything. D'Alembert's famous *Discours préliminaire* stressed the "order and connection of human knowledge," the inner order and harmony of the sciences, and the possibility of making a uniform system of encyclopedic knowledge. It was not clear, however, how he supposed the tree of knowledge to grow—did logical division or historical development control the branching? His ideals were those of a geometer, believing in a regular universe and operating with axioms and principles. "If somebody could survey the whole universe from one standpoint, it would be seen as consisting of one single fact and one single truth."[8] Alphabetical order does not necessarily invalidate coherence, since cross-references could link the various subject matters from a synthetic point of view. How well these combinations in the end corresponded to the tree of knowledge was problematic.

Diderot's article "Encyclopédie" (1755) gave a truer picture of the purposes of the *Encyclopédie*. It responded to accusations that the first volumes did not have the coherence promised in d'Alembert's essay: "It is impossible to improve the arbitrariness of this great original profusion. The universe presents us only with individual things, infinite in number and with almost no fixed and determined division [among them]; none of them can be called the first or the last; everything is connected to everything else by insensible gradations." Later

7. Giorgio Tonelli, "The problem of the classification of the sciences in Kant's time," *Rivista critica di storia della filosofia, 38* (1975), 243–94.

8. D'Alembert, *Inledning*, 57.

Diderot insisted on the value of irregular articles. "The formation of an encyclopedia is like the foundation of a great city." He compared encyclopedic order with a machine, the parts of which fit together, but which can also be assembled in a completely new way.[9] Rather than being the organized accumulation of the total stock of human knowledge, the cross-referenced *Encyclopédie* would be an open-ended conversation among members of a *cité scientifique*.

Diderot's dynamism would become even more pointed in the *Rève d'Alembert*, in which he defended poetry against geometry.[10] That did not mean total hostility toward all "systems," only metaphysical and deductive ones; acceptable systems should act as heuristic devices and hypotheses and as pedagogical tools. When Diderot defended this openness in contrast to closed systems, he also defended the alphabetical arrangement, which predominated thereafter. This arrangement represented an important step toward secularization of knowledge and agnosticism toward taxonomies of knowledge. To quote a recent interpretation: "As the zero degree of taxonomy, alphabetical order authorizes all reading strategies; in this respect it could be considered an emblem of the Enlightenment."[11]

The encyclopedic business thrived on this tension between knowledge and information. The first edition of *Encyclopedia Britannica* (1768–71) was more weighted toward information, whereas the success of the Brockhaus *Conversationslexicon* provoked charges of superficiality.[12] Hegel used the same word as Kant did for natural history works, *Aggregaten,* to describe the state of contemporary

9. Diderot, "Encyclopédie," *Oeuvres complètes,* 7 (Paris: Hermann, 1976), 174–262, quotations on 210, 218.

10. John Pappas, "L'esprit de finesse contre l'esprit de géométrie: Un débat entre Diderot et d'Alembert," *Studies on Voltaire and the eighteenth century, 89* (1972), 1229–53.

11. Charles Porset, "Figures de l'encyclopédie," in Christiane Mervaud and Sylvain Menant, eds., *Le siècle de Voltaire. Hommage à René Pomeau* (Oxford: Voltaire Foundation, 1987), 719–33, on 719.

12. See also the mild contempt in the commentary of the Swedish classicist Adolf Törneros (ca. 1820): "Lexica, e.g., Conversations-Lexica, are like grocery stores [krambodar], where everybody can fetch his temporary or minute [timmeliga eller minuteliga] needs, like snuff for two pence." Aphorism 84, in *Brev och dagboksanteckningar*, ed. Nils Afzelius (Stockholm: A. Bonnier, 1962), 27.

encyclopedias. To Hegel, an encyclopedia should be a summarized introduction to scientific knowledge. An heroic attempt to combine these different traditions, literally identified with the heritage from Francis Bacon and Plato, is found in the "General introduction" to the *Encyclopedia metropolitana* (1817). The objective is no less than a reconstruction of "the circle of knowledge in its harmony" to provide that "unity of design and of elucidation, the want of which we have most deeply felt in other works of similar kind, where the desired information is divided into innumerable fragments scattered over many volumes, like a mirror broken on the ground, presenting instead of one, a thousand images, but none entire."[13] But the future belonged to fragments and to Brockhaus's *Aggregaten.*

Whatever its organization, an encyclopedia required supplements to stay current.[14] A popular work could be updated, commented on, and augmented; but in principle an encyclopedia should stand unaltered. The editors of the *Encyclopédie* failed in their initial plans to translate Chambers' *Cyclopedia.* Too much had to be altered or added. A two-volume *Supplement* to Chambers' work was published in London in 1753. About the same time a new series of supplements began to complete the already published sixty-four volumes of Zedler's *Lexicon,* and Felice's *Encyclopédie, ou Dictionaire universel raisoné des connaissances humaines,* received six volumes of supplements in 1775–6. Four supplementary volumes completed d'Alembert's and Diderot's great work in 1776–7. This constant adding made an encyclopedia something like a journal. Diderot wrote to the point. "An encyclopedia is a rapid and disinterested exposition of mankind's discoveries in all places, of all types, and in all times, without any judgment about people. Journals are only a passing account of books and authors."[15] Still, the comparison was possible,

13. Samuel Taylor Coleridge, "On the science of method," *Encyclopædia metropolitana* (London: B. Fellowes, 1817–45), *1,* 42.

14. See examples listed in Giorgio Tonelli, *A short-title list of subject dictionaries of the sixteenth, seventeenth, and eighteenth centuries as aids in the history of ideas,* Warburg Institute, Surveys, 4 (London: Warburg Institute, 1971).

15. Diderot, "Encyclopédie," 241.

and in a sense the encyclopedia and the newspaper were getting closer to one another.

The increase of information that perplexed encyclopedists also vexed lexicographers. In his *Plan for an English dictionary* (1747), Dr. Johnson declared it to be his purpose to "fix" the language once and for all, because "all change is of itself evil." By the time his *Dictionary* appeared in 1755, Johnson had shifted his position. He now thought that "fixing" the language was as futile an effort as "embalming" it.[16] In the interim he had put in much hard work to collect words and to define them, work similar to what natural historians tried to do in the species jungle. He too had to deal with the problematic relation between words and things. "I am not yet so lost in lexicography as to forget that words are the daughters of earth, and that things are the sons of heaven." Thus he inclined to accept Locke's criticism of essentialism, although as a lexicographer he could not be happy about the instability of meanings. "Definition is, indeed, not the province of man; everything is set above or below our faculties," he sighed on another occasion.[17] The lexicographer had to make the best of the situation, and describe words, not define them; and so Johnson proceeded.

Emanuel Swedenborg also had something to say about the growing awareness of never-ending knowledge. In an undated note from the middle of the century he wrote: "I was allowed to enter a library where there was a great number of books....In the inner parts of still more libraries were books written by adherents of the old churches and still farther were books from the oldest....The libraries had many departments according to the training of the students. There were also many other libraries in heaven, but not public ones. Further on there were still more libraries likewise divided into

16. Cf. James H. Sledd and Gwin J. Kolb, *Dr. Johnson's dictionary* (Chicago: University of Chicago Press, 1935); Tom McArthur, *Worlds of reference* (London: Cambridge University Press, 1986).

17. Johnson, quoted in James McLaverty, "From definitions to explanations. Locke's influence on Johnson's Dictionary," *Journal of the history of ideas* (1986), 277–94, on 291. "Preface to the dictionary," in Howard D. Weinbrot, ed., *New aspects of lexicography* (London and Amsterdam; Carbondale: Southern Illinois University Press, 1972), 73–94.

departments. They were as many as there are research areas."[18] We meet the famous spiritualist and one-time scientist, the inspiration of Borges' story of the libraries of Babel and of Eco's monastery; we recognize the principle of plenitude, and also of order. The multitude is overwhelming, but chaotic only to those who have not found—like Swedenborg and his fellow countryman Linnæus—the general patterns of reality.

The relation between order and plenitude is a constant problem in librarianship. A survey of German *Sachkatalogen* from the 17th to the 19th centuries shows a change from hierarchical systems to alphabetism. Gabriel Naudé, who introduced the concepts of systematic and alphabetic catalogues in the early 1600s, compared a library to an army. Strict order must rule in both. During the second half of the 18th century the Göttingen school of librarianship dominated. Its head, professor of medicine Georg Matthiae, stressed the necessity of arranging the library into "classes" according to a "natural" system. The systematic catalogue should be given priority over an alphabetic one since the coherence of books provides the basis of a scientific arrangement; considerations of size, form, and so on, mark the simple and unscientific librarian. Some disagreed: "Pure arrangement of books by their sciences is almost impossible and by no means necessary. The place a book stands is of no importance."[19] The librarian's main problem was to find room for new books on old shelves; but, like natural historians, librarians wanted to raise their professional status by invoking order, classification, and science. The reverse analogy also held. Linnæus once wrote: "Gardens are like living libraries of plants." The garden and the library shared the same problem: growing knowledge had made systems cumbersome.[20]

18. Emanuel Swedenborg, *Diarium spirituale*, ed. J.F.I. Tafel (London: W. Newbery, 1844), no. 5999.

19. Quoted in Harald Kleinschmidt, "Von System zur Ordnung. Bermerkungen zu Bewertungen von Sachkatalogen vornehmlich im 18. und 19. Jahrhundert," *Libri, 37:2* (1987), 113.

20. Linnæus, *Demonstrationes plantarum in horto upsaliensi* (Uppsala, 1753), sect. 1. In what follows, I treat questions similar to those addressed by Michel Foucault in *Les mots et les choses* (Paris: Gallimard, 1966) and by Wolf Lepenies in *Das Ende der Naturgeschichte* (Munich: C. Hanser, 1977): both contrast the classifying *episteme* of natural history with the new biology. The emphasis here is on the dynamics of fact accumulation and on the interplay between two models of knowledge

THE CHALLENGE OF PLENITUDE

The plenitude of being was a constant cause for religious as well as scientific amazement. John Ray's famous *The wisdom of God manifested in the works of creation* (1691) opens by quoting Psalm 104:24—"How manifold are thy works, O Lord! In wisdom hast thou made them all"—and continues by changing the exclamation into a question. Ray called attention to the number of stars and planets, then calculated the number of animals. His estimate that there are 2,000 insect species in Britain alone, and 20,000 in the whole world, must have seemed very bold at the time, and likewise the figure 40,000 to 45,000 for all plants and animals together. He also discussed the proportions between the number of species, showing that the less perfect genera contain more species than the more perfect, a matter he went on to explain with many insights into the interplay of nature. To him, as to many of his contemporaries, the microscope had proved the plenitude of creation. We should note a typical 17th-century conceit: the greater the number of variations, the greater the glory of Man, since the Lord made them for human pleasure. As Ray put it: "another reason why so many kinds of creatures were made, might be to exercise the contemplative Faculty of Man; which is nothing so much pleas'd as in the variety of objects."[21]

Compare Ray's numbers with the estimates given by Pieter van Musschenbroek half a century later. Following Hermann Boerhaave's lead, Musschenbroek reckoned the number of plant species at 13,000. He allowed 5 insect species to each plant. This made 65,000 insects

handling—the encyclopedia and taxonomy in natural history. Recently Vernon Pratt has argued that the 18th-century obsession with taxonomy should be interpreted as a fascination with the order of nature as such rather than with animals and plants per se: "System-building in the 18th century," in J.D. North and J.J. Roche, eds., *The light of nature* (Dordrecht, 1985). As Pratt himself admits, he risks overemphasizing the point. The system-builders—John Ray, Tournefort, Michel Adanson, and Linnæus—were indeed trying "to articulate the order of nature," but we should not undervalue the fact that they were scientists working in the Baconian tradition. Certainly they would not recognize themselves in the mirror of modern historiography.

21. John Ray, *The wisdom of God manifested in the works of the Creation* (London: Samuel Smith, 1691), chap. 1. Cf. William T. Stearn, "John Wilkins, John Ray and Carl Linnæus," Royal Society of London, *Notes and records, 40:2* (1986), 101–23.

and (he guessed) 72,500 species in the entire animal kingdom—or, rather, twice that many, since each animal serves as food for another. Suddenly, the number has reached 145,000, and, again, the figures should be doubled since so little of the earth had been explored. We have in all 291,000 species, magnificent proof of the skill of the creator. Musschenbroek referred to Antony van Leeuwenhoek, to Henry Baker's book on microscopy, and to René Antoine Réaumur's entomology. It may be objected that Musschenbroek, a professor of physics, reasoned too much like a mathematician to win the confidence of the natural historian. Still, both had recourse to physicotheology.[22]

At first, Linnæus probably did not estimate nature's multitude as extravagantly as had Ray. He inherited the task of describing the 6,000 species in Caspar Bauhin's *Pinax* (1623) from his predecessors, Olof Rudbeck, father and son, who planned to publish pictures of every known plant. (Their project was stopped by the great fire in Uppsala in 1702.) Anyone shouldering such a task would try to keep the numbers down. According to Linnæus, botanists should be able to remember at least all genera. Hence this second Adam, as Albrecht von Haller called him, must have believed it possible to survey nature taxonomically. In his *Oratio de telluris habitabilis incremento* (1743), he supposed that in the beginning the world was a paradise island containing all species.[23] The only thing that had happened since then was the constant increase of individuals—and land.

The number of new discoveries during the 1740s forced Linnæus to change his mind. He had to fit these novelties into the context of his already published textbooks. The characters of the fresh-water polyps and the corallines, and later different fungi, seemed to contradict the taxonomical borderlines of nature. Critics labeled the

22. Pieter [Petrus] van Musschenbroek, *Oratio de sapientia divina* (Leiden: S. Luchtmans, 1744). Cf. Maria Roosenbom, "Petrus van Musschenbroek's 'Oratio de sapientia divina,'" in P. Smith and J.Ch.V. ter Laage, eds., *Essays in biohistory [presented to] Frans Verdoorn* (Utrecht: International Association for Plant Taxonomy, 1970), 177–94.

23. Tore Frängsmyr, "Linnæus as a geologist," in Tore Frängsmyr, ed., *Linnæus, The man and his work* (Berkeley: University of California Press, 1983), 110–55. Formally the *Oratio* is an exposition of sect. 134 of *Fundamenta botanica* (1736) on the fixity of species. "We count today as many species as were created in the beginning."

sexual system scholastic and artificial and contrary to the notion of nature's great chain. As an answer, Linnæus sketched an evolutionary hypothesis in *Genera plantarum* (1764), which also expressed his views about the natural system of plants. The more Linnæus peeked into the secret depths of nature, the more he found intricate connections. He now preferred the concept of *mappa naturae* over the simple hierarchical arrangement of the *scala naturae*.[24]

Still more material flooded into Linnæus' study. The *Systema naturae* expanded from twelve folio pages in the first edition to almost 2,000 pages in the twelfth. He classified some 7,700 plants, 6,200 animals, and 450 mineral species. Altogether Linnæus' nature contained no more than 15,000 species, but that, as he knew, was but a beginning. *Systema naturae* goes from God to gravel, aiming at nothing less than a complete inventory or encyclopedia of the world. To Linnæus, nature's multitude must have been both a wonderful ongoing revelation and a terrible curse. He must have been one of the greatest optimists of all time. His library, now in the Linnæan Society of London, is filled with volumes of additions scrawled in the margins of his books.

Did Linnæus ever indicate the possible "size" of nature's kingdoms? Perhaps. In the fifth edition of *Genera plantarum* (1754), he introduced the idea of an alphabet or letters of the botanical language. These 26 letters combined two at a time were to cover all parts of the flower with the help of four "certain and mechanical principles," namely number, position, figure, and relative position. By a combination of these characters, botanists could give a good description of the genera, which in difficult cases should be checked by the concept of "habitus," the over-all impression of the plant. The largest possible number of genera is 26 × 26 × 4 or 2,684. Linnæus' evolutionary hypothesis of 1764 seems to end up in a similar number. At the beginning God created the "natural orders," which Linnæus elsewhere counted as 58. The next level, of the genera, came into existence by "mixing" the orders two by two: 58 × 58 makes 3,364

24. See Gunnar Broberg, "Linné's systematics and the new natural history discoveries," in John Weinstock, ed., *Contemporary perspectives on Linnæus* (Lanham: University Press of America, 1985), 153–81.

natural genera. The third step, another multiplication by 58, gives us the number of possible species "mixed" by nature. There is also a fourth step, when "casus" or change mixes the species, which constantly happens and to such an extent that the botanist should not bother about all the resultant "varieties." Thus, the successive unfolding of types is caused by God, nature, and chance. The orders and the genera deserve special study. To keep up with the rest will be almost impossible. Here Linnæus seems to have retreated from his earlier optimism about human possibilities.[25] If his reasoning seems odd, consider this from his fellow Swede Christopher Polhem, the great technical innovator. A descriptive name can be constructed by using letters in a prescribed order. For instance, an initial "p" could mean man, a "b" animal, "k" grass, while vowels might stand for the different senses, "a" for sight, "e" for hearing, "i" for smell. Further letters could indicate medical effects. With this system, Polhem guaranteed 1,406,250 possibilities; "more differences should not exist." If they do, it would be easy to add one or more variables or letters.[26]

Linnæus' critic, Buffon, had to face the same problems. At the outset of his series on birds, he admitted a problem in acquiring all existing species. He and his collaborators worked for twenty years, constantly enlarging the collections of the Cabinet du Roi but never approaching completeness. Still, he could offer three times as many species of birds as were found in the tenth edition of *Systema naturae* (1758). (These figures depended on species definition and are not strictly comparable.) There are two competing strains in Buffon's writing: his early fascination with individuality in nature and his later encyclopedic ambitions. These tensions are even more obvious in the writings of Linnæus' second major French critic, Michel Adanson. Adanson faulted Linnæus for, among other things, grossly

25. *Genera plantarum*, 6th ed. (Stockholm: L. Salvius, 1764), unpaginated. Cf. C.E.B. Bremekamp, "Linné's views on the hierarchy of the taxonomic groups," *Acta botanica nederlandica, 2* (1953), 242–53. Linnæus' cortex-medulla theory probably put limits to possible change, since he considered the medulla as solely responsible for producing new properties.

26. Christopher Polhem, *Efterlämnade skrifter*, 4 vols. (Uppsala: Almqvist & Wiksell, 1947–54), *4*, 333–51.

underestimating the number of species of *coquillages* in Senegal.[27] We will return to Adanson. Here he exemplifies awareness of nature's tropical abundance, while Buffon represents the contention that this abundance cannot be counted.

Linnæus and Buffon both received credit in an unusually unbiased way in Haller's preface to the German translation of the *Histoire naturelle* (1750). In view of his old quarrel with Linnæus about who was the better naturalist and who had collected more plants, this sympathetic judgment might seem surprising. It may be explained by the message of the essay, that is, the blessings of hypotheses for the progress of science. "Laws in botany are arbitrary," Haller stated, "but they have done us an unbelievable service. Now we can distinguish ten thousand plants more easily than the Ancients did their six hundred." Linnæus' new theory performed the greatest service. At present, botany was advancing its lead over all sciences. "Not only is it nearest to completion, and has little by little determined the nature of almost all its classes and resemblances; it has [also] spread its laws throughout the entire Kingdom of Nature. Zoologists and mineralogists have received their laws from it and accepted them as the Romans did from the Areopagus." This last allusion is not a poor piece of flattery. Haller certified that botany would soon be finished, that all the work of collecting material and its description would be brought to their happy end. Success was guaranteed by the model he gave for natural history research: just as a land surveyor begins a map by determining some locations without the positions of places in between, the natural historian starts off with an initial sketch, which later knowledge will fill. The analogy is typical for an era of geographical exploration. It hardly implies the notion of an infinite or a dynamic nature, but rather only the hope of completing the process of exploration.[28]

27. Michel Adanson, *Histoire naturelle du Sénégal* (Paris: C.-J.-B. Bauche, 1757), xxvii.

28. Haller, quoted in John Lyon and Philip R. Sloan, eds., *From natural history to the history of nature. Readings from Buffon and his critics* (Notre Dame: University of Notre Dame Press, 1981), 297ff.

A similar promise of prompt success (if only governments and princes would offer enough financial support) appeared in an article with the upbeat title, "Have we still to hope for a complete system of nature?" The author, entomologist Johan Samuel Schröter, did so hope, provided natural historians adopted his rules. In Schröter's scheme, only one person should be responsible for each major natural group and should be endowed with dictatorial powers to suppress quarrels among naturalists. Schröter held that only a group of specialists in natural history, and not a single person alone, could fulfill the Linnæan task. Other necessary preconditions were "ability, leisure, and money."[29]

The most ambitious attempt of the 18th century to secure exact figures for the contents of the natural world, however, was made by Eberhard Zimmermann in his book, *Geographische Geschichte des Menschen und der allgemein vierfüssigen Thiere* (1778–83). Zimmermann combined macroscopic information from the explorers with inspiration from the microscopists' overfilled world. As professor of mathematics—he taught Gauss at Braunschweig University—he excelled in calculations about the size of the known world in relation to the number of known animals.[30] These ratios show an impressive knowledge of zoological detail but also have strong similarities with the more unrealistic parts of political arithmetic as well as with Musschenbroek's calculations. Zimmermann vigorously pursued the Linnæan project, and equally vigorously attacked Buffon's "complete ataxia." But Zimmermann's calculations made the works of the Lord too manifold for the grasp of His creatures.

Zimmermann began small, with minerals, whose maximum number he put at 750. The variety of plants is much greater; recent estimates reached to almost 30,000 known species, only a fraction of the total quantity, which Zimmermann by various calculations raised

29. Johan Samuel Schröter, "Haben wir noch ein vollständiges System der Natur zu hoffen," *Berlinische Sammlungen, 2:1* (1770), 249–71, *3:1* (1770), 353–75, quotation on 263.

30. On Zimmermann, see F.S. Bodenheimer, "Zimmermann's Specimen zoologiae geographiae quadrupedum," *Archives internationales d'histoire des sciences,* Oct–Dec 1955, 251–7.

to 175,000. He showed even more interest in the number of animals. Erxleben, Schreiber, Johann Friedrich Blumenbach, and others had set the quantity of known species at 10,000. Following Roesel von Rosenhoff and Linnæus (*Pandora insectorum,* 1758), Zimmermann proposed that there must be at least five insect species specialized on each plant species. Immediately we have $175{,}000 \times 5$ or 875,000 insect species of which only 51,000 had been described; entomologists had work to do. Nature abounds with parasitic and intestinal worms, with polyzoa and infusoria, and the sea is crammed with life to an extent almost impossible to grasp. "Who can follow the flying fish?" Probably the variety of sea organisms is two or three times that of land organisms. Zimmermann provided for invisible organisms, small and pellucid, on which the insects feed, and also for atmospherical fauna. Bearing all this in mind, he reached seven million species of animals, perhaps more. Then there are the innumerable specimens of each species. In the universe our planet is just a small spot unknown even to the inhabitants of Jupiter. And yet it has such multitude and magnificence! Zimmermann ended his essay with "a paean of thanks" to the Creator, the traditional natural-theological *envoi.* Following the natural-philosophical tradition as well, he offered calculations of the number of mammals undiscovered in different parts of the world, but despite the obvious interest of these estimates for the history of biogeography we leave them out.[31]

This account of Zimmermann's work sums up several of the features of late Enlightenment natural history. He enthusiastically accepted the Linnæan project of species-hunting. Several times he appealed to princes for support for its completion. He was also intoxicated by the idea of nature's plenitude. As Ray did a century earlier, he compared the plenitude of this world with the plenitude of the universe. But we must note the enormous difference in the numbers they gave. And Zimmermann turned the whole architecture of the chain of being upside down. Whereas Ray (and Linnæus in *Oratio,* 1743) claimed that the ratios between nature's kingdoms were such

31. Eberhard August Wilhelm Zimmermann, *Geographische Geschichte des Menschen und der allgemein vierfüssigen Thiere,* 3 vols., *3* (Leipzig: Weygand, 1783), 48.

that the lowest—the mineral—was more numerous than the vegetable, the vegetable more than the animal, Zimmermann proudly formulated a different and "remarkable law: the total of types of organized bodies increases with the degree of sensation and life."[32] Throughout Zimmermann supported the idea of the chain of being, but obviously that notion was collapsing—so to speak—under its own weight. Zimmermann's discussion reflected many of the components of 18th-century natural history, but his figures neither fit the Linnæan project nor supported the single-line chain of being. His account did not visibly change the direction of natural history, but it forced two themes to the surface: the dramatic growth of numbers and the pessimistic prospects for complete knowledge.

So much for external problems; the Linnæan system has internal ones as well. It was fundamentally mathematical in a very simple way, based on the numerical principles of the sexual system as well as on zoological taxonomy. The naturalist counted stamens and pistils, teeth and nipples, toes, scales, antennas, everything. Among other advantages, mathematics afforded brevity and exactness. Linnæus was in every respect an economical man, who had to budget both time and space for his enormous undertaking to map the whole natural world. Without the Linnæan reforms the natural historian would not have been successful in surveying nature; with them, however, nature might appear to be one long list of numbers.

It was reasonable to question the validity of the criteria. Buffon did, in his famous criticism of Linnæus in the "Premier discours" of the *Histoire naturelle* in Buffon (1749).[33] There Buffon argued that species as well as higher taxa are simply constructions of the taxonomist, while nature by contrast only consists of individuals. For Buffon, the counting of stamens and pistils has nothing to do with the true study of nature. He found no room for mathematics in natural history, the more remarkable considering his background in

32. Ibid., 22.

33. Buffon, "De la manière d'etudier et traiter l'histoire naturelle," in Buffon, *Oeuvres philosophiques*, ed. Jean Piveteau (Paris: PUF, 1954), 7. See also Wolf Lepenies, *Das Ende der Naturgeschichte*, and Philip R. Sloan, "The Buffon-Linnæus controversy," *Isis*, 57 (1976), 356–75.

that discipline. Nor did he use numbers to reason about nature's continuity. Indeed, he did not stress continuity in his writings as strongly as is generally thought. He of course made references to the great chain of being; but he rejected the idea of animal reason and opposed linking human and animal via the ape as a link to man. At bottom, however, what makes Linnæan taxonomy impossible in Buffon's eyes is that the number of natural "groups" must be infinitely great, since the only natural entity is the individual. The only way to treat nature, the only "method" or plan in the *Histoire naturelle*, is to accept subjectivity and to arrange descriptions according to how we normally get to know the animals.

In many ways Buffon adhered to the observer tradition, as did the master entomologist Réaumur, who advocated a similar concentration on the individual in contrast to more or less complete taxonomic surveys. Since we never will know everything, Réaumur wrote, we should not regret that some thousands of insects do not figure in our inventories. Instead, we should concentrate on the truly interesting wasps and butterflies. Observations rather than coherent systems are what matter.[34]

In another line of attack, Linnæan mathematics was met with more mathematics. Adanson attacked the essentialistic and scholastic elements in Linnæan taxonomy and recommended an overall statistical evaluation of the plant's character, a method appropriate for modern computers.[35] It is uncertain how far he himself used his method. It is not only extremely time consuming, it also stands in strong contrast to his predilection for completeness and universality. Thus Adanson's suggestions, although in many ways interesting, were a dead end in practice. Practical taxonomy could function only in connection with some sort of essentialism, while nominalism seemed to belong more to philosophy.

34. René-Antoine Ferchault de Réaumur, *Mémoires pour servir à l'histoire des insectes, 1* (Paris: Imprimerie Royale, 1734), preface. In reference to the contrast between "observers" and "taxonomists," see Jacques Roger, "The living world," in G. Rousseau and R. Porter, eds., *The ferment of knowledge* (Cambridge: Cambridge University Press, 1980), 255–83.

35. See Peter H.A. Sneath, "Mathematics and classification from Adanson to the present," in *Adanson. The bicentennial of Michel Adanson's 'Famille des plantes'*, 2 vols. (Pittsburgh: Hunt Botanical Library, 1963–4), 2, 471–98.

The philosophical core of these objections and of 18th-century natural history in general was the idea of the great chain of being. In his thematic study, Lovejoy singles out the species problem, the interest in missing links, and the studies of the microscopists, all of which involved philosophical problems such as continuity and causality. By the end of the century, the idea had suffered many changes. Lovejoy emphasizes what he calls the temporalization of the chain, brought about by the study of fossils.[36] More might well have been said about *continuity*. In general, both philosophers and natural historians adhered to the Cartesian-Leibnizian idea of continuity without any leaps or bounds.[37] Action by contact keeps the world together and makes it go around. The Newtonian model of a void universe operating by distant forces did not appeal to natural historians. If there is continuity along the whole chain, then it should be possible to use the same classificatory devices and characters to judge all organisms. But, as Charles Bonnet objected, humans lack the right perspective and must satisfy themselves with "classifying classes," with specializing within fairly broad limits.[38] Linnæus could stretch his scheme through botany only with difficulty—the sexual system was hardly useful for the cryptogams—and new principles were invoked for the other kingdoms of nature. Natural historians could see only fragments of the great chain before it twisted or disappeared from sight.

The related problem of *plenitude*, so beautifully demonstrated by the observations of the microscopists, also meant continuity along the

36. Arthur O. Lovejoy, *The great chain of being* (Cambridge, Mass.: Harvard University Press, 1936); chaps. 6–9. See also A. Vartanian, "Trembley's polyp, La Mettrie, and eighteenth-century French materialism," *Journal of the history of ideas, 11* (1950), 259–86; Herbert Thienemann, "Der Stufenfolge der Dinge," *Zoologische Annalen, 6* (1910); Philip C. Ritterbush, *Overtures to biology. The speculations of eighteenth-century naturalists* (New Haven: Yale University Press, 1964); William T. Bynum, "The great chain of being forty years after," *History of science, 13:1* (1975), 1–28.

37. Francesca Rigotti, "Biology and society in the age of Enlightenment," *Journal of the history of ideas, 47* (1986), 215–34.

38. See Lorin Anderson, *Charles Bonnet and the order of the known* (Dordrecht and Boston: D. Reidel, 1982).

chain. Its main impact on classification, however, was the implication that God created an unlimited number of individuals rather than "types" or species. The vast numbers that came into being through the laws of generation were studied by Leeuwenhoek and other animal demographers.[39] How to impose limits on this continuity? The question was similar to Zeno's paradox and, if taken seriously, could have led to taxonomy's suicide. The 18th-century natural historian had to deny the physicotheological premises that underpinned his work. To be sure, some very straightforward observations could strengthen him in his denial. As Voltaire wondered shrewdly, why is there no link between animals with two feet and with four? Johann Hermann, in his admirable *Tabula affinitatum animalium* (1783) (affinity had become the new catchword for genetic, evolutionary, or taxonomic relationship), emphasized symmetry in the animal world. To him, the chain was not linear but rather like a *net* with many intermediary and vacant steps. The number of possible types between two species of bug became almost impossible to estimate. Working with only ten variables, he found 10,172,640 possible varieties. Where, then, is the species limit? His conclusion seems to be that, apart from the possibility of an innumerable quantity of species on other planets, about which we know nothing, the Creator must have condensed his unlimited multiplying ability into a limited number of multifaceted species.[40] As a solution of the plenitude dilemma,

39. See Clifford Dobell, *Antony van Leeuwenhoek and his "little animals"* (London and New York: Staples Press, 1932); also Marjorie Hope Nicolson, *The microscope and English imagination* (Northampton, Mass.: Smith College, 1935). Leeuwenhoek calculated the number of sperm in the milt of a single cod to be no less than 150,000,000,000, a figure 13 times greater than the possible number of humans on the earth; a second calculation raised the proportion to 30:1. These investigations made a great impression on Réaumur, Buffon, and Bonnet. Cf. Frank G. Egerton, "Leeuwenhoek as a founder of animal demography," *Journal of the history of ideas, 1* (1968), 1–22. Such counting was connected with utilitarian ambitions: for Wolff's extensive experiments on increasing grain number, see Günter Mühlpfordt, "Physiologie, Biologie und Agronomie im nationalistischen Wissenschaftssystem Christian Wolffs," Martin-Luther-Universität Halle-Wittemberg, *Wissenschaftliche Beiträge*, 1932:6, 472–91. On checks on the growth of any one species, see Clarence Glacken, *Traces on the Rhodian shore* (Berkeley: University of California Press, 1967), esp. 508–12.

40. Johann Hermann, *Tabula affinitatum animalium* (Strasbourg: J.G. Treuttel, 1783), 19–24.

Hermann's suggestions seemed good enough. But they were purely defensive and conjectural.

Kant's criticism of Linnæan natural history also deals with the problems of continuity and plenitude. Either we work from logical entities as Linnæus did in *Systema naturae*, or we work with categories like time and space, that is, with *Naturbeschreibung* or *Naturgeschichte* based on geography. The Linnæan method did not attend to geographical dispersion. Kant sharpened his view: "The systems of Nature so far proposed should more correctly be called Aggregates of Nature; for a system presupposes the idea of the *whole*, from which the multiplicity of things can be derived. Really we have no *Systema naturae*. In our present systems, things are just placed together."[41] This criticism, delivered in a lecture in 1756, was sooner or later bound to affect the ethos of natural history.

We conclude this section with two examples of pessimism, or rather modesty, on the part of late 18th-century epistemologists. The Swedish-Finnish chemist Jacob Gadolin observed that scholars stumble and fumble through the unknown regions of knowledge hoping to find a chain like the Linnæan method to follow. "If every country had its Linnæus, there would be no end of discoveries since Nature is an unfathomable treasure." So far optimism—but as Adam after the Fall was to experience, "the more we open our eyes the more we see that we are naked."[42] Gadolin combined epistemological skepticism with traditional biblical pessimism. Again in a combination of religion and chemistry, Joseph Priestley confessed that for himself he found it impossible "to produce a work that shall be anything like *complete*." "In completing one discovery we never fail to get an imperfect knowledge of others, of which we could have no idea before; so we cannot solve one doubt without creating several new ones." "The greater is the circle of light, the greater is the boundary

41. Immanuel Kant, "Physikalische Geographie" (1756), in *Gesammelte Schriften*, 9 (Berlin: G. Reimer, 1923), 159ff. Cf. James L. Larson, "Linné and Kant," in Ragnar Granit, ed., *Utur stubbotan rot* (Stockholm: Norstedt, 1978), 99–107; Lepenies, *Das Ende der Naturgeschichte*.

42. Jacob Gadolin, *Tal om gränsorne imellan det vi uti naturen känne, och det vi ännu icke hunnit fatta* (Stockholm, 1761), 17f.

of the darkness by which it is confined."[43] The awareness of this truth is the other side of the belief in progress.

THE BROKEN CIRCLE

The success of natural history was overwhelming. With well-founded enthusiasm, Sir James Edward Smith in 1788 pushed to inaugurate a new society entirely devoted to this study, which came to life as the Linnæan Society of London. He wrote: "He who determines with certainty a single species of the minutest moss or meanest insect, adds so far to the general stock of human knowledge, which is more than can be said of many a celebrated name."[44] Smith had good reasons for thinking that his initiative would be well received, since more and more scholars worked in naming and surveying the living world. Linnæan natural history had conquered rival systems and spread into other areas as well. A common, international scientific language thus developed. Internationalism in botany was promoted further by trade in seeds and dried specimens and by the travels of natural historians. Traveling scientists, Linnæan apostles, Blumenbachian headhunters, the Cook company, explorers, adventurers, artists, collectors, all brought home a rich harvest from the field.

Natural history became interesting to the general public also as a part of a new æsthetic appreciation of nature. The growth of gardens and museums demonstrated the institutional strength of natural history. Curators naturally wanted to increase the size of their holdings. Next to salaries, the botanical garden and the library made heaviest demands on university budgets.[45] Visits to famous gardens were

43. Joseph Priestley, *Experiments and observations I–III* (Birmingham, 1790), xviii–xix.

44. J.E. Smith, "Introductory discourse on the rise and progress of natural history," Linnæan Society of London, *Transactions, 1* (1788), 1–55. Cf. Roy Porter, "The new taste for nature in the eighteenth century," *The Linnæan, 4:1* (1988), 14–30; David E. Allen, *The naturalist in Britain: A social history* (London: A. Lane, 1976); Blanche Henrey, *British botanical and horticultural literature before 1800*, 3 vols. (London and New York: Oxford University Press, 1975).

45. See William T. Stearn, "Botanical gardens and botanical literature in the eighteenth century," *Catalogue of botanical books in the collection of Rachel McMasters Hunt, 2* (Pittsburgh: Hunt Botanical Library, 1961), xlii–cxl.

mandatory for the traveler. At the end of the 18th century, Kew Garden had grown to an important promoter of the new tastes in England. To Kew and to the President of the Royal Society, Joseph Banks, came endless material from the colonies.[46] Museums, no longer chambers of curiosities but institutions serving the scientific public, were entering the modern era. In France the Jardin du Roi was reorganized after the Revolution into the Muséum d'Histoire Naturelle according to a modified Linnæan plan.[47]

Among the modifications was abandonment of the hope of a complete inventory of nature. Rather than showing the full divine order, the naturalist collectors wanted to outdo competitors in the size of their herbaria. The competitive element in collecting became increasingly obvious, but it was a race without a finish line. When Carl Peter Thunberg, successor to Linnæus at Uppsala, had his Botanicum built around 1800, his architect provided herbarium space for only 15,000 specimens. Soon the windows in the beautiful main hall would be blocked with the cases of specimens that marked Thunberg's proficiency in the collecting game. In writing he devoted himself to mere description of small units and did not try to put together the big jigsaw puzzle. Thunberg's admirable industry is obvious from his series of dissertations on *Nova plantarum genera* (fifteen parts) and his *Museum naturalium academiae upsaliensis* (twenty-nine parts including appendices). His colleague, the entomologist Schönherr, wrote sixteen volumes totaling 5,000 pages on the insect-group *cucurlionides*. The dream for completeness survived only in completeness in detail. Its form could be labeled "descriptionism," a pure registration of facts.[48]

46. Lucile H. Brockway, *Science and colonial expansion. The role of the British Royal Botanic Gardens* (New York: Academic Press, 1979).

47. Friedrich Klemm, *Geschichte der naturwissenschaftlichen und technischen Museen*, Deutsches Museum, Abhandlungen und Berichte, 41:2 (Munich: VDI-Verlag, 1973); Yngve Löwegren, *Naturaliekabinett i Sverige under 1700-talet* (Uppsala: Almqvist & Wiksell, 1932).

48. For descriptionism, see J.L. Heilbron, "Fin-de-siècle physics," in C.G. Bernhard et al., eds., *Science, technology and society in the time of Alfred Nobel* (Oxford: Pergamon Press, 1982), 51–73.

In the early 19th century, probably only Robert Brown could handle the whole of botany. The last person able to treat all of nature was Alexander von Humboldt, who discovered several hundred plant species along the slopes of Chimborazo in a few months (1800).[49] The mass of little details struck some people as ridiculous. The philosopher Lichtenberg predicted the advent of "a Linnæus of sand."[50] Coleridge complained that despite all the work by the botanists there was "little more than an enormous nomenclature, a huge catalogue, *bien arrangé*, yearly and monthly augmented in various editions, each with its own scheme of technical memory and its own conveniences of references!"[51]

The new situation called forth a variety of means of publication: a growing number of sumptuous books for the connoisseur, textbooks for the tyros, handbooks for the collectors, journals for specialists. It also produced a number of bibliographies by Haller, by Boehmer, by

49. Alexander von Humboldt, *Aus meinem Leben* (Munich: Beck, 1987), 170ff., letter to Wildenow, February 21, 1801: he was now completely convinced "that we do not know more than three-fifths of the plant species." According to J.C. Loudon, the total number of vegetable species known in 1820 was 44,000, of which 38,000 had been described. "It is now (1829) believed that there may be from 100,000 to 200,000 species of plants. Such is the progress of discovery and of ideas," he added proudly. At the end of the 19th century, M.A.J. Möbius counted 415,600 species so far described, which should be compared to the 4,162 of *Systema naturae* (1758). Compare the growth of knowledge in an unusually well researched animal group: in 1758, 564 species of birds were known (*Systema naturae*, 10th ed.); in 1760, 1,500 (Brisson, *Ornithologia*); in 1790, 2,951 (Latham, *Index ornithologicus*); in 1812, 3,779 (Illiger, *Tabellarische übersicht*); in 1841, 6000 (Gray, *British Museum Catalogue*); and in 1909, 18,939 species (Sharpe, *Hand-list of the genera and species of birds*); statistics from Erwin Stresemann, *Die Entwicklung der Ornithologie* (Berlin: F.W. Peters, 1951), 271. The Natural History Museum of Sweden today holds at least 15,000,000 specimens, and in 1980 the British Museum's entomological department alone housed 22,500,000. But according to a recent estimate, there might be 30,000,000 to 40,000,000 species of insects in the world, a multitude nobody can survey; see E.O. Wilson, *Science,* 13 December 1985. The question of nature's multitude has become an issue in the debate over conservation. Cf. S.J. Gould: "I'm just very confused by the whole debate. How can you be confident about an issue when the base level datum, which is the number of species out there that are threatened, is so unknown that estimates are ranging from two million to 30 or 40 million?" (interview in *Dialogue*, 1987:4, 16f.). For more recent estimates, see *Biodiversity*, 29 Feb 1988.

50. Dietrich von Engelhardt, *Historisches Bewusstsein in der Naturwissenschaften: Von der Aufklärung bis zum Positivismus* (Munich: Alber, 1979), 79.

51. Coleridge, "On the science of method," 15. Quoted in Collison, *Encyclopedias*, 261.

Jonas Dryander (on Bank's library), by Joseph Reuss, and so on. A new genre arose, the natural-history *dictionary.* With its alphabetical arrangement, it aped the encyclopedias: facts, in the Lockean sense, would speak for themselves. Considering their number, these dictionaries must have meant good business for the publishers; considering the size of some of them, they did have something to summarize. They are a proof of the success *and* the failure of 18th-century natural history.

The compilers of dictionaries had two banks to rob: *Systema naturae* and Buffon's *Histoire naturelle*, changed into smaller coin and arranged alphabetically. The reader got Linnæus' exactness and Buffon's fine writing, both science and literature, in such extravagant works as Houttyun's *Natuurlyke historie of uitvoerige beschryving der dieren, planten en mineralen*, published in thirty-seven volumes (1761–85), and Valmont de Bomare's *Dictionaire raisoné universel d'histoire naturelle* (1760–), in fifteen volumes, at least five editions, and a Danish translation (*Den almindelige naturhistorie i form af et dictionaire* (1767–). There was also Friedrich Heinrich Martini's *Allgemeine Geschichte der Natur in alfabetischer Ordnung* (1774–), which built on Bomare's *Dictionaire* and invited its readers to participate in making a new edition by filling out pages left blank for the purpose. The new edition would not be complete, of course; in natural history, Martini said, "complete is a harmonious sound without meaning." There were still more dictionaries: the anonymous *Manuel du naturaliste* (Paris, 1770); W.F. Martyn's *A new dictionary of natural history*; Philip Miller's very handsome *The gardener's and botanist's dictionary* (1731, 8th edition 1768); and Lamarck's botanical chunk of the *Encyclopédie méthodique.* A somewhat earlier, similar trend has been noted in geography.[52]

Both Linnæus and Adanson had considered the possibility of making dictionaries. Through his patron Carl Gustav Tessin, Linnæus had the help of a copyist to compile a *Dictionnaire portatif d'histoire*

52. Numa Broc, *La géographie des philosophes géographes et voyageurs français au XVIIIe siècle* (Lille: Université de Lille III, 1972), 347ff.

naturelle. He began in 1757, but abandoned the project two or three years later. It fell victim to his fear that nothing can be omitted. Nothing, in any event, is left of the manuscript.[53] In a *Memorial* printed in 1775, Adanson summarized what he had accomplished so far—the description of 300 species of mammals, 2,000 of birds, and so on, in all some 17,000 animals, 10,000 plants, and 8,000 to 10,000 minerals. He proposed grand and encyclopedic ventures, a new nomenclature, a universal language, a plan to cover all natural history in three "orbes," the first to contain 40,000 species. Nothing, however, came of these plans; and to cope with economic realities he had to work in the dictionary business. To the *Supplément à l'Encyclopédie* Adanson contributed about 450 articles, and among his huge piles of manuscripts there is much material intended for alphabetical publication.[54] In their unsuccessful way both Adanson and Linnæus tried to balance two interests: scientific order and communicable information.

There were other strategies to make natural history scientifically meaningful. Georg Forster returning from the Cook expedition wanted to combine Linnæan formalism and Buffon's individualism, as in his famous monograph on the breadfruit tree.[55] Robinet preferred to concentrate not on the species level but on the type.[56] That was also the way chosen by the idealistic morphologists, by Goethe, Etienne Geoffroy Saint-Hilaire, and Sir Richard Owen, and in Sweden by the mycologist Elias Magnus Fries and the algologist Agardh. Or one could adjust the size of taxonomic groups to make them easier to handle. That seems to have been the reason why Alphonse-Laurent de Jussieu recognized one hundred families of plants, none of which had more than one hundred genera.[57] Then

53. Linnæus, *Svenska arbeten*, ed. Evald Ährling, 2 vols. (Stockholm, 1878–80) *1*, 20, 25, 26, 33, 45.

54. See *Adanson. The bicentennial*, 64–74; also "Plan & tableau de mes ouvrages," in François Rozier, *Observations sur la physique, 5* (Paris, 1775), 257–74.

55. Cf. Ludwig Uhlig, *Georg Forster. Einheit und Mannigfaltigkeit in seiner geistlichen Welt* (Tübingen: M. Niemeyer, 1965), 45–55.

56. On Robinet, see Lovejoy, *The great chain of being*, chap. 9.

57. Cf. Frans A. Stafleu, *Linnæus and the Linnæans* (Utrecht: A. Oosthoek, 1971), 325–32.

there was Humboldtian science, which took not the species or genus as subject of study but the plant community as such.[58]

The widening scope of the natural history project split it into such specialties as ornithology, entomology, and mycology.[59] The new public for natural history, professors as well as amateurs, wanted information to suit diversified tastes. Thus, as a result of the widening fragmentation of the subject and the dissolution of the chain of being, a new science was invented. "Biology" was inaugurated in the very year 1800 by Jean Baptiste Lamarck, a critic of the concept of the great chain but also one of the most prominent pursuers of the natural history project.

Finally, we must note a change of metaphors. Both natural history and encyclopedism had experienced the increase of *quantity*, of boundless, unmeasurable information. The encyclopedic enterprise, aiming at order, totality, and coherence, had exchanged the *esprit systématique* for the *esprit de système*. In natural history, epistemological criticism and awareness of nature's richness had made the great chain of being succumb to its own weight. The metaphors of "chain" and "scale," both very popular during the 18th century, gave way to metaphors less apt for classification, such as "map" or "net," used by, among others, Linnæus, Haller, and Hermann.[60] The geographical imagery recurs in Adanson's "orbe." In a similar manner, d'Alembert used the older metaphors "circle of knowledge" and "tree" in elucidating the concept of encyclopedia, while Diderot preferred "growing city" and "machine." The critics (Kant, Hegel, Törneros) spoke of the modern encyclopedia as an "aggregate" or a "grocery store." There is a clear progression from symbols appropriate to a traditional classification to symbols suited for an open-ended collection. The systematic "encyclopedia" in the old sense of "whole circle of knowledge" was breaking up. This broken circle is one aspect of the

58. Malcolm Nicolson, "Alexander von Humboldt, Humboldtian science and the origin of the study of vegetation," *History of science, 25* (1987), 167–94.

59. On specialization in this period, see Paul Farber, *The emergence of ornithology as a scientific discipline* (Dordrecht: D. Reidel, 1982), chaps. 7–8.

60. See P.F. Stevens, "Metaphors and typology in the development of botanical systematics, 1690–1960, or the art of putting new wine in old bottles," *Taxon, 33:2* (1984), 174ff.

critical work of the Enlightenment. It might be possible to argue that such a shift reflects a general development during the second half of the century from a "geometrical" to an "arithmetical" mentality, when systems and stable structures mattered less than quick, irregular information.

3 Systematics and the Geometrical Spirit

By John E. Lesch

During the latter 18th century European thinkers embraced a systematic model of order with an enthusiasm and conviction unprecedented before and unmatched since. Systematics—the classification of objects into groups according to degrees of identity and difference, and rationalized description and nomenclature—largely constituted the scientific study of the three kingdoms of nature: animal, vegetable and mineral. Drawing upon a century and a half of groping and hesitant effort by naturalists, Linnæus had succeeded in standardizing names and the process of naming, clarifying and simplifying the criteria of classification, and including the known objects of each kingdom in a simple, coherent, and comprehensive arrangement. Systematics dominated the study of the three kingdoms in the latter 18th century, to the disadvantage of anatomical, physiological, and microscopical approaches. Although most obviously successful in natural history, the systematic model may be found in many other domains. Late Enlightenment thinkers classified, or sought standards of description and naming, for objects as diverse as chemicals, diseases, the vital properties or anatomical elements of organisms, algebraic curves and surfaces, machines, and medicinal drugs. Whether we view these activities as the result of a diffusion of the model from one or a few sources—most probably natural history—or as the synchronous expressions of an underlying disposition, the systematic model formed a major modality of late Enlightenment thought.

The chapter begins by identifying the fundamental characteristics of the model by examination of one of its earliest and most prestigious exemplars, systematic botany, and determining in what sense the model participated in a more general geometrical spirit. It then surveys the model's manifestations in other areas of thought, emphasizing mineralogy, chemistry, and medicine, and ends by suggesting that the success of the systematic model, and its decline in the early 19th century, must be understood in relation to its coherence with other elements of late Enlightenment culture.

THE SYSTEMATIC MODEL

Systematic botany emerged as a fully developed science in the first half of the 18th century, as the culmination of a gradual development of concepts and methods over the preceding two hundred years. The stimulus of Renaissance humanism and naturalism in art, the growing popularity of natural history as a leisure pastime, and increasing exploration of the various habitats of European flora as well as the findings of European explorers in Asia, Africa, and America brought rapid increases in the numbers of known plants. A six-to-tenfold increase can be recorded between 1550 and 1623, when 6,000 species were listed in a leading textbook. By 1690, the number had more than tripled again, rising to 19,000.[1]

From around 1600 problems of botanical nomenclature and classification became acute. Writers on botany recognized that creation of a standardized, uniform, and accurate system of naming, describing, and grouping plants was fundamental to the construction of a science of botany. Until such a system existed, the rapidly expanding knowledge of plants would yield more embarrassment than glory. By the time that Linnæus began his scientific career in the 1730s, many of the materials and methods of systematic botany were already in existence. It was only through Linnæus' work, however, that an authoritative system of nearly universal appeal emerged.[2] In Linnæus' *Systema naturae* (1735–67, thirteen editions), *Critica botanica* (1737), *Philosophia botanica* (1751), *Species plantarum* (1753), and

1. A.G. Morton, *History of botanical science* (London and New York: Academic Press, 1981), 115–286. On numbers of plant species, see also Gunnar Broberg, chap. 2 in this volume.

2. W.T. Stearn, "Introduction" in Carl Linnæus, *Species plantarum. A facsimile of the first edition 1753*, 2 vols. (London: The Ray Society, 1957), *1*, 1–176, esp. 3.

other writings, botanists were at last in possession of a comprehensive, rationalized language upon which to found their science.[3]

The result was the relatively sudden appearance, around the middle of the 18th century, of a dogmatic confidence that the correct approach to botany had been found. Although particular points of Linnæus' classificatory schemes were questioned or modified, his methods of nomenclature and description had been generally adopted by botanists by the time of his death in 1778. Linnæus' work not only consolidated earlier developments in systematic botany, it also defined its future tasks. The search for new species and genera, their naming and systematic placement, became the main aim of botany, to the relative neglect of plant anatomy and physiology.

Linnæus' botany was a project of rationalization, an effort to create a set of concepts and procedures that would bring uniformity, consistency, and coherence to the representation of a clearly defined domain of natural objects. "Minerals grow, plants grow and live, animals grow, live, and have feeling," wrote Linnæus in the *Systema naturae*, thus marking off in epigrammatic style the three kingdoms to be organized.[4] The essential tasks of the organizing system were to describe, to name, and to classify every object within its kingdom by uniform and consistent principles. In the *Systema naturae* Linnæus wrote that "the first step in wisdom is to know the things themselves; this notion consists in having a true idea of the objects; objects are distinguished and known by classifying them methodically and giving them names. Therefore, classification and name-giving will be the foundation of our science."[5]

3. Frans A. Stafleu, *Taxonomic literature: A selective guide to botanical publications with dates, commentaries, and types* (Utrecht: International Bureau for Plant Taxonomy and Nomenclature, 1967), 275–90; William T. Stearn, "Linnæan classification, nomenclature, and method," in Wilfrid Blunt, ed., *The compleat naturalist: A life of Linnæus* (New York: Viking Press, 1971), 242–49; Gunnar Eriksson, "Linnæus the botanist," in Tore Frängsmyr, ed., *Linnæus: The man and his work* (Berkeley: University of California Press, 1983), 63–109.

4. Carolus Linnæus, *Systema naturae 1735. Facsimile of the first edition*, ed. and transl. M.S.J. Engel-Ledeboer and H. Engel (Nieuwkoop, Holland: B. de Graaf, 1964), 19 (quoted); Julius Sachs, *History of botany 1530–1860* (Oxford: Clarendon Press, 1890), 7–9; Stearn, "Introduction," 3.

5. Linnæus, *Systema naturae*, 19, 22–4. In practice Linnæus sometimes allowed his sense of affinities to overcome the rigid application of his criteria. See Stearn,

In the *Critica botanica* and *Philosophia botanica*, Linnæus laid down in aphoristic form detailed rules for describing, naming, and classifying plants. His binomial nomenclature for designating genus and species placed the plant within a logically integrated system. Binomials had been used in a haphazard way before Linnæus to define more or less isolated groups, but in the Linnæan system they became instruments of a rationalizing project to impose unity, consistency, and logical order on the whole field of botany.

Also central to Linnæus' botanical work was his sexual system of classification, the fundamental features of which were abstraction, numeration, and artificiality. Abstraction, because Linnæus proceeded by setting aside as irrelevant all but a few select qualities of the plant. The essence of plants, Linnæus said, consists in sexual reproduction or fructification. The essence of fructification consists in the flower and fruit, and the essence of the flower consists in the stamen and pistil. The essence of the stamen consists in anthers, the essence of the pistil consists in the stigma. Thus did Linnæus justify the narrowing of classificatory criteria to a part of a part of the plant. As W.T. Stearn and James L. Larson have remarked, Linnæus' thinking was informed by Aristotelian logic, and Linnæus drew an analogy between logical and natural forms. Yet it is founded also on a close study of particulars. Linnæus went well beyond his predecessors in distinguishing and naming the organs of fructification.[6]

Having distilled the essence of the plant by abstraction, Linnæus classified it by numeration. The sexual system, in Stearn's words, is

---

"Linnæan classification," 244. On consistency as a criterion of systems, see Michel Adanson, *Familles des plantes*, 2 vols. (Paris: Vincent, 1763), *1*, xli. Daudin contrasts Linnæus' truly systematic approach, in which the number of categories is universally and necessarily limited to five (class, order, genus, species, variety), with "a simple 'synopsis,' in which the number of successive dichotomies vary at will from one section to another," Henri Daudin, *De Linné à Jussieu: Méthodes de classification et idée de série en botanique et en zoologie* (Paris: Librairie Felix Alcan, 1926), 38.

6. Linnæus, *Systema naturae*, 22–4; J.L. Larson, *Reason and experience: The representation of natural order in the work of Carl von Linné* (Berkeley: University of California Press, 1971), 2–3, 149; Stearn, "Introduction," 2–3. See also V.H. Heywood, "Linnæus—the conflict between science and scholasticism," in John Weinstock, ed., *Contemporary perspectives on Linnæus* (New York: University Press of America, 1985), 1–15.

"a basically simple but ingenious arithmetical system, whereby the genera are grouped into twenty-four classes according to the number of stamens (together with their relative lengths, their distinctness or fusion, their occurrence in the same flower as the pistil or their separation in unisexual flowers, or their apparent absence), while division into orders within each class is determined by number of pistils"[7] (fig. 3.1). Such a system gives a privileged place to the numerical or spatial qualities of its objects, and to the visual sense. In the *Systema naturae* Linnæus refers to his science as one of "describing and picturing," and defines a naturalist as one who distinguishes and names the parts of natural bodies by sight.[8]

Abstraction and numeration gave the sexual system a clarity and simplicity that translated into ease of use and certainty of identification. These advantages came at a price, however, namely the divergence that opened up between the classifications so constituted and those based on a less sharply defined but more intuitively satisfying grouping of the vegetable kingdom that resulted from simultaneous consideration of multiple characteristics. Julius von Sachs argued that the major aims of systematic botany from the late 16th century to Linnæus were to arrive at a natural grouping of plants and to identify the groups by a few easily recognizable marks. Linnæus was, in Sachs' view, the first to recognize that these two aims were incompatible. Linnæus sketched the elements of a natural system—an outline that became the basis of later attempts by others—but devoted his major effort to a classification, based on the sexual system, that was avowedly artificial. Even his use of the parts of the flower, justified by the flower's functional importance, contained an irony, since just those aspects of the parts used as criteria—number and connection—are functionally irrelevant.[9]

Paying the price of artificiality proved a good investment for systematic botany. In the *Systema naturae* Linnæus compared tables of

7. Stearn, "Introduction," 22–6.

8. Linnæus, *Systema naturae*, 19.

9. Linnæus, *Systema naturae*, 22–4; Stearn, "Introduction," 24–6; Sachs, *History of botany*, 7–9, 82–3. Linnæus considered his genera to be less artificial or conventional than his classes or orders. Cf. Daudin, *De Linné à Jussieu*, 34–48.

classification to maps. They were like maps in abstracting from a complex reality, in representing degrees of proximity and distance of a sort, and, above all, in their utility. The sexual system's clarity, simplicity, and ease of use opened up botany to the contributions of amateurs. The very step that constituted systematic botany as a science made it more, not less, accessible to popular participation. In this case, at least, it appears that the simplification effected by a successful effort of rationalization facilitated a real, although subordinate, role for the nonexpert.[10]

In all of this the formal qualities of Linnæus' science are evident. Linnæus' formalism is expressed in his insistence on explicit rules, especially in the *Critica botanica* and *Philosophia botanica*, and their consistent application in the systematic works. It is found in the analogy of logical and natural forms that is implicit in the very idea of a system of nature and explicit in the parallel Linnæus drew between the hierarchies of logical and systematic categories. It is embodied in the a priori habit of mind that led Linnæus to embrace an artificial system placing clarity of concept over empirical intuition, and that may have led him to the sexual system in the first place. It is, finally, most obvious in his presentation, which is methodical, concise, impersonal, and—to some—arid. The formality of the Linnæan system was facilitated and reinforced by its atemporality. Time did not appear in the Linnæan scheme because within it species were constant, while each natural group had as its basis a common type, which, like a Platonic form, was beyond the reach of temporal change.[11]

Linnæus' system was universal by intention and also in effect. In ways not yet systematically studied, botanists in the 18th century

10. Linnæus, *Systema naturae*, 19; Stearn, "Introduction," 3; Blunt, *The compleat naturalist*, 183–92; and Stearn, "Linnæan classification," 244.

11. Linnæus, *Systema naturae*, 18, 22; Sachs, *History of botany*, 7–9; Stearn, "Introduction," 2–3, 17, 24–26; Larson, *Reason and experience*, 149–50. The epistemological issues raised by the analogy of logical and natural categories, and by the problematic status of systematics in relation to mathematical physics, were taken up by Kant in the *Critique of judgment*. See Ernst Cassirer, "Das Problem der Klassifikation und der Systematik der Naturformen" in Cassirer, *Das Erkenntnis problem in der Philosophie und Wissenschaft der neueren Zeit von Hegels Tod bis zur Gegenwart (1832–1932), 4* (Stuttgart: W. Kohlhammer, 1957), 127–44.

formed an international community, corresponding, exchanging specimens and visits, and struggling toward a common scientific language. Publishing in Latin, and developing his system through an extensive network of international correspondents, Linnæus achieved widespread acceptance. By his death in 1778, the binomial system of nomenclature was universal and the sexual system in general use.[12]

Related to the universalism of Linnæus' system was its ease of generalization. It was readily—almost promiscuously—transferable from one domain of objects to another. Nowhere is this better illustrated than in Linnæus himself, whom Sachs rightly compared to a "classifying, coordinating, and subordinating machine." Besides plants, he provided systematic arrangements for animals, minerals, and diseases. In the *Bibliotheca botanica* (1736) he classified botanists, in the process nicely revealing the relative value and place he assigned to the various components of the science (fig. 3.2).[13]

The universality of these qualities of the systematic model may be seen in counterpoint in the work of one of Linnæus' major critics, the French botanist Michel Adanson. Confronted for the first time by a tropical flora during his six-year service in Senegal with the Compagnie des Indes, Adanson brought back to France a conviction of the inadequacy of the Linnean and every other system. In place of "systems," Linnæan or otherwise, which based classification on one or a few parts of the plant defined at the outset, Adanson proposed the "natural method." The botanist was first to consider "the ensemble of all the parts of plants," including roots, twigs, and fruits as well as flowers, insofar as these could be studied in their number, figure, situation, relative proportion, and symmetry. Affinities were to be determined by subsets of common features arrived at empirically and open to revision by new experience. In his *Familles des plantes* (1763), Adanson elaborated the principles of the natural method and applied it to the formation of fifty-eight families of plants.[14]

12. Stearn, "Introduction," 2–3.

13. Sachs, *History of botany*, 89–91; Sten Lindroth, "The two faces of Linnæus," in T. Frängsmyr, ed., *Linnæus*, 1–62 (22–4); Linnæus, *Bibliotheca botanica* (Amsterdam, 1736).

14. Michel Adanson, *Familles des plantes, 1*, esp. clv. On Adanson, see Morton, *History of botanical science*, 301–14; Frans A. Stafleu, "Adanson and the 'Familles

Adanson gave the first full theoretical statement and justification of the method by which a natural arrangement of plants could be constructed. Despite his thoroughness, however, Adanson was not successful in his aim to displace the Linnæan system. That achievement fell to his younger contemporary and associate Antoine-Laurent de Jussieu, who benefited not only from the teachings of his uncle Bernard and of Adanson, but also from detailed study of the many new plants yielded by voyages of exploration, and from his adoption of the popular Linnean binomials. It was largely through A.-L. de Jussieu's *Genera plantarum* (1789) and other writings that the natural method gained ascendency in systematic botany in the early 19th century.[15]

Although the differences between natural and artificial systems loomed large to their protagonists, on most points the two approaches shared the general features of the systematic model. Adanson's project was one of rationalization that stressed comprehensiveness, internal consistency, and disciplined use of language. Adanson insisted that his method was comprehensive—or, as he put it, "universal"—in that he strove to include all plants, tropical as well as temperate and glacial. Consistency, too, was to be sought in steady adherence to the more empirical rules of the natural method. Even Adanson's rejection of binomial nomenclature resulted from his commitment to principles of the priority and stability of names, in opposition to the precedence given by Linnæus to their significance.[16]

des plantes,'" in *Adanson: The bicentennial of Michel Adanson's 'Familles des plantes',* 2 vols. (Pittsburgh: The Hunt Botanical Library, Carnegie Institute of Technology, 1963), *1*, 123–264, and *Linnæus and the Linnæans: The spreading of their ideas in systematic botany, 1735–1789* (Utrecht: Oosthoek, 1971), 310–39. An instructive comparison of "system" and "method" is given by Michel Foucault, *The order of things: An archeology of the human sciences* (New York: Vintage Press, 1973; translation of *Les mots et les choses*, 1966), 128–45.

15. Linnæus' attempt at a natural arrangement—the *Methodi naturalis fragmenta*—first appeared as part of his *Classes plantarum* (1738). See Morton, *History of botanical science*, 301–14; Stafleu, *Linnæus*, 310–39; and Stearn, "Linnæan classification," 244. Stafleu notes common features in the views of Linnæus and advocates of the natural method. See "Adanson and the 'Familles des plantes,'" esp. 155, 166, 167, 229, 236.

16. Adanson, *Familles des plantes, 1,* clii–cliii, cxci–cxcii. On Adanson and nomenclature, see Stafleu, *Linnæus*, 311; and "Adanson and the 'Familles des plantes,'" 187.

Adanson, like Linnæus, abstracted from the totality of the plant's features criteria for defining degrees of identity and difference. The two botanists differed not on the goal of abstraction but on the tactics used to obtain it. Where Linnæus' procedure was a priori and essentialist, at least at the level of classes and orders, Adanson's was inductive or "experimental," first looking at all features, then narrowing to a subset. The two naturalists also converged in their emphasis on numerically or spatially definable traits and the associated precedence of the visual sense. Here Adanson was, if anything, more explicit than Linnæus, remarking that botany "distinguishes plants only by their relations of quantity, whether numerical or discrete, or continuous, which gives us the extent of their surface or their size, their figure, their solidity."[17]

On one point—simplicity, or ease of mastery and use—Linnæus' system had a seeming advantage. Adanson conceded as much and also that artificial systems had increased the popularity of botany and the numbers of botanical publications. But he complained that the result was merely superficiality, the spoiling of the most penetrating minds, and the production of quantities of catalogues on Linnæan principles. Most important, Adanson regarded artificial systems as incapable of bringing a permanent end to the confusion and uncertainty of botany. Certainty, stability—in short, a true system based on true principles—would come to botany only through the slower, more empirical, but ultimately more reliable procedures of the natural method.[18]

In its insistence on explicit rules, its positing of a hierarchy of systematic categories implicitly analogous to logical ones, and its methodical, precise, and impersonal form of presentation, Adanson's natural method can be described as formalistic. The categories of Adanson's method, like those of Linnæus' system, are untouched by

17. Morton, *History of botanical science*, 303, 306–8; Adanson, *Familles des plantes*, *1*, cc–cci (quote).

18. Adanson, *Familles des plantes*, *1*, xli–xlii, clii–cliv, cxci–cxcv. Adanson sounded an Enlightenment theme when he appealed to nature as a source of truth in opposition to "the old prejudice in favor of systems and the ideas on which they are based." Ibid., clvii.

time or change.[19] Where Adanson's formalism departed from that of Linnæus was in his clear break with essentialism and with Linnæus' a priori habits of mind. Like Linnæus, Adanson intended his method to be universally adopted. But the success of the Linnæan system, Adanson's rejection of binomials, and his lack of an institutional teaching position doomed the *Familles des plantes* to an oblivion from which it had to be rescued by Henri Baillon in the following century. Only through its adoption and extension by A.-L. de Jussieu did Adanson's method achieve a delayed acceptance and play a role in the formation of modern systematics.[20]

A collaborator of Diderot and d'Alembert, described by Stafleu as *plus encyclopédiste que les encyclopédistes*, Adanson conceived a vast work that would extend the natural method to other parts of natural history, physics, chemistry, ethnology, philology, and related subjects. Although never brought to fruition, the project expresses Adanson's conviction of the compatibility of his method with the systematic model.[21]

In one respect, at least, Adanson's natural method had an advantage over artificial systems that added a significant quality to the systematic model. Having identified the general features of a family, the Adansonian botanist could predict that other members of the group as yet unknown would have numbers of features in common. No such predictive quality could be relied upon for artificial systems. Besides its value for systematics, predictiveness enhanced the utility of the method, particularly at a time when voyages of exploration were bringing new plants to Europe in ever greater numbers. For example, plants of medicinal value might be found among the newly discovered specimens belonging to families already known to include medically useful plants.[22]

19. Morton, *History of botanical science*, 309–11.

20. Stafleu, *Linnæus*, 311, and "Adanson and the 'Familles des plantes,'" 126–7, 187.

21. Stafleu, "Adanson and the 'Familles des plantes,'" 136, 197; Jean-Paul Nicolas, "Adanson, the Man," in *Adanson: The bicentennial, 1*, 1–121, esp. 35–6, 65–78.

22. Adanson, *Familles des plantes, 1*, lxxiii–lxxx, cxcv–cxcvi; Morton, *History of botanical science*, 306; Stafleu, *Linnæus*, 330; and "Adanson and the 'Familles des plantes,'" 161, 196. When, in the first decades of the 19th century, Paris pharmacists turned to the chemical analysis of medicinal plants, they incorporated the bo-

The Model and the Geometrical Spirit

The systematic model was one expression of the geometrical spirit of the late Enlightenment. In its most general sense the geometrical spirit may be identified with analysis, a term and concept placed by Ernst Cassirer at the center of Enlightenment thought. In its first meaning analysis is indeed mathematical—or "geometrical" in 18th-century usage—referring especially to algebra. Enlightenment thinkers, however, allowed the term a wider formulation that gave it nearly unlimited applicability. In this formulation, "analysis" refers to a double movement of analysis and synthesis by which the phenomena of a field are reduced to their elements, and then restructured into a true whole that can be known by reason because, in Cassirer's words, reason "can reproduce it in its totality and in the ordered sequence of its individual elements." As Cassirer remarks further, for the Enlightenment "to 'know' a manifold of experience is to place its component parts in such a relationship to one another that, starting from a given point, we can run through them according to a constant and general rule." For Michel Foucault, too, analysis represents a universal method of classical thought that includes but is not confined to its mathematical expression in algebra. Systematics, in this view, is made possible by analysis in the form of a system of signs, and is a nonmathematical expression of the quest for a mathesis or universal science of measurement and order.[23] The kinship of the systematic model with mathematics is therefore not a question of derivation or a direct modeling of systematics on one or another field of mathematics, but of the sharing of a generalized method of analysis and of the qualities that make such a method possible.

The most obvious of these are abstraction and numeration. Just as the mathematical sciences confined their treatment of the physical

---

tanists' prediction into their research program. See John E. Lesch, *Science and medicine in France: The emergence of experimental physiology 1790–1855* (Cambridge, Mass.: Harvard University Press, 1984), 125–44.

23. Ernst Cassirer, *The philosophy of the Enlightenment*, transl. Fritz C.A. Koelln and James P. Pettegrove (Boston: Beacon Press, 1955; original German edition 1932), 13-6, 23-4; and Foucault, *The order of things*, 46–77.

world to its numerable or measurable qualities of extension and motion, so did Linnæan and Adansonian systematics reduce the plant to the number, form, connection, and spatial arrangement of its characters. The formal qualities of Linnæus' and Adanson's approaches—their insistence on explicit rules, their methodical, impersonal, and economical presentation, and, in the case of Linnæus, a priori thinking subordinating the empirical to the conceptual clarity of the artificial—all have clear analogues in mathematics, as do the certainty and simplicity each system offered its users.

Adanson grasped the connection clearly. Insisting that botany was a science not merely of names but also of facts, he added that "we even believe that we find in it an immediate relationship with geometry." This relationship consisted not only in the exclusive use of characters subject to number and measure but also in the botanist's ability to pose questions analogous to "the most sublime geometry" in difficulty and instructiveness. "Find the most sensible point that establishes the line of separation or of definition between the family of the Scabiosa and that of the honeysuckle," or "Find a known genus of plants (natural or artificial, it does not matter) which occupies an accurate middle point between the family of dogbane and that of borage." Properly constructed, such questions would yield conclusions on the possibilities "that would be as evident and as well-demonstrated as the truths of the best geometry." In this way, too, the botanist would be able to estimate how many families or genera were lacking between two distant families or genera whose intermediaries were unknown, "presumptions which, if they would not have all the precision of mathematics, nevertheless would yield large views, and would furnish new means of extending our knowledge in botany." While admitting that absolute perfection in a botanical system—which he equated with "the necessary exactitude, which characterizes mathematical perfection"—was not possible, he nevertheless credited the families determined by the natural method with bringing to botany all the certitude, stability, comprehensiveness, concision, ease of use, and utility of which it was capable.[24]

24. Adanson, *Familles des plantes, 1*, cxci–cxcii, cc–cci. Underlying Adanson's presumption of the existence of unknown intermediaries between known families

Another who saw systematics as an extension of the mathematical way of thinking to the ordering of empirical objects was Johann Heinrich Lambert. Much of his life work was an effort to reshape science and philosophy in the image of mathematics and to assure for those fields of knowledge the exactness and certitude of their exemplar. Lambert treated the problem of order and its measure not in relation to natural history or any domain of particulars, but abstractly as a problem of knowledge in general.[25] Mathematics, he pointed out, had been most successfully applied where the objects of knowledge could be construed as homogeneous entities, because only such entities could be added, subtracted, and related to one another as more and less. In its formation of generic names and the abstract ideas derived from them, language often lost sight of the homogeneities on which the names and ideas were originally based. The result was conceptual confusion and failure in communication. Where homogeneities could be defined, however, "mathematics shows us that. . .they present ideas that are simple, very knowable, and exempt from logomachy. And this is what is necessary for a clear and well-arranged system."[26]

According to Lambert, the degree of order of a given arrangement may often be calculated as a proper fraction. In the simplest case, a linear succession of objects, each of which is assigned a rank or value, absolute order is rank order. The degree of disorder that results from displacement of one object from this absolute order is taken as the product of the number of places the object is displaced and the value

---

and genera was his commitment to the idea of a progression that connects "in a continuous series the families that resemble one another the most, and in each family, the genera that have the most general relationships." Cf. Adanson, *Familles des plantes, 1,* clxxxviii; Stafleu, *Linnæus,* 328; and "Adanson and the 'Familles des plantes,'" 194. On the idea of series in botany and zoology, see Daudin, *De Linné à Jussieu,* esp. 79–187.

25. Johann Heinrich Lambert, "Essai de taxéometrie, ou sur la mesure de l'ordre," Akademie der Wissenschaften, Berlin, *Nouveau mémoires,* 1770, 327–42 and 1773, 347–68. See Christoph J. Scriba, s.v. "Lambert, Johann Heinrich" *Dictionary of scientific biography (DSB),* 7, 595–600; *Colloque international et interdisciplinaire Jean-Henri Lambert* (Paris: Editions Ophyrs, 1979); and Gereon Wolters, *Basis und Deduktion: Studien zur Entstehung und Bedeutung der Theorie der axiomatischen Methode bei J.H. Lambert (1728–1777)* (Berlin: Walter de Gruyter, 1980).

26. Lambert, "Essai de taxéometrie," 327–9.

of the object displaced. Similar calculations of degrees of order may be carried out in the more complex case of systems of classification. Lambert gives the example of a well-arranged library in which the books are classified first according to the sciences, next according to their age, their format, their binding, and so on. If each book satisfies all conditions, the library will be absolutely well-arranged. Its order will then be unity. It cannot be greater, but can admit of fractions.[27] Suppose that there are n books and that each book must satisfy three conditions, a, b, c. The product $n(a + b + c) = 1$. But suppose an arrangement in which all books satisfy a, while of the other two conditions m books satisfy b and c, p books satisfy only b, q books satisfy only c, and r books satisfy neither b nor c. Then the degree of order of this arrangement would be expressed as the fraction:

$$\frac{m(a + b + c) + p(a + b) + q(a + c) + ra}{n(a + b + c)}$$

What such a calculation measures is the degree to which a given classificatory system conforms to a set of explicitly defined, consistent criteria. It does not touch the fit of the scheme with nature, and Lambert, who was no naturalist, did not discuss the specific problems of botany or any other particular field. Nevertheless, Lambert's conviction that systematics could be treated as an extension of mathematics, and his implicit commitment to the wide applicability of the systematic model, do indicate once more the confluence of that model and the late Enlightenment's ambition to establish a general science of order.

Considered in a wider perspective, mathematics and the systematic model may also be associated as instruments of a movement of rationalization that was pervasive in the late Enlightenment. This movement found literary expression in numerous encyclopedias and dictionaries, of which the *Encyclopédie* was only the most famous.[28] It appeared in government in forms as diverse as attempts to monitor

27. Ibid., 329–37.
28. Stearn, "Introduction," 11–2.

and improve public health through statistics, formulation of the metric system, and the shaping of armies as tools of the absolutist state. It entered the empirical sciences as an effort to classify their objects and reform their nomenclatures. Often its actions, especially in the sciences, were driven by pedagogy, as professors sought to order their subjects for presentation to students. Everywhere rationalization harnessed the geometrical spirit that, in different ways, informed both mathematics and the systematic model.

The development of systematic botany and of the fields for which it came to serve as a model was conditioned by material and social factors that remain to be investigated in detail. By the mid-18th century botany had long since achieved its intellectual independence of medicine, although—as the case of Linnæus indicates—important institutional links were preserved. Exploration and empire now provided the richest opportunities and resources for botanists and zoologists. Like early Christians spreading their creed over Roman roads, 18th-century naturalists were moved by their own sort of zeal to the far-flung corners of European colonial and commercial empires, and brought or sent back the specimens that gave substance to the systematists' projects. Adanson's experience in Senegal is an excellent case in point. Mineralogy owed much to the increasing need for technical expertise in mining and metallurgy, and chemistry was stimulated by its ever-closer association with pharmacy and industry. Nosology, of course, was an integral part of medical theory. The stimulus afforded by the high popularity of natural history among the educated classes must not be overlooked. Linnæus' productive stay in Holland from 1735 to 1738 was largely supported by the patronage of a wealthy banker, George Clifford. The mineralogist, Jean Baptiste Louis Romé de l'Isle, too, long derived his sole financial support from wealthy patrons with amateur interests in natural history. Another mineralogist, René Just Haüy, was himself an amateur botanist. And the medical profession, historically sensitive to areas of science invested with prestige by the lay public, could not fail to be affected by the popular prestige accorded the natural history disciplines.[29]

29. Stearn, "Introduction," 8–10; Blunt, *The compleat naturalist*, 102–8, 116–8.

The need for rationalization was felt most acutely in fields untouched by the great synthesis of Newton and the continental mathematicians. The mathematicians had provided a new basis for sciences—including astronomy, optics, and mechanics—that had already acquired mathematical form. In other fields, for which the conceptual structure was less secure or coherent, the order of the day was the gradual establishment of foundations. For some of these, such as electricity and magnetism, the path led from discursive theorizing and experimental manipulation to mathematization.[30] For others, including chemistry and medicine as well as natural history, the systematic model of order was decisive. Where electricity and magnetism had to deal with specialized physical phenomena, the latter fields had to contend with an ever increasing quantity of diverse specimens.[31]

MINERALOGY

The pervasiveness of the model's influence between 1760 and 1810 is well exemplified in mineralogy, chemistry, and medicine. Spurred by the practical needs of mining and metallurgy and the curiosity of naturalists, mineralogy was an active field in the latter 18th century. Linnæus included a scheme for the mineral kingdom in the *Systema naturae*. His example was decisive both in his implicit commitment to the existence of mineral species, and in his use of crystalline form as classificatory criterion. The formal parallel between crystalline forms and the sexual parts of plants was strengthened by the analogy he perceived between the chemical formation of crystals

30. J.L. Heilbron, *Electricity in the 17th and 18th centuries: A study of early modern physics* (Berkeley: University of California Press, 1979).

31. "If the multiplicity of objects to be described has been the true source of this useful method, and if it was on account of this multiplicity that Linnæus believed he had to imagine a new and short method to characterize these objects, should not all the natural sciences, in which the facts and observations multiply in such a way that they require the presentation of immense details, follow the same path and adopt the same descriptive system?" Antoine Fourcroy, s.v. "Caractères," *Encyclopédie méthodique: Chimie, pharmacie et metallurgie*, 6 vols. (Paris, 1786–1815), *2* (1792), 784–5.

and the reproduction of living things, an analogy that assured constancy of species in the mineral as in the plant or animal kingdoms. Linnæus distinguished each mineral genus by a basic geometrical figure, and the species within each genus by truncation of the edges or angles of the generic figure. In keeping with his general practice, he assigned each species a genus-species binomial.

Linnæus' system entirely subordinated physical and chemical properties of minerals to the geometrical form of the crystal. In part this may be attributed to the weakness of contemporary chemical analysis and to the absence of quantitative techniques for the measurement of physical properties like hardness. Far more determinative, however, was the compatibility of crystalline form with the requirements of Linnæus' systematic model. Just as the sexual system abstracted from the plant just those visible qualities that could be expressed in numbers or spatial relationships, so did crystalline form abstract from the mineral visible external characters that could be numerically or geometrically defined.[32]

Linnæus' mineral scheme did not enjoy the success of the sexual system. Mauskopf has identified three distinct approaches to mineral classification in the last quarter of the 18th century, based on chemical analysis, groups of external characteristics, and crystal form. All three recognized the need to know chemical composition. Given the state of chemistry at the time, however, chemical criteria were very difficult to apply. The other two approaches looked for characteristics other than the chemical that would still express degrees of essential identities and differences.[33]

In 1774 Abraham Werner, a professor of mineralogy at the mining school of Freiberg, published a work entitled *On the external characters of minerals.* Aiming to produce a practical handbook for the miner and naturalist, Werner made use of readily accessible mineral characteristics like color, shape, hardness, and texture. He did not

32. John G. Burke, *Origins of the science of crystals* (Berkeley and Los Angeles: University of California Press, 1966), esp. 52–77; Seymour H. Mauskopf, "Crystals and compounds: Molecular structure and composition in 19th-century French science," American Philosophical Society, *Transactions, 66:3* (1976), 7–20.

33. Mauskopf, "Crystals," 14.

group his species in higher categories, but did restate the Linnæan concept of primary forms of crystals, and pointed out that certain forms were related and could be derived from one another by truncation.[34]

More theoretical, and more directly in the Linnæan tradition, was the work of the French crystallographers Romé de l'Isle and Haüy. Both came to mineral classification by way of crystallography. In his *Essai de cristallographie* of 1772 and his revised and expanded *Cristallographie* of 1784, Romé de l'Isle attempted a comprehensive classification of crystals based on the theory that there were a limited number of primitive crystalline forms. In his theory, the diversity of forms observed in nature arose from variations on the primitive ones induced by different conditions of solution or by varying proportions of the constituent chemical principles. The *Cristallographie* incorporated steps toward a quantitative science of crystals, including use of the contact goniometer and statement of the fundamental law of constant interfacial angles.[35]

Linnæus had not drawn a hard-and-fast line between the kingdoms of living things and minerals. He had therefore not felt a need to justify mineral taxonomy. As the line between organic and inorganic was more and more sharply drawn, however, the need for justification and explicit methodological discussion became inescapable. Prompted by an essay of Louis Daubenton that denied the existence of mineral species, Romé de l'Isle made explicit his commitment to their reality, distinctness, and fixity. In *Des caractères extérieurs des mineraux* (1784), he argued that invariable laws of chemical affinity assured the same fixity for mineral species that reproduction did for organic ones. In practice, however, he relied on more accessible external features—crystal form, hardness, density—that he presumed to be the direct expression of uniform chemical composition.[36]

34. Ibid.; Burke, *Origins*, 59–62.

35. Mauskopf, "Crystals," 9–11; Burke, *Origins*, 62–7, 69–77; R. Hooykaas, s.v. "Romé de l'Isle," *DSB, 11*, 520–4. See also R. Hooykaas, *La naissance de la cristallographie en France au XVIIIe siècle* (Paris: Palais de la Découverte, 1953).

36. Mauskopf, "Crystals," 16–8; Burke, *Origins*, 71–7.

Haüy is said to have come to crystallography from botany, seeking a mineral analogue to botanical regularities of form. If so, the botanical inspiration did not immediately extend to questions of mineral taxonomy, for in the *Essai d'une théorie sur la structure des cristaux* of 1784 Haüy denied the relevance of crystal form to mineral classification. The *Essai* introduced the concept of the *molécule constituente* (later termed the *molécule intégrante*), a theoretical entity understood as the smallest molecule of a crystal that displayed a characteristic chemical composition and geometrical form. In his emphasis on the geometry of the *molécule intégrante* and its relationship to the geometry of macroscopic crystals, and in his insistence on the agreement of theoretical and measured results, Haüy took a decisive step in the mathematization of crystallography and attracted the patronage of Laplace. He also provided an original basis for mineral taxonomy.[37]

In a paper of 1793, Haüy defined the mineral species in chemical terms, remarking that just as in botany it is reproduction that assures uniformity in the species, so in mineralogy it is the nature and proportions of the combined chemical principles that guarantee specific identity. In this sense, Haüy argued, chemistry is well suited to accomplish one of the two main purposes of method, that is, classification. For the other purpose—the ready recognition and naming of bodies—chemistry is ill adapted, however, if only because chemical analysis often requires long and laborious procedures that use up all or part of the specimen. Seldom can species be grouped into genera by a single character that is easy to recognize. Thus classification in mineralogy does not compare favorably with that in botany, "where the characters, always drawn from the figure of the organs, that is to say, from a modification that is plainly visible (*qui parle aux yeux*) follow a simple, uniform course, and have the merit of offering a picture in which a small number of colors suffice to give a rich and varied expression." In botany, unlike mineralogy, the same means serve the ends of both classification and recognition. A useful

37. Mauskopf, "Crystals," 18–9; Burke, *Origins*, 108–13; R. Hooykaas, s.v. "Haüy, René Just," in *DSB, 6*, 181–2; Roger Hahn, chap. 13 in this volume.

mineralogical method therefore involves much more groping than is the case in botany.[38]

In Haüy's major work of the early 19th century, the *molécule intégrante* figures in the chemical composition and geometrical form that defined mineral species. The *molécule intégrante* was, however, a theoretical entity. In practice the results of crystallography and chemistry sometimes diverged. Bodies grouped together by crystallography might be separated by chemistry, and vice versa. Some of this dissonance might be accounted for, Haüy argued, by the imperfection of current chemical analysis or by impurities in the mineral samples. So far as species determination was concerned, however, crystal structure offered the more certain guide: "Only for geometry are all minerals pure." Crystal form, more accessible and less ambiguous than chemical composition, dominated Haüy's determination of species, while his genera, orders, and classes were decided by chemical composition[39] (fig. 3.3).

The practical difficulties resulting from the use of both chemical and crystallographic criteria in mineral classification are reflected in Haüy's struggles with nomenclature. His ideal was a binomial based on the new chemistry by Lavoisier and his circle. The state of the art did not make the ideal possible. In practice, therefore, the unavoidable mix of chemical and crystallographical criteria blurred the clarity of Haüy's nomenclature, which remained only partially rationalized.[40]

38. René Just Haüy, "Mémoire sur les méthodes minerologiques," *Annales de chimie, 18* (1793), 225–40, on 237 (quote).

39. Haüy, *Traité de mineralogie*, 5 vols. (Paris: Conseil des Mines, 1801), and *Tableau comparatif des résultats de la crystallographie et de l'analyse chimique relativement à la classification des mineraux* (Paris: Courcier, 1809), i–xxxv, on xv. In the *Tableau comparatif*, Haüy once more showed that he had a close eye on the practice of botanists, calling crystals "the flowers of minerals," comparing crystalline structure to the organization of living things, and citing A.-L. de Jussieu as a model in the presentation of classificatory results. He also justified measures taken to enhance the unity and simplicity of his systematic methods by appeal to *l'esprit géométrique*. *Tableau comparatif*, xvii–xix, xxv. For a more detailed discussion of Haüy's double method of classification, see R. Hooykaas, "The species concept in 18th-century mineralogy," *Archives internationales d'histoire des sciences, 18* (1952), 45–55.

40. R. Hooykaas, "Haüy," *DSB*, 181–2.

CHEMISTRY

In mineralogy the two modalities of the geometrical spirit—systematics and mathematics—intersected at Haüy's theory of crystal structure. A similar convergence and intersection may be seen in chemistry, though with a difference. Whereas in mineralogy the point of contact between systematics and mathematics lay in criteria of classification, in chemistry it was most pronounced in the establishment of rationalized nomenclature. The chemists' reform of their system of naming was inspired by algebra, legitimized by philosophy, and modeled on botany.

By the third quarter of the 18th century, pressure was mounting for reforms in the language of chemistry. A growing list of substances, of which the newly isolated atmospheric gases formed only a part, raised the problem of how names should be formulated. Criticism of the dominant phlogiston theory highlighted the potential theoretical content of chemical names. The views of the philosopher Condillac, for whom a science was a well-made language based on the natural order of mental processes and on exact correspondences between words and things, were gaining increasing currency among French savants.

Only after Linnæan systematics became available as a model and was so perceived by chemists was substantial progress made. The initiative came from Linnæus' student, the professor of chemistry at the University of Uppsala, Torbern Bergman. Bergman was disturbed by the lack of system and order in chemical names. A name might be based on the appearance or properties of a substance, its place of discovery or occurrence, the name of its discoverer, or its alchemical association with the planets. As new substances became known to chemists, they were assigned names ad hoc. The resulting confusion and imprecision of language made it difficult for aspiring chemists to master their subject and for established chemists to communicate with colleagues. In the 1770s Bergman set out to formulate a binomial system that would do for chemistry what Linnæus had done for botany.[41]

41. Maurice P. Crosland, *Historical studies in the language of chemistry* (Cam-

Bergman's project gained a positive reception from the French chemist Guyton de Morveau. In a paper of 1782, Guyton cited the rapid increase in the number of known substances in the preceding twenty years as a major motive for reform. Another motive came from theoretical changes in chemistry as a result of Lavoisier's studies of combustion and the overthrow of the phlogiston theory. In the 1780s Lavoisier's theory was accepted by major scientists, including the physicist Laplace and the chemists Berthollet, Guyton de Morveau, Antoine Fourcroy, and Joseph Black. With the ensuing controversies, the problem of nomenclature became still more acute, since the terminology embodied theoretical views.[42]

In France, Lavoisier, Guyton de Morveau, Berthollet, and Fourcroy joined to suggest appropriate reforms. One of the central pieces of the resulting *Méthode de nomenclature chimique*, which appeared in 1787, was an article by Lavoisier explaining the principles on which the proposed reforms were based. Lavoisier's interest in precise language and nomenclature was based in part on his impression of the contrast between the systematic, logical exposition of mathematical physics and the confusion and disorder of chemistry. He had also been impressed with Condillac's writings. Citing Condillac, Lavoisier emphasized the need for control of chemical reasoning by consistent reference to observation and experiment, a requirement closely connected with the reform of nomenclature. In Lavoisier's view there was to be an exact correspondence between a fact, the idea of the fact, and the word used to express the idea.[43]

Lavoisier and his collaborators tried to arrive at a list of simple substances, that is, of bodies they could not decompose by any existing means of chemical analysis. The total of these substances came to

---

bridge, Mass.: Harvard University Press, 1962), 139–52; Douglas McKie, *Antoine Lavoisier: Scientist, economist, social reformer* (New York: Schuman, 1952), 263.

42. Crosland, *Historical studies*, 153–76.

43. Guyton de Morveau et al., *Méthode de nomenclature chimique* (Paris, 1787), 1–25; Antoine-Laurent Lavoisier, *Elements of chemistry, in a new systematic order, containing all the modern discoveries* (New York: Dover, 1965), xiii–xxxvi; Crosland, *Historical studies*, 153–92. On Condillac, see Isabel Knight, *The geometric spirit: The Abbé de Condillac and the French Enlightenment* (New Haven: Yale University Press, 1968), and Robin Rider, chap. 4 in this volume.

fifty-five. Most already had well-known names, which the reformers decided to keep unless they gave rise to confusion. If so, or if the substance was new, a new name would be given, usually derived from Greek and expressing the substance's most general properties. An example was hydrogen, so called because it was one of the constituents of water. To deal with the great many bodies composed of two simple substances, it was necessary to establish a classification. Here the binomial nomenclature modeled on botany took effect. The acids, which Lavoisier thought of as composed of oxygen plus one other simple substance, are a good example. *Sulfuric acid* is the combination of sulfur with oxygen. The similar acid containing less oxygen was called *sulfurous acid*. The metallic calces, which Lavoisier had shown to be compounds of metals and oxygen, had the generic name *oxide* and specific names derived from the names of the metals. The reforming chemists listed the simple substances, and gave a classification of their compounds with examples, in an expansive *tableau de la nomenclature chimique*. The result was a revolution in the language of chemistry that made the chemical name of a substance a direct expression of its elementary composition.[44]

Underlying Lavoisier's theory of acids was his prior commitment to the existence of a systematic order for chemicals analogous to those already established for the plant and animal kingdoms. Fourcroy made this commitment explicit. Writing in the *Encyclopédie méthodique*, he praised the Linnæan method for establishing the characters by which natural objects are recognized and described, for expressing these characters in concise phrases in which words represent precise ideas, and for reducing the description (*tableau*) of immense numbers of objects to a single comprehensive framework. Fourcroy recalled that when he began to teach chemistry his mind was full of the language and descriptions of Linnæus, and at the same time weighed down with the immense quantity of chemical properties and experiments that he found in the existing literature. The reform of chemical nomenclature in which Fourcroy had

44. *Méthode de nomenclature chimique*, 26–100 (table follows p. 100); Lavoisier, *Elements*, xxv–xxviii.

participated had been conceived and executed in a spirit "analogous to that which had directed Linnæus," and "we therefore find in modern chemical nomenclature a course similar to that adopted in natural history." Chemical compounds could be arranged in classes, orders, genera, and kinds (*sortes*) on the basis of their principal and common properties characterized simply and concisely. In the article "*axiomes chimiques*," Fourcroy presented such a scheme, which included 34 genera of salts estimated to total some 240 species. Evidently prompted by Lavoisier's naming of oxygen, Fourcroy suggested that the three species making up the genus alkalis are all formed by combinations of nitrogen with other substances, and that nitrogen or *azote* should therefore be called *alcaligène*. Lavoisier and Fourcroy were soon proved wrong in their theories of acids and alkalis, but both theories testify to the strength of the impulse to establish clear generic categories for chemistry.[45]

MEDICINE

If the role of the systematic model in chemistry was centered on the problem of nomenclature, its influence in medicine was most clearly perceptible in questions of definition, description, and classification of disease. The desire to bring order to a large and ill-integrated body of knowledge, disillusionment with the results of mechanistic or chemical physiology in practical therapeutics, and a need to overcome the opposite extreme of radical skepticism, were among the factors behind the development of nosology, as medical systematics came to be called.

45. *Encyclopédie méthodique, chimie, pharmacie, metallurgie, 2*, articles "Alkalis," "Axiomes chimiques," and "Caractères," on 20–9, 455–89, 784–5, resp. The formlessness that Fourcroy found in chemistry when he began to teach the subject in the 1770s had also been noted by J.H. Lambert, who pointed to chemistry textbooks as examples of "a very inferior degree of order" in the arrangement of the different parts of a science. Lambert, "Essai de taxéometrie," 337. The importance of generic categories in shaping chemical perception and research is discussed for a particular case in John E. Lesch, "Conceptual change in an empirical science: The discovery of the first alkaloids," *Historical studies in the physical sciences, 11* (1981), 305–28.

The beginnings of the systematic model in medicine appear early in the writings of the English physician of the late 17th century, Thomas Sydenham. A friend of Locke and admirer of Francis Bacon, Sydenham called for the setting aside of hypotheses and philosophical systems in favor of "a natural description or history of all diseases." He took the significant step of asserting that as there were species of plants, so too there were species of disease. This implied that diseases were distinct entities, not merely disturbances blending into one another, and that these distinct entities could be systematically grouped or classified. In practice Sydenham concentrated on exact definition and description. He gave concise accounts of smallpox, dysentery, cholera, plague, and other diseases. He was the first to differentiate between measles and scarlet fever, and his description of gout made possible its separation from rheumatism. For Sydenham specificity of disease implied specificity of remedy, and it was this idea, as much as the example of cinchona bark for malaria, that was behind his doctrine of specific medicines.[46]

The 18th century continued Sydenham's botanical approach while shifting its emphasis from description of individual species to comprehensive classifications using higher taxonomic categories. The first such effort was made by François Boissier de Sauvages, a professor at the medical school of Montpellier familiar with contemporary botany. Sauvages' *Treatise on the classes of diseases* of 1731 arranged its objects in classes, orders, and genera as well as species. His major work carried the instructive title *Methodical nosology, in which diseases are arranged by classes according to the system of Sydenham and the order of the botanists.* Like Sydenham, Sauvages distrusted contemporary physiological theories and rejected the idea that a classification of diseases could be based on knowledge of their

46. Knud Faber, *Nosography in modern internal medicine* (New York: Hoeber, 1923), 5–49; Lester S. King, *The medical world of the 18th century* (Huntington, N.Y.: Krieger), 193–226; Michel Foucault, *The birth of the clinic. An archeology of medical perception* (New York: Pantheon Books, 1973), 3–21; and Sergio Moravia, "Philosophie et médecine en France à la fin du XVIIIe siècle," *Studies on Voltaire and the 18th century, 89* (1972), 1089–1151, esp. 1129–38. On Sydenham, see also *The works of Thomas Sydenham*, 2 vols. (London: The Sydenham Society, 1848–50), *1*, 3–24.

underlying causes. He insisted that classification be based instead on study of directly observable symptoms. Species definition depended largely on designation of the various circumstances in which symptoms might appear. Species proliferated accordingly, finally reaching 2,400, divided into 315 genera, 44 orders, and 10 classes.[47]

Sauvages set the pattern for the nosology of the latter 18th century. Linnæus had been a medical student when Sauvages' first book appeared in 1731. Subsequently the two men corresponded, became friends, and influenced one another. When he became professor of medicine at Uppsala, Linnæus based his lectures on Sauvages' nosological system. In 1763, the year that Sauvages' *Methodical nosology* appeared, Linnæus published his own work on *Genera morborum*. Given the long-standing botanical association of nosology, the increasing success and prestige of Linnæus' botanical system after midcentury no doubt strengthened the appeal of the nosological approach to medicine. There is a clustering of nosological treatises in the 1760s and 1770s. Vogel published at Göttingen in 1764; Cullen, at Edinburgh in 1772; Macbride, at Dublin in 1775; Sagar, at Vienna in 1776; and Vitel, at Lyons in 1778. All of these works followed Sauvages in departing from Sydenham's original emphasis within the systematic program. Rather than seeking new, more accurate descriptions of disease, the nosologists of the latter 18th century took existing descriptions and tried to catalogue and group them, usually on the basis of symptoms.[48]

The nosologists' insistence on observable symptoms as criteria of classification reveals in several ways the kinship of their enterprise with contemporaneous systematic endeavors in other fields. It is analogous to the use of visible external characters in botany and

47. François Boissier de Sauvages, *Nouvelles classes de maladies, qui dans un ordre semblable à celui des botanistes, comprennent les genres et les especes de toutes les maladies, avec leurs signes et leurs indications* (Avignon, 1731), and *Nosologia methodica sistens morborum classes, genera et species juxta Sydenhami mentem et botanicorum ordinem*, 2 vols. (Amsterdam: De Tournes, 1763). See also King, *The medical world*, 205–14.

48. Linnæus, *Genera morborum, in auditorum usum* (Uppsala, 1763); Fredrik Berg, "Linnés systema morborum," *Uppsala Universitets Årsskrift*, 1957:3, 1–132; King, *The medical world*, 198–204; Faber, *Nosography*, 25–6.

mineralogy and appears consonant with their empiricist philosophical posture. It is also artificial, not in the sense of limiting the numbers of external characteristics to be considered, but in the sense of ruling out, at least for the moment, consideration of the structural and functional processes underlying disease.

The Paris clinical school of the early 19th century would challenge what it took to be the superficiality of this approach and insist that diseases be studied in anatomical depth, in the structural changes they produced in the organs and tissues of the body. The transition can be seen in the work of the Paris physician Philippe Pinel. Pinel's *Nosographie philosophique, ou la méthode de l'analyse appliquée à la médicine*, first appeared in 1798, and went through six editions in two decades (fig. 3.4). On the surface it is a work of the 18th century. To be sure, Pinel was critical of the "overloaded tables" and "arbitrary and vacillating" classifications of his predecessors. He granted, however, that there was an "absolute necessity" for some such method to save physicians from uncertainty, perplexities, risk, and precipitous decisions, and to save patients from mistakes. Pinel accepted the validity of the systematic model for medicine, and attempted to describe and classify the full range of known diseases.

In making his nosography "philosophical" and in citing the "method of analysis," he referred to the same views of Condillac that had helped motivate Lavoisier's reform of chemical nomenclature and Haüy's revision of the language of mineralogy. He declared that he would replace the earlier medical motto, "Given a disease, find the remedy," with a new motto of his own: "Given a disease, determine its true character, and the rank that it must occupy in a nosological table." Viewed more closely, however, Pinel's approach does not exactly correspond to that of his predecessors:

> We must make every effort [he wrote] to introduce into medicine the method now followed in all the other parts of natural history, that is, a severe exactitude in descriptions, precision and uniformity in nomenclature, a wise reserve in rising to general views without giving reality to abstract terms, and a simple, regular classification founded invariably on the relation of the structure or organic functions of the parts.

The last crucial phrase separates Pinel's work from the nosologies of the 18th century. For Pinel a valid classification had to be based not only on symptoms but also on pathological anatomy. The class of inflammations—to give only one example—Pinel divided into orders on the basis not of symptoms but of the kind of membrane that was attacked. The result was a system greatly simplified by comparison with earlier nosologies. It included 5 classes, 80 genera, and fewer than 200 species of disease. As Pinel's *Nosography* went through successive editions, the classificatory scheme itself became less and less prominent. With the last edition of 1818, the 18th-century tradition of nosology came to an end.[49]

The varied expressions of the systematic model in mineralogy, chemistry, and medicine only begin to indicate the omnipresence and diversity of its applications in the late Enlightenment. Some idea of this variety may be gained from a brief look at several of the model's less predictable incarnations, in mathematics, physics, mechanics, the theory of machines, physiology, anatomy, and materia medica.

VARIA

If the geometrical spirit is found in systematics, the systematic spirit is also found in geometry, more precisely in analytic geometry. The field that joined algebraic analysis and geometry for the benefit of each was the subject of intense activity in the first half of the 18th century. The study of curves was taken up first primarily as a means to determine the roots of equations. Increasingly curves came to be of interest in themselves, however; and by midcentury they had become a candidate for systematization. Leonhard Euler's *Introductio in analysin infinitorum* (1748) and Gabriel Cramer's *Introduction à l'analyse des lignes courbes algébriques* (1750) were largely successful, though different, efforts to meet the need for systematic order acknowledged by both.[50]

49. Phillippe Pinel, *Nosographie philosophique, ou la méthode de l'analyse appliquée à la médecine*, 6th ed. (Paris: J.A. Brosson, 1818), iv–xviii; Faber, *Nosography*, 28–30; King, *The medical world*, 224–6.

50. Pierre Speziali, *Gabriel Cramer (1704–1752) et ses correspondants* (Paris: Palais de la Découverte, 1958), 14–6; Phillip S. Jones, s.v. "Cramer, Gabriel," *DSB, 3*,

Cramer's treatise, described by Speziali as "a true encyclopedia of algebraic curves," was well received in the latter 18th century and remained a classic text well into the 19th. Although Cramer conceded that the ancients had formulated useful particular propositions about curves, he gave modern mathematics high marks for its method, its art of deducing from a single universal principle a great number of truths, submitting them to general rules, and connecting them so as to stimulate new discoveries. Algebra, the "universal key of mathematics," provided the mathematician with "an ingenious means of reducing problems to the simplest and easiest calculation that the question proposed can admit."[51] And the key to algebra was the curve. A proper distribution of curves into orders, classes, genera, and species would serve mathematicians as a "well-arranged arsenal," enabling them to choose without hesitation the arms that might serve in the resolution of a proposed problem. Cramer credited Newton, whose *Enumeratio linearum tertii ordinis* had first appeared in 1704, with opening the way to such a classification.[52]

Cramer defines the order of algebraic curves "according to the degrees of their equations....Thus one may form, for each order of lines, a *general equation* that represents all the possible lines of this order." To do so, Cramer makes use of J.P. de Gua de Malves' analytic triangle, a modified version of Newton's parallelogram arrangement of the terms of algebraic equations. The analytic triangle gives the different terms of a general equation of a given degree on a horizontal line, starting with a degree zero (a) at the bottom, then degree one (by, cx), degree two (dy^2, exy, fx^2) and so on.[53]

459–62; Pierre Speziali, "Leonard Euler et Gabriel Cramer," in *Leonhard Euler 1707–1783. Beiträge zu Leben und Werk* (Basel: Birkhäser, 1983), 421–34; Carl B. Boyer, *History of analytic geometry* (New York: Scripta Mathematica, 1956), 180–91.

51. Speziali, *Cramer*, 17–8; Boyer, *History of analytic geometry*, 193–6; Gabriel Cramer, *Introduction à l'analyse des lignes courbes algébriques* (Geneva: Les Frères Cramer & Cl. Philibert, 1750), vi–vii; Moritz Cantor, *Vorlesungen über Geschichte der Mathematik*, 2d ed., 4 vols. (Leipzig: Teubner, 1907–13), *3*, 605–9, 823–41.

52. Cramer, *Introduction*, viii–ix; W.W. Rouse Ball, "On Newton's classification of cubic curves," London Mathematical Society, *Proceedings, 22* (1890–91), 104–43; Boyer, *History of analytic geometry*, 138–40, 146–7; and D.T. Whiteside, ed., *The mathematical papers of Isaac Newton*, 8 vols. (Cambridge: Cambridge University Press, 1967–81), *1*, 155–212; *2*, 10–88; *4*, 346–405; and *7*, 565–655.

53. Using the triangle Cramer derives the formula $v^2/2 + 3v/2$ for the number of

The subdivision of the first five orders of curves into their classes and genera is accomplished via the number, nature, and position of the infinite branches of each type of curve. The second order contains three classes: the ellipse, of which the circle is one species; the hyperbola; and the parabola. In essentials, Cramer follows Newton's *Enumeratio* in his division of third-order curves into four classes and fourteen genera. His fourth order has nine classes, based on the number and the hyperbolic or parabolic character of the infinite branches of its curves. For example, the eighth class contains the curves that have six hyperbolic branches. This class contains three genera: those curves that have only two, nonparallel asymptotes; those that have three asymptotes, of which two are parallel; and those that have three nonparallel asymptotes. The fifth order has eleven classes, defined by procedures similar to those for the fourth order (fig. 3.5).[54]

Cramer's classification, although comprehensive for its subject, was confined to plane curves. In the 1770s several mathematicians, including Euler, Joseph Louis Lagrange, and Gaspard Monge, began to take analytic geometry into three dimensions. From the outset of his epoch-making work in this field, Monge had as one of his concerns the groupings of its objects. In his *Feuilles d'analyse* (1795 and 1801) he discussed some twenty families of surfaces defined by their mode of generation.[55]

The *Feuilles d'analyse* first appeared as notes to Monge's course at the Ecole polytechnique, where as professor he was responsible for solid analytic geometry. Finding himself without a satisfactory

---

coefficients of the general equation of degree v, and concludes from this that a curve of order v can be made to pass through $v^2/2 + 3v/2$ points. The demonstration leads him into presentation of a rule for solving v linear equations in v unknowns. This rule, together with a paradox developed out of the same formula, had its own subsequent history. See Cramer, *Introduction*, xi, xiii–xiv, 52–79, esp. 52–60; Jones, "Cramer," 460–1.

54. Cramer, *Introduction*, 352–99.

55. Gaspard Monge, *Feuilles d'analyse appliquée à la géométrie* (Paris, 1795); René Taton, "Une correspondance mathématique inédite de Monge," *Revue scientifique, 85* (1947), 963–89, esp. 979–82; Taton, s.v. "Monge, Gaspard," *DSB, 9*, 469–78; and *L'Oeuvre scientifique de Monge* (Paris: Presses Universitaires de France, 1951), 209–20; Boyer, *History of analytic geometry*, 204–25.

textbook, he was compelled to put his results in systematic order. Similar pedagogical challenges created by the new Ecoles helped motivate other efforts at systematization and rationalization during the Revolution and the First Empire. In these efforts, the full systematic model in its Linnæan form, with logically nested categories—especially the canonical five, class, order, genus, species, variety—and binomial nomenclature, was frequently replaced by other arrangements more or less improvised for the matter at hand. Natural history categories might or might not be used, rationalized nomenclature might or might not be a goal, and other qualities of the full model such as numeration or formalism might be present in different degrees. What these undertakings did consistently have in common with the full systematic model were its emphasis on rationalization and method, its aim for comprehensiveness, and its promise of utility.

The work of Etienne Barruel, an examiner in physics at the Ecole polytechnique, is a case in point. As its title advertises, Barruel's textbook, *Physics reduced to systematic tables* (1799), offered students a complete summary of current physics in tabular form. Apart from brief prefatory remarks, the entire volume consists of thirty-eight tables, most in large fold-out format.[56]

Barruel defined physics "properly speaking" as the science that considers the properties of natural bodies, in contrast to natural history, which studies their varieties, organization, and so on, and chemistry, which considers their combination. "In the methodical order that I have adopted," Barruel wrote, "a science that has for its object the properties of bodies, cannot be subject to any other division than that of these same properties." Accordingly the first table, labeled *tableau général de la physique*, lists twenty-one fundamental physical properties, divided into two major groups. The first group comprises properties "that affect bodies in a constant manner" (extension, impenetrability, mobility, inertia, gravity). The second is constituted by properties "that affect bodies in a variable manner" (porosity, sonority, affinity, caloricity, electricity, elasticity, solidity,

56. Etienne Barruel, *La physique réduite en tableaux raisonnée, ou programme du cours de physique fait à l'Ecole polytechnique* (Paris: Baudouin, an VII).

liquidity, gaseousness, capillarity, hygrometricity, meteoricity, crystallizability, light, magnetism, galvanism). Barruel subdivides the first group into "constant and essential" properties such as extension and impenetrability, and "constant and non-essential" properties, made up of gravity alone. He subdivides the second major group into "variable properties that belong to all bodies in general," such as porosity and affinity, and "variable properties that belong only to certain bodies," for example hygrometricity and light.[57]

The *tableau général* characterizes each property in one or two lines. For example, hygrometricity is the property by virtue of which liquids capable of wetting bodies enter their pores. Light is the property by virtue of which bodies excite a vivid impression in the organ of sight. At least one additional table is devoted to each property, elaborating on such points as its manner and circumstances of appearance, the laws to which it is subject, its relation to the senses, or instruments that may make use of it. Light, to give a single example, is the subject of seven tables. The initial division is made according to whether light reaches the eye directly, after reflection, or after refraction. Among the later subdivisions under refraction is a table that considers light in relation to instruments including telescopes and microscopes, subdivided in turn into their kinds.[58]

Except for its initial division of physics according to the properties of bodies, Barruel's text does not have the form of a classification of objects of the same general type (plants, animals, minerals, chemicals, algebraic curves, etc.) according to a determined set of categories. Instead it offers an arrangement of the different aspects of knowledge about what Barruel took to be single things (here properties, such as light) in a table formed by successive dichotomizations. No rule limits the number of dichotomies, and the kinds of criteria by which they are made may vary substantially within the same table. Barruel

57. Barruel, *La physique*, 3, and table 1. In a second edition of 1806, Barruel added an introduction that expanded on the characterizations of properties. A revised *tableau général* redefined the major divisions of properties as physical and chemical. He also revised the content of the tables and changed their order, although the total number remained the same.

58. Barruel, *La physique*, table 1, 27–33.

does not attempt a rationalized system of names, and indeed he derides reliance on "a simple nomenclature" for its aridity. Barruel's was nevertheless a project of systematization and rationalization, and he remarked that "it is to be desired that the elements of all sciences were accompanied by similar tables."[59]

One who heeded Barruel's call was Gaspard Prony. In 1800, when he published his *Mécanique philosophique*, Prony could draw upon five years of teaching experience at Ecole polytechnique. The title deliberately imitated Antoine Fourcroy's *Philosophie chimique* (1792), for, like Fourcroy, Prony intended to present a synoptic view of his science systematically organized on the basis of an analysis of its elements. Every even-numbered page contained formulas, definitions, and brief discussion. Each facing odd-numbered page was divided into four columns, the first defining the letters in the formulas, the second listing items defined in the text, the third and fourth stating theorems and problems. As far as possible, Prony eliminated demonstrations and "intermediate calculations" in favor of a concise presentation of results.[60]

Prony's procedure embodied a double movement of analysis and synthesis. He began with a complex mass of material—his lectures, current knowledge of mechanics in all its detail—and analyzed it into its *matériaux primitifs*, its simplest, most fundamental propositions or elements. Out of these he then constructed the complex structure of the science of mechanics, but in such a way that its components were grouped so that their relations were transparent. In this effort of

59. Barruel, *La physique*, 3–4.

60. In five major divisions of the text he covered preliminary notions, including elementary concepts of mechanics; the mechanics of solid bodies, statics and dynamics; the mechanics of fluid bodies, hydrostatics and hydrodynamics; the application of mechanics to machines and engineering problems; and "transcendental mechanics," general propositions such as the principle of virtual velocities or the principle of least action. Gaspard Clair François Marie Riche, baron de Prony, *Mécanique philosophique, ou analyse raisonnée des diverse parties de la science d'équilibre et du mouvement* (Paris: Imprimerie de la République, an VIII). Fourcroy's work, *Philosophie chimique, ou vérités fondamentales de la chimie moderne, disposées dans un nouvel ordre* (Paris, 1792), was based on Fourcroy's article, "Axiomes chimiques" for the *Encyclopédie méthodique*, which in turn was directly inspired by Linnæan systematics. Cf. W.A. Smeaton, s.v. "Fourcroy, Antoine François de," *DSB, 5*, 89–93.

systematization, in his implicit commitment to a sensationalist theory of knowledge, and in his remarks on science as a well-made language, Prony revealed his links with the encyclopedic tradition and the philosophy of Condillac as well as the kinship of his work with other systematic endeavors.[61]

Mechanics of a more immediately practical sort was the subject of Jean-Nicolas-Pierre Hachette's *Traité élémentaire des machines* (1811). A colleague of Barruel and Prony and student of Monge, Hachette taught a course on machines for engineers at the Ecole polytechnique. The *Traité* was based on these lectures and on the ideas of Monge, who treated the theory of machines as a branch of descriptive geometry.[62]

Hachette limited his treatment to machines that transform motion of one type into motion of another type. The types could be exhaustively enumerated: continuous circular, alternating circular, continuous rectilinear, and alternating rectilinear. These four types of motion make six when taken two at a time; and to these six combinations may be added the four that result when each movement reproduces itself. Hachette concluded, therefore, that there are ten different series of elementary machines.[63] Part of his table of elementary machines is reproduced here (fig. 3.6). Each machine is represented by a picture in a small box. The table presents the ten series of machines in numbered horizontal ranks; capital letters placed above the vertical columns allow a brief designation of each box. So, for example, the box 3^{a}D designates a machine (the windmill) of the third series, which changes continuous rectilinear movement (wind) into continuous circular movement (the mill). Two series, the second and eighth, are empty, for no known machines changed continuous rectilinear into alternating rectilinear motion, or alternating rectilinear motion into itself.[64]

61. Prony, *Mécanique philosophique*, i–iv, citing Barruel's *La physique* as a model.

62. Jean-Nicolas-Pierre Hachette, *Traité élémentaire des machines* (Paris: J. Klosterman fils, 1811), x; Taton, "Monge," 477.

63. Hachette counts each pair of different motions as defining a single kind of machine, even though in these six cases the order, and thus the source of motion and the receiver of motion, may be reversed, and although he concedes that this reversal rarely occurs in the same machine. Hachette, *Traité*, 4–5, 7–8.

64. Hachette, *Traité*, 6–8, 261–9.

The inclusion of two empty series in the table reveals in a graphic way the a priori and arithmetically determined character of Hachette's systematization. He identifies an exhaustive set of possibilities, subdivides them, and distributes existing machines into the preexisting categories. Once again, mathematics intersects the systematic model in criteria of classification.

Prompted in part by the same pedagogical needs that moved Monge, Prony, and Hachette, physiologists and anatomists analyzed and reanalyzed the human body into its functional and structural elements, classified these elements, and provided them with rationalized nomenclature. Xavier Bichat, who taught surgery at the Hôtel-Dieu in Paris, ordered the vital properties into which he analyzed the phenomena peculiar to living things into classes, genera, species, and varieties, and associated the divisions with the elementary tissues yielded by his anatomical analysis. Bichat published his work in a series of textbooks. So, too, did the professors of medicine, François Chaussier and Charles Louis Dumas, who were determined to provide their students with a rationalized nomenclature and classification of the parts of the human body.[65]

Physicians and pharmacists disturbed by radical criticism of traditional materia medica and drug therapy turned to the systematic model as the key to rationalization of knowledge of medicines. Pharmacists had an especially strong motive for reform, since the legitimacy of the body of esoteric knowledge on which their claim to a professional status comparable to that of physicians and surgeons was being threatened. From the 1780s on, systematic arrangements of the materia medica proliferated, especially in France, where differing

65. William R. Albury, "Experiment and explanation in the physiology of Bichat and Magendie," *Studies in history of biology, 1* (1977), 47–131; Lesch, *Science and medicine in France*, 62–8; François Chaussier, *Tableau synoptique des muscles de l'homme* (Paris: T. Barrois le jeune, 1797), and *Exposition sommaire de la structure et des differentes parties de l'encéphale ou cerveau suivant la méthode adoptée à l'Ecole, de médecine de Paris* (Paris: Théophile Barrois, 1807), and *Planches anatomiques à l'usage des jeunes gens qui se destinent à l'étude de la chirurgie, de la peinture et de la sculpture...avec des notes et explications suivant la nomenclature méthodique de l'anatomie et des tables synonymiques*, 2d ed. (Paris: Pancoucke, 1823); Charles Louis Dumas, *Système méthodique de nomenclature et de classification des muscles du corps humain* (Montpellier: Bonnariq, Avignon et Migueyron, an V).

versions were published by Antoine Fourcroy, Xavier Bichat, C.J.A. Schwilgué, Jean Alibert, and Jean Baptiste Barbier.[66]

CONCLUSIONS

The diversity of its applications places in clear relief both the ease with which the systematic model could be brought to bear on different subject matters and its prestige and pervasiveness among late Enlightenment thinkers as a way of ordering knowledge or experience. The systematic model rode a wave of enthusiasm that carried it above and beyond its connections to any particular field of inquiry. Its successes may be readily catalogued. They are not so easily explained, although several conclusions may be ventured on the basis of the preceding analysis.

In several cases, the adoption of rationalized nomenclature or the systematic grouping of the objects of a field in hierarchical categories was the result of a direct modeling on systematic botany, and not simply the expression of a more fundamental impulse of which botany itself was one manifestation. As Foucault pointed out, botany enjoyed epistemological precedence in the classical age, if only because the externality of the significant characters of plants lent itself most easily to specification of degrees of identity and difference according to visible marks. The role of the botanical model is also suggested by the temporal priority of systematic botany, which had emerged full-fledged by the middle of the 18th century. Tentative efforts for parallel reforms in medicine, mineralogy, and chemistry are visible before consolidation of Linnæan botany, but it was only after this consolidation that substantial movement took place in other fields. The serious botanical interests evident in the early careers of such key figures as Sauvages in medicine and Haüy in mineralogy, and the involvement of Linnæus himself in the systematics of minerals and diseases, lend further support to the view that systematic botany served as a model for other fields.

66. Lesch, *Science and medicine in France*, 130–5, 145–8.

Utilization of the model was not mechanical and uniform. Each field seized on those aspects that promised to meet its own most pressing requirements. For chemists it was above all the possibility of a rationalized and radically simplified nomenclature that appealed. For mineralogists and physicians it was the concept of species, the emphasis on exact description, and the promise of a comprehensive ordering of data that counted most. For all of these fields, and for others such as analytic geometry, mechanics, the theory of machines, physiology, anatomy, and materia medica, the rationalization made possible by the systematic model was in part motivated by, and often served the needs of teaching.

The place and importance of the systematic model in the late Enlightenment was enhanced by its resonance with other elements of the time. The kinship and—on occasion—the intersection of the model with mathematics as modalities of the geometrical spirit have been mentioned. Systematics and mathematics were also associated in the prevailing epistemology, best represented by Condillac. Condillac's name and ideas reappeared again and again in applications of the systematic model to fields as diverse as chemistry, mineralogy, nosology, and anatomy. Although his inspiration came from algebra, Condillac's stress on science as a well-made language helped to motivate and sanction attempts to construct rationalized systems of nomenclature. His emphasis on the epistemological priority of sensations validated the use of externally observable features as criteria of classification. And his notion of analysis underlined the importance of seeking the elementary constituents out of which any larger whole was constituted.

In their efforts to systematize and rationalize all knowledge within their chosen domains, proponents of the systematic model moved easily with the broader currents of the encyclopedic movement. As W.T. Stearn has noted, Linnæus was a born encyclopedist in the extent of his knowledge of particulars, in the clarity and concision of his expression, in his industry, perseverance, and talent for methodological organization, and in his attention to utility. Adanson not only contributed to the *Encyclopédie*, but conceived vast encyclopedic ambitions of his own. Systematic efforts in particular subject areas were easily subsumed under the more comprehensive task of organizing all knowledge, as the articles on special fields within the *Encyclopédie méthodique* make plain.[67]

By the time that Bichat published his books on physiology and anatomy around 1800, the model's prestige and influence were already beginning their decline. Bichat himself was moving away from his taxonomy of vital properties and toward a more strictly experimental approach to physiology. While they continued to do systematics, botanists and zoologists—the latter led by Cuvier—turned increasingly to more natural systems that embraced structural and functional aspects of the organism, and to anatomical and physiological studies in their own right. While classification and nomenclature of minerals continued to be active issues well into the 19th century, the center of interest in mineralogy shifted to chemical composition and physical properties. Inorganic chemists, with rational nomenclature firmly established, moved on to new problems: proximate and elementary analysis, atomic theory, electrochemistry, and so on. Nosology was displaced by pathological anatomy.

No doubt the model proved less appropriate and productive outside botany and zoology than within them. This is clear enough in retrospect, since evolutionary community of descent provides a natural basis for hierarchical grouping of plants and animals that does not exist for minerals, chemical compounds, or diseases. To some extent the model was eclipsed by its very success. At least this was the case in botany and chemistry, where the establishment of rationalized nomenclature and classification enabled botanists and inorganic chemists to carry out more effective studies of other problems in their fields. Where these foundations were not in place, the systematic model could find new life. The appearance of burgeoning numbers of new organic compounds from the 1820s provoked fresh preoccupation with problems of naming and grouping in chemistry. To these problems the model could and did respond, with substantial consequences for chemical theory.[68]

67. Stearn, "Introduction," 11–2.

68. N.W. Fisher, "Organic classification before Kekulé," *Ambix, 20* (1973), 106–31; and *21* (1974), 29–52.

That is not the whole story, however. The very extravagance with which the systematic model was embraced between 1750 and 1810 suggests that more was involved in its success than its appropriateness for particular sciences at a determinate stage of their development, and more involved in its demise than the evolution of those sciences to a higher stage. The resonances of the systematic model with applied mathematics as a modality of the geometrical spirit, with sensationalist epistemology, with the encyclopedic movement, and with the rationalizing endeavors of the absolutist state, suggest the coherence of systematics with general and characteristic features of the culture of the late 18th century.

4 Measure of Ideas, Rule of Language: Mathematics and Language in the 18th Century

By Robin E. Rider

In the 18th century mathematics both shone as an example and served as an instrument of enlightenment: it was at once rational, universal, certain, precise, unambiguous. Many 18th-century authors hoped that the same could shortly be said of common language. Toward that end they sought to infuse language with rationality and to render it universal; to guarantee its certainty, increase its precision, strip it of ambiguity. Some sought to convert into mechanical processes both the derivation of words and their transmission. Throughout their struggles with the inadequacy of existing languages and the complexity of creating new ones, language reformers were guided by mathematics.

By the second half of the 18th century the close connection between the content of thought and its form of expression was a familiar philosophical refrain.[1] In his *Diversions of Purley*, for example, John Horne Tooke carefully re-examined a century's worth of views on the intimate connection between language and thought, then pushed the philosophical commonplace one step further. For Tooke, the operations of the mind were none other than the operations of language—hence, language *is* thought.[2]

1. Many authors echoed John Locke: "Though therefore it be the Mind that makes the Collection, 'tis the Name which is, as it were, the Knot, that ties them fast together." *An essay concerning human understanding* (London: Basset, 1690), 205. Cf. Hans Aarsleff, *Study of language in England, 1780–1860* (Princeton: Princeton University Press, 1967), 51; Murray Cohen, *Sensible words. Linguistic practice in England, 1640–1785* (Baltimore: Johns Hopkins University Press, 1977).

2. John Horne Tooke, *The diversions of Purley*, 2d ed., 2 vols. (London: Johnson and Hunter, 1798–1815), *1*.

Growing appreciation of the link between language and thought and renewed concern for the quality of thought gave rise to new laments about the inadequacies of existing languages and their unhappy consequences. Authors in the Age of Enlightenment followed a century-old lead in condemning the arbitrary assignment of words to things. Tooke, for instance, recalled Francis Bacon, who had likened words to a Tartar's bow: they "shoot back upon the understanding of the wisest, and mightily entangle and pervert the judgment."[3]

Language studies in the 18th century also addressed questions of syntax—the structure of phrases and sentences—and what was called universal or general grammar. The latter was defined in the *Encyclopédie* as the "invariable and universal rule that must serve as the foundation for the particular construction of any given language." General grammar thus rested on the premise that all languages follow "the laws of logical analysis of thought; and these laws are invariably the same everywhere and always." Or, more pointedly, "a sane logic is the foundation of grammar."[4]

AN INSTRUMENT OF ANALYSIS

Armed with the view of language as a species of logic, and mindful of the power of language to deceive or mislead, writers on language in the latter part of the 18th century looked for ways to harness language in the service of enlightenment. For many,

3. Likewise John Wilkins, with others in the early Royal Society, had called for the "unmasking [of] many wild errors, that shelter themselves under the disguise of affected phrases." Bacon and Wilkins as quoted in Tooke, *Diversions of Purley, 1*, 31–3.

4. *Encyclopédie, 7* (1757), s.v. "Grammaire." This view derived from a precept underlying the Port-Royal grammar and logic texts of the late 17th century—that language reflects the processes of rational thought. On the shift in emphasis from the lexical to the syntactic, see Cohen, *Sensible words*, 26, 41, 51–2, 60, and Stephen K. Land, *From signs to propositions. The concept of form in eighteenth-century semantic theory* (London: Longman, 1974). On the evolution of general grammar in the 18th century, see Aarsleff, *The study of language in England*, 14–5; James Knowlson, *Universal language schemes in England and France 1600–1800* (Toronto: University of Toronto Press, 1975), 147; and Cohen, *Sensible words*, 101–2.

mathematics, the embodiment of enlightened values, served as a guide. Mathematics was eminently rational to 18th-century eyes; its symbols and results were truly international; its reasoning was at once powerful and certain. More than that, at the heart of 18th-century mathematics stood analysis, and the method of analysis enjoyed privileged status among the *philosophes.* Small wonder that, in an age that prized the rational and the universal, mathematics—that sharp and sure instrument of reason—offered inspiration and example to the reformers of language.

The privileged status accorded to the philosophical method of analysis during the Enlightenment helped to focus attention upon its manifestation in mathematics, and in particular upon algebra as a powerful instrument of analysis,[5] especially in the hands of a master. Few could rival the mastery of Leonhard Euler. In Condorcet's view, Euler "sensed that algebraic analysis was the most comprehensive and certain instrument one can employ in all sciences, and he sought to render its usage general. This revolution...earned [Euler] the unique honor of having as many disciples as Europe has mathematicians."[6] The power of the Eulerian formulation of infinitesimal analysis provided dramatic confirmation of the mathematical potential of algebra and lent credence to the view of algebra as the fundamental language of mathematics.

Algebraic analysis found a vigorous advocate in Etienne Bonnot de Condillac, whose treatises on philosophical analysis, language, and mathematics explicitly invoked mathematics as a model for the perfection of language. Condillac's treatise on logic rejected prevailing approaches to logic and deprecated the grand philosophical *systèmes* of the 17th century. What was needed for building knowledge, Condillac argued, was not "to imagine ourselves a system," but instead to

5. See Robin E. Rider, *Mathematics in the Enlightenment: A study of algebra, 1685–1800* (Ph.D. dissertation, University of California, Berkeley, 1980; *DAI*, 42/01A, 351), esp. chaps. 3 and 9. Cf. John Fauvel and Jeremy Gray, eds., *The history of mathematics: A reader* (London: Macmillan, 1987) on style and formalism in 18th-century mathematics.

6. Marie Jean Antoine Nicolas Caritat, marquis de Condorcet, "Discours sur les sciences mathématiques," in Condorcet, *Oeuvres*, ed. A. Condorcet O'Connor and M.F. Arago, 12 vols. (Paris, 1847–49), *1*, 453–81, on 467.

"attend to what nature teaches us." And what nature teaches us is to analyze: "analysis is the only method by which accurate knowledge is to be acquired."[7]

Condillac offered graphic examples of analysis at work. When you order a copy of a sample dress, seamstresses "will naturally perceive that it is necessary to take [the sample] apart and remake the pattern of each part." When you throw open a window and look at a landscape, you take the scene apart by focusing on first one object, then another. "We make this decomposition only because an instant is not sufficient for us to study all those objects. But we only decompose in order to recompose." Analysis for Condillac is thus nothing more than observing "in a successive order the qualities of an object, so as to give them in the mind the simultaneous order in which they exist."[8]

Condillac threw open a window on our thought processes themselves. "To see distinctly all that at once offers itself in my mind, I must decompose it, as I have decomposed what offered itself to my sight; I must analyse my thoughts." In his enthusiasm, he contended that analysis, which some called the method of invention, "has made all discoveries": "by the medium of analysis we become capable of creating arts and sciences." He thus echoed Voltaire's dictum—"Let us make an exact analysis of things"—which served as a motto for the Enlightenment. It was clear to Condillac that seamstresses knew more than philosophers, who used synthesis, not analysis, to wrap up the truth "in a heap of vague notions, opinions, errors" and who "strayed prodigiously, when they forsook analysis" for the "tenebrous method" of synthesis.[9]

In his depiction of language as an instrument of reason, Condillac also renewed the critique of existing languages and their unhappy effects on thought. In an oft-quoted passage, he attacked philoso-

7. Etienne Bonnot de Condillac, *The logic of Condillac*, edited and with a preface by Daniel N. Robinson (Washington: University Publications of America, 1977), 7–8, based on Joseph Neef's English translation (published in 1809) of *La logique* (1780), part of the course of study Condillac devised for Ferdinand, duke (and later) prince of Parma, whom Condillac served as tutor during the years 1758–67.

8. Ibid., 13, 10, 11.

9. Ibid., 11, 13, 67, 76, 73.

phers as "subtle, singular, mysterious, visionary, unintelligible, [who] often seemed to be afraid of not being obscure enough, and affected to cover with a veil their real or pretended knowledge. Therefore, the language of philosophy has been nothing else but a jargon of gibberish for many centuries past." The tirade continued, "The art of misapplying words was for us the art of reasoning: arbitrary, frivolous, ridiculous, absurd, it had all the vices of disordered imaginations."[10]

Analysis, that lever of the mind, offered Condillac a solution to the problem of language: "To speak so that we may be understood, we must conceive and express our ideas in the analytical order which decomposes and recomposes each thought." The result will necessarily be "clearness and precision." Condillac then took the argument one step further: "The art of reasoning is in truth only a well constructed language." He rejected the remedy, often proposed, of "seeking in words their essential qualities"; instead, he exhorted us to seek in words only "what we placed in them, the relation of things to us, and those relations which they bear to each other." Nature displays to us a natural order; by our analysis of nature we perceive and recreate that order in our mind. Language should reflect a natural taxonomy that corresponds to the "order of our ideas." Condillac concluded that a well-constructed language is nothing other than an analytical method.[11]

Condillac was captivated by the power and achievement of the 18th-century brand of mathematical analysis and set out to facilitate analytical thought in metaphysics by studying algebra—the fundamental language of mathematical analysis. In his treatise on logic, Condillac insisted that no distinction need be drawn between mathematical analysis and logical or metaphysical analysis, "because in all of them [analysis] leads from the known to the unknown, by reasoning; that is, by a series of judgments which are included in each other." For Condillac, the advantage of mathematical analysis consists strictly in

10. Ibid., 64, 69.

11. Ibid., 46, 12, 67, 66. On the connection between language and taxonomy, see M.M. Slaughter, *Universal languages and scientific taxonomy in the seventeenth century* (Cambridge: Cambridge University Press, 1982).

the fact "that it speaks there the most simple language"—and that most simple language is algebra.[12]

For Condillac the mathematical advances of the late 17th and especially the 18th century offered a powerful argument in favor of algebra as the language of mathematics. Algebra constituted "very striking proof that the progress of the sciences depends solely on the progress of languages; and that correct languages alone could give analysis that degree of simplicity and precision of which it is susceptible." Further, he called attention to the central role played by Euler and Joseph Louis Lagrange in putting mathematical analysis in algebraic form: they "are great mathematicians, because they are great analysists [*sic*]. They excellently write algebra, a language in which good writers are most scarce, because it is the most correct." Condillac put the point forcefully. Algebra "is a language, and cannot be any thing else." It is "a language which could not be badly constructed"; this in turn means that mathematicians who couched their analysis in algebraic language are able thereby to "speak with precision."[13]

Condillac maintained that languages are in and of themselves analytical instruments, which ought to lay bare the relations and analogies of our ideas. Most languages, in Condillac's view, were too blunt, too imprecise. But in algebra, conceived as the rules by which equations and their components could be manipulated and transformed, the language (and the analytical method) corresponded exactly to the analogies existing among mathematical ideas. As he explained it, "One goes from identity to identity until the conclusion that resolves the question, in other words, from identical equation to identical equation until the final equation." Condillac's explanation of the chain of identities sounds a good deal like a passage from d'Alembert's *Preliminary discourse* to the *Encyclopédie*: "Thus, the chain of connection of several geometrical [mathematical] truths can be regarded as more or less different and more or less complicated

12. Condillac, *Logic*, 77, 80. Cf. *La langue des calculs*, which appeared first as vol. 23 of his *Oeuvres*, 23 vols. (Paris, an VI [1798]), and was reprinted in *Oeuvres philosophiques*, ed. Georges Le Roy, 3 vols. (Paris, 1947–51).

13. Condillac, *Logic*, 82, 75, 82, 62; cf. *La langue des calculs.*

translations of the same proposition and often of the same hypothesis." D'Alembert was describing a series of mathematical propositions deduced one from the other: "It is almost as if one were trying to express this proposition by means of a language whose nature was being imperceptibly altered, so that the proposition was successively expressed in different ways representing the different states through which the language had passed. Each of these states would be recognized in the one immediately neighboring it; but in a more remote state we would no longer make it out, although it would still be dependent upon those states which preceded it and designed to transmit the same ideas."[14]

Condillac's enthusiasm for algebra went beyond a recommendation to study its method: he provided a sort of elementary algebra textbook in the work *La langue des calculs*. In line with the interest, common during the second half of the 18th century, in the origin of language, Condillac attempted in this work to trace the origin of algebra, that most excellent of languages. He traced the development of calculating from finger-counting to words to letters and symbols. Familiar numerical expressions like 1, 10, and 100 were not introduced until well into the book, since, according to Condillac, the use of numerals followed the use of letters and symbols in the historical development of calculating. The remainder of *La langue des calculs* explained the basic operations of calculating with letters and numbers. Condillac confined his exposition to the canons by which algebraic and arithmetic quantities were to be expressed and manipulated. In stopping short of the solution of polynomial equations or their application in solving problems, he showed himself to be concerned primarily with algebra as the expression of relations among mathematical entities and as the means to make those relations reveal new facts.

Condillac's intent in writing *La langue des calculs* was not to add another volume to the lengthening shelf of elementary algebra textbooks in the 18th century, nor can his rational reconstruction of the

14. Condillac, quoted in Knowlson, *Universal language schemes*, 171. D'Alembert, *Preliminary discourse to the Encyclopedia of Diderot*, transl. Richard N. Schwab with Walter E. Rex (Indianapolis, 1963), 28.

birth of calculating be considered a serious addition to the historiography of mathematics. Nor should his work be read as a recommendation that all thought be "mathematicized" or "algebraicized." In our attempts to reform languages, we should not wield recklessly the instrument of algebra, but instead should attend to the lessons of mathematics as we devise new tools. Condillac called attention to algebra precisely because the chain of reasoning in a series of identities "is more easily perceived, when we express ourselves with algebraical signs." He issued a call to action: "If, therefore, any sciences are not very exact, it is not because we do not talk algebra in them; it is, because their languages are not correct....All the sciences would be exact, if we knew how to speak the language of every one."[15] His work thus represents an important example of the confluence of two deep philosophical currents in the 18th century: advocacy of the method of analysis in all domains of thought and the desire to perfect language. In both aspects Condillac was strongly influenced by what he knew of mathematics.

ARTIFICIAL LANGUAGES

In the last third of the 18th century, perceived flaws in existing languages called forth a spate of proposals for artificial ones free from flaws—and the very profusion of existing languages, like antiquated social structures and inconsistent systems of weights and measures, cried out for rationalization. To answer this call, several authors proposed a "pasigraphy," from the Greek terms for "universal" and "writing"—a set of rational, universal symbols each person could read in his or her own language. Consider the artificial language proposed by György Kalmár in his *Praecepta grammatica atque specimina linguae philosophicae sive universalis, ad omnevitae genus adcommodatae.*[16] The fact that Kalmár thought it necessary to publish his proposal in Latin, German, and Italian editions in the space of just two

15. Condillac, *Logic*, 81, 72.

16. Published in Berlin and Leipzig in 1772. Cf. Kalmár, *Precetti di grammatica* (Rome, 1773) and *Grammaticalische Regeln* (Vienna, 1774).

years testified to the need, as he saw it, for a universal mode of communication. Clearly, his native Hungarian would not suffice; nor, he thought, would other existing languages. All were crippled by grammatical irregularities and orthographic confusion. For his new language, Kalmár constructed both a general, rational grammar and a new set of 400 primitive characters. Kalmár's familiarity with Hungarian as well as other languages helped him to construct a general language capable of accommodating "the details, and even the anomalies, of all existing languages."[17]

Two more examples, taken from opposing camps in the political turmoil of the 1790s, illustrate further the broad appeal of the rationalization of grammar and the invention of a language all nations might share. In 1795 Jean Delormel presented to the National Convention in Paris his project for a new and universal language. Shortly thereafter Joseph de Maimieux, a nobleman who had fled to Germany and no friend of the Convention, published his own pasigraphy. Both Delormel and de Maimieux intended their schemes to further the objectives of rationalization and universal communication; both drew inspiration and justification from the *esprit géometrique.*

Delormel recognized that "extraordinary epochs" offer the opportunity, impossible in normal circumstances, to realize "interesting projects." He presented his own proposal during just such an epoch, at a moment ripe for disseminating "the principles of equality." Delormel's proposal included a binomial classification of substantives by genera and species. In tune with the call for rationalization, Delormel coupled the taxonomy with a system for the regular formation of derivative words; no longer would language need to submit to the tyranny and caprice of usage. As a reviewer of Delormel's scheme commented in 1797, "changes on words are to be rung with all the regularity of a multiplication-table."[18] Such comments doubtless

17. As noted by the polymath Johann Heinrich Lambert, himself the inventor of a symbolic logic, in a letter to de Francheville (1 Mar 1772), quoted by Kalmár, *Precetti*, 110.

18. Jean Delormel, *Projet d'une langue universelle, présentée à la Convention nationale* (Paris: Chez l'auteur, an III), 1, 3. The critic in the *Monthly review*, Sep–Dec 1797, 562, is quoted by Knowlson, *Universal language schemes*, 150.

pleased Delormel, who considered the analogy to numeration to be a prime reason for the simplicity of his scheme.

His system prescribed ten vowels, according to the spirit of the day, and twice that number of consonants. The thicket of synonyms was cleared away, replaced by seven degrees of comparison. Delormel measured the advantages to be gained: where ordinary dictionaries contained 30,000 words, he claimed that one-tenth that number would suffice in his project. More than that, he proclaimed that his new language would promote the "central unity" of the Republic and, by uniting savants of different nations, would spur the progress of science. "Enlightenment brings together men of all sorts, and this language, by facilitating communication, will propagate enlightenment."[19]

Joseph de Maimieux also billed his pasigraphy as offering all the advantages of a rational scheme of knowledge expressed in a universal form. And, he claimed, an enthusiastic audience awaited. One exuberant disciple would later honor the all-encompassing nature of de Maimieux's scheme by dubbing him a second Leibniz. De Maimieux likened pasigraphy to numerals in arithmetic, lines of music, and "characters of chemistry"—"equally intelligible from Petersburg to Malta, Madrid to Peru, London and Paris to Philadelphia or the isle de Bourbon."[20] He devised twelve characters, some of which were mirror images of one another; twelve grammatical rules universally applicable and permitting no exceptions; and three sets of tables. These sets of tables corresponded to the three species of pasigraphic words: those of three, four, or five characters respectively. Words of three characters constituted what de Maimieux called the *Indicule*. In the *Indicule*, the first character specified the relevant column (out of twelve columns); the second specified the *tranche* (six for each column); and the third, the line (one of six) within a given *tranche*. The *Indicule*, together with the *Petit*

19. Delormel, *Projet*, 4, 7, 9–10, 48, 50.

20. Joseph de Maimieux, *Pasigraphie...ou...premiers élémens du nouvel art-science d'écrire et d'imprimer* (Paris, 1797), 2, 20, which had a German translation (Paris, 1797) and new French edition as *Pasigraphie et pasilalie* (Paris, 1801). See *Biographie universelle*, 85 vols. (Paris: Michaud frères, 1811–62), s.v. "Maimieux."

Nomenclateur (for words of four characters) and the *Grand Nomenclateur* (for words of five characters), made up the second part of de Maimieux's pasigraphy. Each page of the tables contained scores of French words, arrayed according to de Maimieux's outline of knowledge.

He claimed much for this scheme. Unlike the "alphabetic chaos" of standard dictionaries, "the pasigraphic order is a natural order." Every word in the system could express "thought, state, action, or passion by means of a progressive analytic development, but without any analytic *appareil*." The twelve grammatical rules governed declension, modification, conjugation, and enunciation, and, de Maimieux puffed, yielded up great logical and grammatical riches. The tables of the *Indicule, Petit Nomenclateur*, and *Grand Nomenclateur* also provided a *mappemonde intellectuel* of visual, analytic, and mnemonic convenience.[21]

Although the printer might have complained about having to cast a new font of type (which might never be used again), de Maimieux was fortified by enthusiasm. The relative complexity of concepts would be evident at a glance, measured by the number of characters used in a given word; nature and knowledge could be surveyed with ease from an armchair. De Maimieux proclaimed his lowered expectations, at least by comparison to those held by Wilkins and others in the creation of universal characters in the late 17th century. De Maimieux aimed, not at a representation of truth, but at a handy chart to facilitate communication across linguistic boundaries.

The coordinates in this *mappemonde* covered varied lexical terrain. Column 9 of the *Indicule* concerned simple aspects of "science, grammaire, calcul." At the 6th *tranche*, line 1, de Maimieux put "Plus, au plus, de plus"; four lines down he placed "Beaucoup, bien, très, fort."[22] Cadre 6, column 6 of the *Petit Nomenclateur* listed civil acts, including *privilège, procès-verbal,* and *confronter*; more lively were the inhabitants of cadre 3, column 6, tranche 4: "Rhinoceros, girafe, onagre, zèbre, buffle, cerf, daim, rène, chamois, gazelle,

21. Ibid., 5, 22, 32.
22. Ibid., 2e Partie, 6.

grisbock, chevreuil, cabri, vigogne, musc, élan, original." In the *Grand Nomenclateur* de Maimieux mapped out more complex concepts, with columns for such diverse categories as *Dieu, etre, esprit; astres, signes, élémens; insouciance; actes religieux; meubles*; *arts chymiques*. He paid tribute to the reigning confusion of measures and money by supplying an additional four-page alphabetic, multilingual list of available units.

De Maimieux thus borrowed from the encyclopedic spirit of the age, reckoned with the profusion of measures and tongues, marshaled concepts of universal grammar, and hammered his system into a numerical matrix. By analogy to latitude and longitude, de Maimieux's characters could lead a reader straight to the location of the idea in the "topographic map of the domain of thought."[23] De Maimieux followed this with a *Carte générale pasigraphique* in 1808. Containing some 8,000 words, the tableau of 1808 was nearly as complete as the original version, but multiplied pasigraphic confusion by its incompatibility with its predecessor.[24]

Both de Maimieux and his defenders resorted to mathematical terms in describing the merits of his pasigraphies. De Maimieux saw his system as "a sort of general *glossomètre* which will give the measure of the richness of each language [and] rectify the inexactitudes of translations, in applying to languages a common scale."[25] One of his disciples went so far as to describe algebra as "that pasigraphy of quantities."[26]

De Maimieux's pasigraphy is a multidimensional variant on proposals in the 17th and 18th centuries for numerical dictionaries—polyglot dictionaries with numbered entries. All these proposals built upon what was seen as the universal intelligibility of Arabic numerals.[27] As Robert Boyle had written in the mid-17th century,

23. De Maimieux's preface to his *Epître familière* (1802), quoted by Knowlson, *Universal language schemes*, 153. Cf. ibid., 156, 160, 260, and Armand-Charles-Daniels, comte de Firmas-Périès, *Pasitélégraphie* (Stuttgart: Metzler, 1811), 21.

24. Michaud, *Biographie universelle*, s.v. "Maimieux."

25. Firmas-Périès, *Pasitélégraphie*, 28, quoting de Maimieux.

26. Ibid., 27, quoting Charles Dalberg.

27. See the discussion of universal language proposals leading to that of Leibniz, in Louis Couturat, *La logique de Leibniz d'après des documents inédits* (Hildesheim: Olms, 1969; reprint of 1901 edition), 51–3.

"Since our arithmetical characters are understood by all the nations of Europe the same way, though every several people express that comprehension with its own particular language I conceive no impossibility that opposes the doing that in words, that we see already done in numbers."[28] The idea that mathematical concepts, especially the concepts of number and their representation by Arabic numerals, were universally comprehensible, held persistent appeal to those who aimed at universal, rational languages. In 1801, for example, Zalkind Hourwitz proposed a *Polygraphie* that relied on the assignment of a number to each word in a basic polyglot dictionary. The same number thus served to designate words with the same meaning in several languages.[29] As the subtitle of his book indicates, Hourwitz intended polygraphy to facilitate the art of communication across national boundaries. Not long afterward Jacques de Cambry did likewise, explicitly invoking the blessings of numeracy in his *Manuel interprète de correspondance, ou Vocabulaires polyglottes alphabétiques et numériques en tableaux.*[30] His *mappemonde* encompassed not merely Francophones, but also speakers and writers of Italian, Spanish, German, English, Dutch, and the ever-vexing "Celto-breton."

MECHANICAL COMMUNICATION

In the final decade of the 18th century, French revolutionary expansionism lent special urgency to the problem of long-distance communication.[31] The logistic necessity of swift and sure communi-

28. Boyle to S. Hartlib (1647), in Boyle, *Works, 1,* 22, quoted in Knowlson, *Universal language schemes,* 21.

29. Zalkind Hourwitz, *Polygraphie: ou, L'art de correspondre, à l'aide d'un dictionnaire, dans toutes les langues, même dans celles dont on ne possède pas seulement les lettres alphabétiques* (Paris: Chez l'auteur, an IX). Cf. Knowlson, *Universal language schemes,* 259, on Montmignon, *La clef de toutes les langues. Biographie universelle,* s.v. "Montmignon," refers to the "numérotage des mots."

30. Paris: Mme Johanneau, 1805.

31. A number of solutions to the problem had already been tried. Aeschylus' account of the Persian wars, for example, described signal torches; beacons warned England of the approach of the Spanish Armada; the future King James II devised a system of signal flags for communication at sea. Thomas Northmore, writing in the 1790s, called the telegraph "new only in name": *A quadruplet of inventions* [etc.], 2d ed. (Exeter: Trewman and Son, 1799; 1st ed. 1796), 7.

cation with allied military units on distant fronts called for a language simple enough to be transmitted over many miles. Here was an obvious and practical instance of the importance, as proclaimed by philosophers, for a simple, clear, expressive, unequivocal language. In most of the late 18th-century proposals for an optical or military telegraph, the preferred means of transmission were mechanical; the elements of the message were geometrical or numerical.

The first telegraph to attract international attention was the invention of Claude Chappe. Chappe began by exploring the possibility of electrical telegraphy.[32] In 1790 he abandoned the electrical for the mechanical as a means of transmitting language and built a prototype optical telegraph. By 1791 he had extended its range to 15 kilometers. Revolutionary crowds, however, feared telegraphy as a royalist instrument and tore down Chappe's Parisian stations.[33]

The apparatus consisted of a 1-foot pole atop a tower or tall building, a 14-foot crossbar pivoting at the center, a 6-foot arm at each end of the crossbar, control wires by which an operator could manipulate the arms, and lanterns on the arms for nighttime operation. The crossbar could be positioned either horizontally or vertically; the arms then pivoted around the ends of the crossbar. Each arm could assume seven different positions with respect to the crossbar. Each possible position was assigned a number.

With the help of a friend familiar with diplomatic codes, Chappe formulated a manual of nearly 10,000 words. As in the numerical dictionaries,[34] each word corresponded to a number. The positions of

32. Cf. T.K. Derry and Trevor I. Williams, *A short history of technology from the earliest times to A.D. 1900* (New York: Oxford University Press, 1961), 623–4; Keith Dawson, "Electromagnetic telegraphy: Early ideas, proposals and apparatus," *History of technology, 1,* (1976), 113–42; Shelby T. McCloy, *French inventions of the eighteenth century* (Lexington: University of Kentucky Press, 1952), 41. For an account of the claims of Christoph Ludwig Hoffmann, see Hans Walter Wichert, "Ein Vorschlag zur optischen Telegraphie aus Westfalen ausdem Jahre 1782," *Technikgeschichte, 51:2* (1984), 86–93.

33. K.T. Rowland, *Eighteenth-century inventions* (Newton Abbot: David and Charles; New York: Barnes and Noble, 1974), 154; McCloy, *Inventions,* 42–3.

34. See, for example, Jacques Cambry, *Manuel interprète de correspondance, ou vocabulaires polyglottes alphabétiques et numériques en tableaux* (Paris: Johanneau, 1805), plus other examples cited by Firmas-Périès, *Pasitélégraphie,* 22–4.

the arms signaled first the page number and then the numbered word on that page. Eventually Chappe added two more code books—one containing phrases and another containing place names. Operators would then transmit sets of three signals: the first indicated which code book to use; the second, the page number; and the third, the number of the word on the page. The scheme thus resembled Hourwitz's numerical dictionary and de Maimieux's matrix of knowledge. As a reviewer commented, Chappe had coupled his mechanical apparatus with an analysis of language: "The table of characters. . .is a tachygraphic method. . .fruit of [Chappe's] long and laborious meditations."[35] In the interest of enhancing its tachygraphic possibilities, a member of the Chappe organization later suggested the use of two-person teams, one to dictate in a sort of numerical shorthand what he saw on the distant tower, the other to record the information.

Through the good offices of his brother, a deputy in the Legislative Assembly, Chappe's system came to the attention of the Assembly's committee on public instruction. After the demise of the Assembly, its successor, the National Convention, was persuaded in April 1793 to order a test "in order to determine the utility of the telegraph" and to consider the use in war of such a "rapid messenger of thought." The *commissaires* appointed by the Convention—Lakanal, Arbogast, and Daunou—were joined by "several celebrated savants and artists" on the day of the test.[36] They were apparently impressed: Lakanal's official report on the merit of Chappe's telegraph envisioned its "great utility. . .especially in wars on land and at sea, where prompt communication and rapid awareness of maneuvers can have a great influence on success."[37] The positive report led to Chappe's appointment as official *ingénieur-télégraphe* and, more telling, a high-priority claim on scarce supplies with which to construct a full-scale telegraph system. In September 1794 public enthusiasm was

35. Lakanal, *Rapport sur le thélégraphe* [*sic*] *du citoyen Chappe* (Paris: Imprimerie nationale, n.d.), 2.

36. Lakanal, *Rapport sur le télégraphe* (Paris: Imprimerie nationale, an III), 3; *Rapport sur le thélégraphe* (n.d.), 2.

37. Lakanal, *Rapport sur le thélégraphe* (n.d.), 1–2.

buoyed (and suspicions about Chappe's loyalties presumably quelled) when the newly constructed telegraph line carried to the capital news of the capture of Condé. A year later the Directory authorized continued support of Chappe's venture, and in 1798 a second line (connecting Paris and Strasbourg) was completed. Eventually the system consisted of eight principal lines and covered some 3,000 miles.[38]

News of the successful deployment of Chappe's system spread quickly. The first printed version of the "alphabet Chappe" appeared anonymously in Leipzig in 1794; periodicals carried accounts of the *bullétin télégraphique* reporting the capture of Condé; a second edition of Chappe's *Beschreibung und Abbildung des Telegraphen* appeared in 1795. The need for an optical telegraph was widely felt. From 1794, for example, committees of the Patriotic Society of Hamburg had discussed the development of a telegraph system to replace the cumbersome apparatus of messenger boats, observers, and towers used to regulate shipping traffic between Hamburg and Cuxhaven. The Society's secretary, Friedrich Johann Lorenz Meyer, also noted the military potential of a telegraph system. On a visit to Paris in 1796, Meyer witnessed a demonstration of Chappe's optical telegraph, and carried the news of its success back to Hamburg, where a select committee was convened to study its merits. Their cost-benefit analysis argued against establishing a Chappe-style system in Hamburg, on grounds both of cost (estimated at 20,000 marks) and of poor visibility in the northern German climate.[39]

38. McCloy, *Inventions,* 44–5; Rowland, *Eighteenth-century inventions*, 154; Derry and Williams, *Short history*, 622; Henri Gachot, *Le télégraphe optique de Claude Chappe. Strasbourg-Metz-Paris et ses embranchements* (Saverne: Imprimerie et Edition Savernoises, 1967), 44.

39. Henry E. Lowood, *Patriotism, profit, and the promotion of science in the German Enlightenment: The economic and scientific societies, 1760–1815* (Ph.D. dissertation, University of California, Berkeley, 1987; *DAI* 49/06A, 1563), 177–8. Cf. Friedrich Johann Lorenz Meyer, *Ueber Telegraphie und über Errichtung e. telegraph. Correspondenz zwischen Hamburg und Kuxhaven* (Hamburg, 1798), as cited in *Gesamtverzeichnis der deutschsprachigen Schrifttums (GV): 1700–1910*, 160 vols. (Munich: Saur, 1979–87), s.v. "Meyer." Meyer also promoted the use in Hamburg of the French metric system: Lowood, *Patriotism, profit, and the promotion of science*, 178–9.

The issue of the optical telegraph also engaged the attention of savants and royalty elsewhere in the German states. In Berlin, the academician and chemist Franz Karl Achard demonstrated his version of the telegraph to Friedrich Wilhelm II in 1795, successfully sending such meaningful messages as "The King is loved by his subjects just as much as he is feared by his enemies." As reward for his efforts, Achard received a substantial bonus of 500 Reichstaler, equal to one-third of his annual salary at the Berlin Academy.[40]

Another savant inspired to imitate or improve upon Chappe's invention was the Swedish academician and natural philosopher Abraham Niklas Edelcrantz. He seized on its military possibilities: "In case of war," a telegraph would afford the possibility of "quick communication for discussion between several armies or divisions of the same army."[41] Edelcrantz began his investigations in September 1794 with variants on Chappe's design of crossbars and pivoting arms. He then switched to a lattice of ten "holes" or windows with shades visible in daylight and lanterns behind them at night. This arrangement permitted 1,024 different signals. The first trials of the Swedish telegraph took place between Stockholm and Drotningholm on 30 October and 1 November 1794. Demonstrations of the system the following year were conducted in "the presence of the king, the regent, and the whole royal court."[42]

Like Chappe, Edelcrantz compiled a codebook, which he called a *telegraphische Chiffren-Tabelle*. It contained short words, syllables, and a few phrases, each corresponding to a three-digit number.[43] Each

40. Karl Hufbauer, *The formation of the German chemical community (1720–1795)* (Berkeley, Los Angeles, London: University of California Press, 1982), 206. See also *Allgemeine deutsche biographie*, s.v. "Achard"; Dieter Herbarth, *Die Entwicklung der optischen Telegrafie in Preussen* (Cologne: Rheinland-Vert., 1978); Volker Aschoff, "Franz Karl Achard und die optische Telegraphie," *Technikgeschichte, 144* (1977), 246–54.

41. Abraham Niklas Edelcrantz, *Abhandlung über Telegraphen und ein Versuch zu einer neuen Einrichtung derselben*, transl. J.J.F. Plagemann (Hamburg: Meyn und Mahncke, 1802), 82 fn. First published in Swedish (1796); translated into French as *Traité des télégraphes*, transl. Hector B. (Paris: C.F. Patris, an IX [1801]).

42. Edelcrantz, *Abhandlung*, 22–3, 25, 26–8; quotation on 87.

43. Throughout his discussion of the *Chiffren-Tabelle* (and also in his assessment of competing systems), Edelcrantz showed himself to be well acquainted with combinatorial mathematics. This was just at the time that Karl Friedrich Hindenburg and his colleagues were issuing their *Sammlung combinatorisch-analytischer Abhand-*

number designated one of the 1024 signals possible with the ten-window lattice. Edelcrantz saw his codebook as the basis for a universal language: "A cipher-table or dictionary explaining signs for all languages should be made to go along with this language-instrument, which can be completely portable."[44]

Thomas Northmore, inspired by what he knew of Polybius' telegraph, also linked his proposal for a Nocturnal or Diurnal Telegraph (which used reflecting lamps moved by a winch) with an account of his new universal character. He proclaimed the "ground-work of the whole super-structure" to be "that if the *same numerical figure* be made to represent the *same word* in all languages, an universal medium is immediately obtained." "Diversity of idioms" would prove no obstacle in Northmore's simplified language, which required as a "constant companion" only a "small pocket numerical dictionary," containing some 5,000 to 6,000 "*select* words."[45]

One of de Maimieux's royalist disciples likewise saw the connection between universal language and telegraphy. The comte de Firmas-Périès, who served in the army of the prince de Condé, recognized the versatility of de Maimieux's pasigraphy and reformulated it for use in telegraphy. Firmas-Périès' mechanical semaphore system, which he labeled "pasitelegraphy," converted de Maimieux's latitude and longitude indicators to arrangements of wooden "hands" on a large clock face operated by chains and pulleys *à la Vaucanson.*[46] To decipher the message, the operator turned to de Maimieux's tables of words (and knowledge).

Reports concerning the French telegraph, including one retrieved from the pocket of a French prisoner in 1794, also inspired English inventors, including John Gamble, chaplain of the staff of H.R.H.

---

lungen, 2 vols. (Leipzig, 1796–1800). Cf. Robin E. Rider, *A bibliography of early modern algebra, 1500–1800* (Berkeley: Office for History of Science and Technology, 1982).

44. Edelcrantz, *Abhandlung*, 86 fn., with a manuscript annotation "Pasiphrasie."

45. Northmore, *A quadruplet of inventions*, 10, 17–8, 23–4. Northmore carried the mathematical imagery further by alluding (on p. 26) to algebraic powers as analogous to the sign he used for plural words. For his account of earlier examples of numerical universal characters, ibid., 27.

46. Firmas-Périès, *Pasitélégraphie*, preface, 348–9, 355; *Biographie universelle*, s.v. "Firmas-Périès." Cf. Knowlson, *Universal language schemes*, 160.

Frederick, Duke of York and Albany. As would Edelcrantz, Gamble analyzed "the different modes which have been, or may be adopted for the purpose of distant communication"—this at the request of the Duke.[47] Gamble recognized the need for a "figurative language" adequate for communication and capable of quick transmission, and settled on a portable apparatus of five shutters ("lever boards"), which in various combinations denoted the letters of the alphabet. The use of lamps behind the shutters looked promising for nighttime transmissions.[48] The Admiralty implemented Gamble's suggestion on a series of "telegraph hills" between London and Deal as early as 1796 and found that a short message could be relayed in a minute's time. Like the other systems of the late 18th century, the Admiralty telegraph was quick (though labor-intensive) and promptly proved its value for the transmission of military intelligence.[49]

The efficiency of mechanical-optical telegraph systems itself came under scrutiny in Edelcrantz' study. He treated the visibility of signals at a distance like a problem in exact experimental physics, and assessed the accuracy of contemporary telescopes, threshold values for the angular diameter of an object visible at a given distance, and differential effects of color and humidity on visibility. The analysis was informed by Edelcrantz' familiarity with contemporary work on experimental optics, meteorology, and other branches of 18th-century physics. Indeed, Edelcrantz saw the possibility that a network of telegraph operators and stations, suitably equipped with an array of scientific instruments, might also advance the cause of science: "Every telegraph is a real Observatory, which can become an

47. From the subtitle of his *Observations on telegraphic experiments* (London, 1795?). Cf. Gamble, *An essay on the different modes of communication by signals* (London: S. Gosnell for W. Miller, 1797), 69, 71.

48. Gamble, *Observations*, 77, 13–5. On the difficulties of using lamps, see Gamble, *An essay*, 114.

49. Derry and Williams, *Short history*, 622–3. Cf. Edelcrantz, *Abhandlung*, 89, and figures 35–42; McCloy, *Inventions*, 48; Tom W. Holmes, *The semaphore: The story of the Admiralty-to-Portsmouth shutter telegraph and semaphore lines 1796 to 1847* (Ilfracombe, Devon, 1983). McCloy also describes a system devised by Richard Lovell Edgeworth as early as 1767 and put into operation in Ireland circa 1797 to contend with the threat of a French invasion. On Gamble's difficulties with the Admiralty, see Gamble, *An essay*, 89–91.

astronomical one, with the appropriate instruments and the understanding of the director, and science would be enriched with new discoveries."[50]

SIGNS AND THOUGHT

The larger problem of the connection between signs and thought once again claimed the attention of philosophers and linguists in the last third of the 18th century. In his history of universal language schemes, Knowlson has identified some of the reasons for this renewed interest. Studies of the origin of languages and of general grammar helped to stimulate thinking on the subject by focusing on the evolution of nomenclature and the logical significance of syntactical structure. What Knowlson calls a "general climate of rationalization," as demonstrated by the design of new systems of schools and of weights and measures, promoted the search for a new and rational language. Political events and trends also played their part. Hopes that a rationally constructed language would break down political and cultural barriers were fed by republican aspirations to extend the realm of *liberté, égalité*, and *fraternité*. The reconstruction of chemical language, and its happy consequences for the science of chemistry, lent an optimistic note. Condillac's views of language also attracted more attention at the end of the century as editors prepared a complete edition of his writings. It would include the first appearance in print of *La langue des calculs*, left unfinished at Condillac's death in 1780.[51]

A series of essays in the memoirs of the Berlin Academy evinces the widespread interest in "the reciprocal influence of reason on language and language on reason" (the phrase comes from the title of Sulzer's memoir of 1767). In his essay of 1774, for example, Dieudonné Thiébault argued in favor of a universal language, rejecting Johann Gottfried Herder's contention that such a language would be nearly useless. Thiébault envisioned a language with a limited number

50. Ibid., 85.

51. Knowlson, *Universal language schemes*, 149, 161–3, 172–3, quotation on 163.

of base words or concepts, a uniform system of derivation, and a regularization of grammatical relations. Such a language would provide, he claimed, "at once the best book of logic and metaphysics." Castillon took up a similar question at the end of the century, in his "Mémoire touchant l'influence des signes sur la formation des idées."[52]

Anne-Robert-Jacques Turgot, well acquainted with attempts to rationalize activities of the French state, also concerned himself with the connections of language, logic, and mathematics. He suggested that "The study of language, if well done, would perhaps be the best of logics," since language served as both the "expression" and the "measure" of thought. Language exemplified Turgot's view of progress as passage through various states of perfection, each state accompanied by new symbols. For Turgot, as his biographer Frank Manuel has explained, the language of mathematics could serve as "an impregnable barrier against retrogression"—"the armor of numbers and equations" in defense of reason. Turgot's protégé Condorcet also envisioned a philosophical language, clad in algebraic notation, applicable to all of knowledge and assured, by virtue of its mathematical foundation, of precision.[53]

Among those who thought that the whole matter of language and its relation to thought bore reexamination were the *idéologues*, who, building upon the ideas of Condillac, Turgot, and Condorcet, launched a thorough review of the question of signs and thought. *Idéologie* (the term was introduced by Destutt de Tracy) tackled the analysis of sensations and ideas, and drew upon the 18th-century discussion of general grammar and logic. The Institut national de France, and in particular its Classe des sciences morales et politiques,

52. Sulzer, in Akademie der Wissenschaften, Berlin, *Mémoires*, 1767, 413–40; Thiébault, "Observations générales sur la grammaire et les langues," ibid., *Nouveaux mémoires*, 1774, 520–2; Castillon, ibid., *Mémoires*, 1799–1800, Classe de philosophie speculative, 3–28.

53. Turgot, *Réflexions sur les langues*, quoted in Frank Manuel, *Prophets of Paris*, 30; ibid., 332–3, quotation from Manuel on 43; Keith Michael Baker, *Condorcet: From natural philosophy to social mathematics* (Chicago: University of Chicago Press, 1975). For Condorcet's view of Condillac, see Baker, "Un 'éloge' officieux de Condorcet: Sa notice historique et critique sur Condillac," *Revue de synthèse, 88:47–8* (1967), 227–51.

proved to be a hotbed of *Idéologie*, and its members debated vigorously the connection between signs and thought. To encourage consideration of this issue, the Classe set as its first prize question, "To determine the influence of signs on the formation of ideas."[54]

Among the entries was an essay by P.F. Lancelin. His work received honorable mention, and in expanded form was subsequently published under the title *Introduction à l'analyse des sciences, ou de la génération, des fondemens, et des instruments de nos connoissances.*[55] As the title page announced, Lancelin aimed at nothing less than the perfection of human reason. He saw language as the lever of the mind and the perfection of language as the best mechanism for assuring the reign of reason and for avoiding error and prejudice.[56] The mechanical metaphor is worth noting. As an Ingénieur-constructeur de la Marine française, Lancelin was a product of the technical training establishment in France, and his "analysis of the sciences" bears the stamp of his training, which saluted the military virtues of mathematics and mechanics. Lancelin's intent was to take apart the human mind in the same way an engineer takes apart a machine.[57]

The influence of Condillac is also evident. Like Condillac, Lancelin reduced the art of thinking to the use and analysis of a well-constructed language—"a language based on good observations, correct ideas, and exact facts"—and sought to reveal the development of the human mind by examining "the exact formation of mathematical languages." Language was an analytical method; properly used, it would enable the mind to soar like "an eagle over the whole globe of human knowledge." For Lancelin, as for Condillac, algebra—the language of mathematics—was just such an analytic method and as such deserved careful attention, since a method that delivers exactitude in one sphere of human knowledge will do so in all.[58]

54. Cf. Knowlson, 174–5.

55. In 3 vols. (Paris: Imprimerie de Bossange, Masson et Besson [etc.], an IX).

56. Lancelin, *Introduction, 1*, xxxv.

57. Ibid., xvi, xviii.

58. Ibid., xxx, vi, xxxviii, xxxvi, xxi, xxii, 284. See, for comparison, the words of J.-J. Rossignol, "Que l'Arithmétique nous fait aller à pied, que la Géométrie nous fait voyager à cheval et que l'Algèbre nous donne des ailes pour planer dans les airs," in his *Collection complète* (1804), 94.

Lancelin shared in the general Enlightenment enthusiasm for mathematics and professed the common faith in the correctness and certainty of mathematical reasoning. But in order to conclude that the method that gives us exact knowledge in the realm of mathematics can do so in other realms of thought, Lancelin had to consider why mathematics works at all. Mathematics deals with what Lancelin called "measurable ideas"—those formed by successive addition or repetition of identical elements. The restriction to measurable ideas clearly enhances the certainty of mathematical reasoning, Lancelin argued, but it is not the sole cause for that certainty. At least as important as a guarantee of correct reasoning is the art of mathematical signs. Or, turning the argument around, differences of opinion outside mathematics are to be blamed as much on the inexactitude of signs as on the inexact formulation of the ideas they represent.[59]

In an attempt to improve signs and thus thought, Lancelin built upon Condillac's contention that propositions, judgments, and equations are all essentially the same—that is, expressions of the formation of complex ideas out of simple elements. He concluded that ordinary sentences can usefully be translated into a sort of algebra. He offered as example the statement "gold is yellow, heavy, meltable, and malleable," which he transformed into the "equation" $O = a + b + c + d$, where O denotes gold, a denotes yellow, b denotes heavy, and so on. By a liberal use of algebra, the language of mathematical analysis, Lancelin thought he could facilitate the application of analysis outside the normal realm of mathematics.[60]

Nor did Lancelin stop there. He advanced an ambitious plan for representing "in a series of synoptic tables the ideas and languages of all peoples." His idea for tables of knowledge and nature was scarcely novel.[61] What is noteworthy, however, is the way the suggestion flowed directly from his ideas about the perfectibility of language and the privileged nature of mathematical reasoning. The whole discussion was embedded in an extravagant reform program that also

59. Lancelin, *Introduction, 1*, 90, xxxv.

60. Ibid., xxxiii, 224. Cf. 285–95, and the plate opposite 280.

61. Ibid., 297, xxxiii; cf. table at end of vol. 2. See also Slaughter, *Universal languages and scientific taxonomy*.

included the perfection of naval engineering, a complete overhaul of the French educational system, a new approach to solar astronomy, and an "experimental physics of the soul," in which the "thinking force" could be measured and expressed in an analytic formula.[62] However ambitious and impractical his plans, it is evident that they were shaped by his faith in mathematics.

A more sober and critical approach was taken by Baron Joseph Marie de Gerando in his prize-winning essay for the competition sponsored by the Institut national. In considering the relation of signs and thought, de Gerando compared analysis and synthesis in both mathematics and metaphysics, argued for the privileged nature of mathematical knowledge, and presented a careful critique of proposals for a philosophical language and an algebra of thought. His essay, published in four thick volumes, constituted a thorough reexamination of ideas on language and mathematics from the mid-17th century to the end of the 18th century.

Like so many of the authors whom he discussed, de Gerando deplored the abuse of words and urged that we hasten to discover a remedy. He thus granted the desirability of reforming language, but did not see flaws in language as the root of all wrong thinking. Instead, he blamed the imperfections of language on imperfect thought.[63]

He then compared the methods of analysis and synthesis in both philosophical and mathematical contexts and decided in favor of analysis (although not with the same determined enthusiasm as had Condillac). De Gerando was especially interested in the "prodigious successes" of algebra, which seemed to offer the greatest hope to those who sought to perfect all abstract sciences, because the use of algebra permits us both to display the relations between quantities and to eliminate extraneous information. His analysis focused on the advantage of algebra over ordinary calculation or over reasoning using words, the praiseworthy character of the signs used in algebra, and

62. Lancelin, *Introduction, 1*, xlv, and the conclusion of section III.

63. De Gerando, *Des signes et de l'art de penser considérés dans leurs rapports mutuels*, 4 vols. (Paris: Goujon fils, Fuchs, Henrichs, an VIII), *4*, 43, 238, 240, 242, 252.

the connection between algebraic signs and the nature of the ideas they represent.[64]

De Gerando also drew attention to a particular feature of algebraic calculations: there is no need at the outset, as there is in arithmetic, to determine an idea before establishing its relations to other quantities. By using signs for unknown and known quantities alike, "we are not stopped by our own ignorance." In citing the virtues of algebraic analysis, he quoted Laplace: "Such is the fecundity of analysis, that it suffices to translate particular truths into this universal language, in order to see how to proceed from these expressions to a host of new and unexpected truths. . . .Thus the geometers of this century, convinced of its superiority, have applied themselves to extending its domain and pushing back its boundaries."[65]

De Gerando's enthusiasm for mathematics in general, and for algebraic analysis in particular, was qualified by his recognition of the privileged character of mathematical ideas. He divided all ideas into four categories, of which the most straightforward and least problematic was the class of "ideas of simple modes." This class is roughly equivalent to Lancelin's category of measurable ideas. The properties and advantages of "la langue du calcul," recalling Condillac's phrase, were to be seen as a "privilege" of such simple ideas: the essential identity of the elements of such ideas means that both the nature of the elements and their mode of combination are already determined. Thus mathematicians reason only on clearly conceived and unequivocal notions.[66]

De Gerando then undertook an evaluation of schemes proposed for philosophical or universal languages in the hopes of demonstrating which reforms are desirable and which are feasible. He defined the desiderata for an artificial language—analogy of signs to ideas, analogies of signs among themselves, simplicity of the whole system as well as of its details, neat and precise distinctions between signs, and "a sufficient abundance" of signs—and described the general categories into which all philosophical languages can be divided.

64. Ibid., *2*, 211–2, quotation on 212. Cf. 218–9.

65. Ibid., *2*, 218–9, 215, 213–4; Laplace quoted in ibid., *4*, 217.

66. Ibid., *2*, 198, 203; *4*, 215.

Using these categories, and working from a thoughtful analysis of mathematical reasoning itself, he set up criteria for the evaluation of universal and philosophical languages. The conclusion was a gloomy one. He found—sometimes by use of numerical assessments of the set of base concepts or the complexity of the character systems—that such language proposals necessarily fell far short of the desired ends. Thus, for each category, including the pasigraphies of the 1790s, he found it impossible to satisfy the necessary conditions with which he began. (De Maimieux's pasigraphy came in for special criticism because it forced the distribution of ideas to conform to a scheme requiring equal numbers of division or subdivisions, regardless of content.) Hence the goal of a truly philosophical language, however desirable, was an impossible dream.[67]

De Gerando also traced the attempts to create an algebra of thought from Ramon Lull and Athanasius Kircher through Leibniz to Lambert and other 18th-century proponents of symbolic logic. Here, too, de Gerando offered a careful and ultimately pessimistic critique, warning that a true algebra of thought will require a large number of primitive ideas to be capable of handling more than very simple and general propositions. Indeed, the number of primitive ideas and symbols required will be prohibitively large. Unlike Lancelin, with his boundless enthusiasm for mathematics, de Gerando cautioned against confusing the "method of reasoning of geometers with the mechanical processes of their *calculs*." Although geometers and metaphysicians share the same method of reasoning, he argued, we cannot assume that the virtues of mathematical algebra—"the simplicity of forms and the speed of execution"—will follow from an uncritical application of its procedures to metaphysics.[68] He thus invoked the privileged character of mathematics in order to discount the likelihood that its procedures can be applied uncritically in realms of thought outside mathematics.

Although mathematics was an instructive and inspirational guide, the path chosen by 18th-century language reformers was steep and

67. Ibid., *4*, 44, 80–1, 352, 353–5, 412, 435–6.
68. Ibid., *4*, 448–9, 450, 451.

rocky. The efforts to construct artificial, universal languages made some progress toward rationalization, especially in replacing idiosyncracies of existing languages by means of new, regular schemes of word derivation. Most of these schemes were informed by at least the rudiments of combinatorics, and it was hoped that by pursuing the analogy to numbers language reformers would approach the precision and transparency characteristic of mathematical language. Establishment of a set of base words and categories capable of expanding to encompass all of human knowledge proved more difficult. Also elusive was the conversion of language into an instrument of discovery as effective and certain as mathematical reasoning.

Mechanicians of language, with their optical telegraph systems, were more successful. Taking inspiration from military need and encouragement from governmental coffers, they took account of combinatorics in constructing signaling apparatus, devised clear and unequivocal codes, and treated as essentially interchangeable words, numbers, and geometrical patterns. Using such systems, trained operators were able to transmit messages rapidly and without distortion over long distances. However, military exigencies, which called for the codebooks to remain secret, blocked hopes that telegraphic cipher-tables might serve as a universal language, at least in times of war.

Whether they met with success, as in the optical telegraph, or failure, as in the universal language schemes, efforts in the 18th century to invoke mathematics in the consideration of language were frequently self-conscious. Condillac, Lancelin, de Gerando, Gamble, Meyer, and Edelcrantz all took seriously the task of evaluating the possibility of achieving their objectives. Condillac, although he did not recommend the algebraicization of all fields of knowledge, still thought it wise to study algebra as the best available example of a well-made language. Lancelin's faith in mathematics, which exceeded Condillac's, led him to argue that it was practical to remake language in the image of mathematics. De Gerando's assessment was more careful and conservative. Edelcrantz and Gamble also looked for mathematical measures of the functioning of telegraph systems and applied considerations drawn from mechanics and experimental physics in their appraisals.

Of the 18th-century attempts to employ mathematics in the construction of language, only the optical telegraph achieved practical success. The quest for a universal or philosophical language modeled on mathematics was an uphill struggle; indeed, in de Gerando's opinion, it was assured of failure. That language reformers were so determined to ground their projects in mathematical terms, whatever the outlook for success, is yet another measure of the pervasive appeal of mathematics in the late Enlightenment.

PART 2

MORE EXACT SCIENCES

5 Late Enlightenment Meteorology

By Theodore S. Feldman

Meteorology in the 18th century meant more than observation and prediction of the weather. The complex and manifold processes occurring in the atmosphere brought it into contact with a large range of topics of natural philosophy—topics differently arranged than they are today, their boundaries less distinct and far more fluid. Meteorology naturally overlapped much of pneumatics, or the study of gases. The pneumatical parts of meteorology included the expansion of air with pressure and heat, evaporation and precipitation, latent heat, the behavior of aqueous vapor in air, and the closely related problems of the production of steam in both air and the vacuum. In the last decades of the century the discoveries of the chemical components of air and water led meteorologists to pneumatic chemistry and the chemical constitution of the atmosphere.

Meteorology also shared topics with electricity and magnetism. Atmospheric electricity, including lightning and the electricity of the air and clouds in fair and stormy weather, was a popular topic and the electrometer a familiar instrument of weather observation. Terrestrial magnetism belonged to meteorology at this time, and, especially in the last third of the 18th century, meteorologists improved the magnetic compass and discovered regularities in the magnetic variation.

Research for this chapter was supported by a Mabelle McLeod Lewis Memorial Fellowship; the National Science Foundation; the University of California, Berkeley; Lewis and Clark College; the American Philosophical Society; and the University of Southern Mississippi. Sources of 18th-century German meteorology are difficult to obtain, and I want especially to thank Henry Lowood for access to his collection on the German patriotic societies and David Cassidy for materials from the *Ephemerides meteorologicae* of the Societas meteorologica palatina. The following abbreviations are used in the notes: *PT*, *Philosophical transactions* of the Royal Society of London; SRM, Société royale de médecine.

These topics, coming from pneumatics, heat, electricity, and magnetism, belonged to natural philosophy or physics proper; a second grouping of subjects—including mechanics, geometrical optics, acoustics, and surveying and cartography—made up applied or "mixed" mathematics, as it was called.[1] Meteorology had much in common with these as well. Barometry had its 17th-century beginnings in hydrostatics, which was part of mechanics. In the late 18th century, when the barometer became useful for the measurement of heights, it was taken up by surveyors. Atmospheric refraction, rainbows, haloes, the color of the sky, and the diminution of light through the atmosphere brought meteorology into contact with geometrical optics and the new science of photometry. Mixed mathematics gave these parts of meteorology a mathematical content throughout the 18th century.

Finally, meteorology was involved with parts of the life sciences, especially agriculture, medicine, and plant and human geography. The influence of the weather on agriculture and human health involved meteorologists in the numerous projects of agricultural and medical reform of the last third of the century. Botanists such as Linnæus and Willdenow began to consider the role of climate in the geographical distribution of plants. Montesquieu's discussion of the role of climate in human geography is well known.[2]

1. Applied mathematics was distinct from natural philosophy for most of the 18th century and was quantitative, where most of natural philosophy was descriptive and nonquantitative. As has been shown in several studies, quantification in physics in part reflected an alteration in the relation between these two fields. See J.L. Heilbron, *Electricity in the 17th and 18th centuries. A study of early modern physics* (Berkeley: University of California Press, 1979), 1–19. Theodore S. Feldman, "Applied mathematics and the quantification of experimental physics: The case of barometric hypsometry," *Historical studies in the physical sciences, 15* (1985), 127–97, 128, lists further studies. The present chapter will not be concerned with this aspect of quantification.

2. For pneumatical meteorology, see Feldman, *The history of meteorology, 1750–1800. A study in the quantification of experimental physics* (Ph.D. dissertation, University of California, Berkeley, 1983; *DAI* 45/03-A, 922), section I, and W. Knowles Middleton, "Chemistry and meteorology, 1700–1825," *Annals of science, 20* (1964), 125–41. For atmospheric electricity, see Heilbron, *Electricity*. For earth magnetism, see D.H. Hall, *History of the earth sciences during the scientific and industrial revolutions* (Amsterdam, New York: Elsevier Scientific Pub. Co., 1976). For barometry, see Feldman, "Applied mathematics." For plant geography, see Adolf Engler, *Die Entwicklung der Pflanzengeographie in den letzten hundert Jahren und weitere Aufga-*

Meteorology not only overlapped these areas of natural philosophy; it enriched them. This was especially true in physics. Both the law of uniform expansion of air with heat and Dalton's law of partial pressures grew out of or were closely connected to meteorological investigations.[3] Capital discoveries in electricity in the mid-18th century were the electrical nature of lightning and the phenomenon of electrostatic induction, which derived in part from the study of the electricity of clouds.[4] A prize question on the magnetic compass led to Coulomb's discovery of the law of magnetic attraction.[5] More important than these, meteorologists made substantial contributions to the design in the 1770s and 1780s of the first precise instruments of physics. These instruments and an emphasis on systematic measurement—which together may be called "exact experimental physics"—were crucial aspects of quantification in late 18th-century natural philosophy. They generated reliable quantitative data that could be used as a foundation for mathematical laws.[6]

This chapter treats quantification in one part of meteorology: weather observation and climatology. The term "climatology" is something of an anachronism; it and its cognates are not to be found in the 18th century but made their appearance in the first years of the 19th.[7] For convenience "climatology" is used—sparingly—to

---

ben derselben (Berlin: W.H. Kuhl, 1899). For Montesquieu and other studies of the role of climate in human geography, see Clarence J. Glacken, *Traces on the Rhodian shore. Nature and culture in western thought from ancient times to the end of the eighteenth century* (Berkeley: University of California Press, 1967), chap. 12.

3. Students of barometric hypsometry—the barometric measurement of heights—investigated the expansion of air with heat in the 1770s. See Feldman, "Applied mathematics," 175–6. The law of partial pressures for mixtures of air and water vapor was discovered in the context of hygrometry in the 1770s and 1780s. Cf. Feldman, *The history of meteorology*, chap. 4.

4. For discussion of induction, see Heilbron, *Electricity*, 374.

5. Heilbron, *Electricity*, 469.

6. For discussions of exact experimental physics, see Feldman, "Applied mathematics," 151–2, 164–77, and "The history of meteorology"; Heilbron, *Electricity*, 75–83; Maurice Daumas, "Precision measurement and physical and chemical research in the 18th century," in Alistair C. Crombie, ed., *Scientific change* (New York: Basic Books, 1963), 418–30.

7. For example, W.A. Lampadius, *Systematischer Grundriss der Atmosphärologie* (Freiburg: Craz und Gerlach, 1806), chap. 4, "Climatologie"; and Leopold von Buch, "Klimatologische Betrachtungen über die Westküste Norwegens," *Annalen der Physik*, *25* (1807), 318–32.

denote the analysis of observations for weather patterns. In climatology the organizational imperative—the need for coordinated observations carried out in different locations according to a common plan—meant that strong meteorological institutions were a prerequisite for quantification. So was exact experimental physics. The two appeared almost simultaneously after 1770. In order to clarify their roles, this chapter begins with an overview of weather observation and climatology prior to 1770, then turns to quantification in the two decades 1770–90.

WEATHER OBSERVATION AND CLIMATOLOGY PRIOR TO 1770

From the invention of meteorological instruments in the middle of the 17th century, natural philosophers recognized that little could be won from individual weather diaries or registers; instead, organized groups of observers were needed. As G.A. Hamberger, Christian Wolff's mentor at the University of Jena, put it:[8]

> It is not enough to examine the state of the air in our own location, but we must direct our attention to the surrounding regions. This may best be done if well-informed persons throughout several provinces and neighboring kingdoms record [weather observations] simultaneously. . . .If many ephemerides of this type, from various locations, are published and compared, they will throw great light on [weather] phenomena.

Short-lived organizations or networks of weather observers had been set up around the middle of the 17th century by Périer, Pascal's brother-in-law; by Robert Hooke at the Royal Society of London; and by Ferdinand II of Tuscany, patron of the Accademia del Cimento in Florence.[9] Little came of their efforts and interest in

8. G.A. Hamberger, *De barometris* (1701), quoted in Gustav Hellman, *Repertorium der deutschen Meteorologie* (Leipzig: W. Engelmann, 1883), 884. As the title of the work suggests, Hamberger was discussing barometric observations in particular.

9. Hellman, "Die Entwicklung der meteorologischen Beobachtungen bis zum Ende des 18. Jahrhunderts," Preussische Akademie der Wissenschaften, Physisch-mathematische Klasse, *Abhandlungen*, 1927, 9; W. Knowles Middleton, *A history of the thermometer and its use in meteorology* (Baltimore: Johns Hopkins Press, 1966), 30–2; H. Howard Frisinger, *The history of meteorology: To 1800* (New York: Science History Publications, 1977), 101–2.

coordinated weather observation declined from about 1660 to the end of the century.

In the early years of the 18th century as few as three or four natural philosophers in all of Europe culled weather observations from newspapers and magazines and from reports of correspondents. After 1715 more energy is evident: at Breslau the physician Johann Kanold published in his quarterly *Breslauer Sammlung* observations he compiled from a dozen locations across Europe.[10] In 1723 the *Philosophical transactions* of the Royal Society of London carried James Jurin's "Invitatio ad observationes meteorologicas."[11] In it Jurin, the Society's secretary, laid out a plan for daily readings of barometer, thermometer, wind strength and direction, precipitation, and the state of the sky. Some fifteen observers, from Bengal and St. Petersburg to Cambridge, Massachusetts, responded with weather diaries, which William Derham, a Fellow of the Society and author of works on physicotheology, edited for the *Philosophical transactions.* This labor made Derham the most prolific meteorologist of the first third of the century.

Although Kanold and Jurin had ambitious hopes for their networks, the results fell short of expectations. Observers had difficulties in obtaining instruments. For example, Derham located few observations of the hard winter of 1709–10 made with instruments.[12] There seem to have been no instruments in the American colonies prior to 1716, when the physician Cadwallader Colden began observing in New York; as late as 1727 neither barometer nor thermometer was to be found in the Boston area, where Isaac Greenwood, first Hollis Professor of Mathematics and Natural Philosophy at Harvard, served as Jurin's delegate on the weather watch.[13] A number of Jurin's

10. *Sammlung von Natur- und Medicin- wie auch hierzu gehörigen Kunst- wie Literatur-Geschichten*, Breslau, 1717–30.

11. James Jurin, "Invitatio ad observationes meteorologicas communi consilio instituendas," *PT, 32:2* (1722–3), 422–7.

12. William Derham, "History of the great frost in the last winter 1708 and 1708–9," *PT, 26* (1708–9), 454–78.

13. Derham, "An abstract of the meteorological diaries communicated to the Royal Society," *PT, 37* (1731–2), 261–79, on 267; J.H. Cassedy, "Meteorology and medicine in colonial America: Beginnings of an experimental approach," *Journal of the history of medicine, 24* (1969), 193–204, on 197–9; Brooke Hindle, *The pursuit of science in revolutionary America, 1735–1789* (Chapel Hill: University of North Caro-

volunteers observed without instruments, as did most of Kanold's—instruments seem to have come late and few to central and eastern Europe, where about half of Kanold's observers were stationed.[14] A late 18th-century meteorologist at Prague reported that meteorological instruments first appeared in Bohemia in 1750 and were still rare two decades later.[15]

Even when instruments were available, their measurements were nearly useless unless the instruments were comparable—that is, unless instrument scales were interconvertible. Comparability of barometric observations posed no difficulty in principle, since the length of a mercury column serves as a natural scale for atmospheric pressure—though the variety of national and even regional units of length confused matters considerably. But the variety of scales for the thermometer greatly diminished the usefulness of temperature readings. Although Jurin attempted to secure comparable observations by asking his volunteers to specify the scale and make of their thermometers, many failed to do so. Derham found reducing their data "a matter so perplexed and difficult, as not to answer the great trouble of it."[16]

Lack of precision and reliability posed further problems. Instruments neither rendered accurate readings, nor could they be depended upon to render the same reading twice in identical situations. Precision and reliability, in fact, did not trouble early 18th-century natural philosophers.[17] Imperfect instruments generated much confusion and waste—as late as 1750 measurements made with

---

lina Press, 1956), 88.

14. Jurin distributed thermometers to his observers (as had Hooke and Ferdinand II), but they were frequently lost or broken during transit. Cf. Louise Diehl Patterson, "Thermometers of the Royal Society, 1663–1768," *American journal of physics, 19* (1951), 523–35; "The Royal Society's standard thermometer, 1663–1709," *Isis, 44* (1953), 51–63; also J.L. Heilbron, *Physics at the Royal Society during Newton's presidency* (Los Angeles: William Andrews Clark Memorial Library, 1983), 105–6.

15. Anton Strnad, "Meteorologische Beobachtungen auf das jahr 1775," Privatgesellschaft zur Aufnahme der Mathematik, der vaterlandischen Geschichte, und der Naturgeschichte, Prague, *Abhandlungen, 2* (1775), 392–9.

16. Derham, "An abstract of the meteorological diaries communicated to the Royal Society," *PT, 38* (1733–4), 101–9, 334–44, and 458–70, on 464.

17. However, see Heilbron, *Electricity,* 81.

inaccurate barometers led eminent natural philosophers to question Boyle's law.[18] As for weather observation, "how many observations have we lost," lamented one late 18th-century meteorologist, "through the imperfection and uncertainty of Mr. Hauksbee's thermometer!"[19] This was just the thermometer Jurin used and supplied to his network.

Observers were scarcely more reliable than their instruments. The discipline of recording daily temperature, pressure, humidity, winds, and cloud cover over a period of years did not come easily. As Derham put it, "these investigations require not only industry and inclination, but also leisure and means and opportunity, which you seldom find together."[20] Rather than follow Jurin's instructions, several of his informants simply submitted registers they had completed in earlier years. Under these circumstances, a consistent collection of observations was unlikely. The labor of reducing the registers delayed publication, making matters worse. In 1732 and 1733, when the project had almost ended, Derham published comparisons of the weather of 1707 at Upminster and Coventry, the weather of 1715–22 at Upminster and Cambridge, New England, and the weather of 1724 in Lund and St. Petersburg.[21] Analyses of the project's last registers, which had been submitted in 1734, did not appear until 1742.

Imprecision and unreliability of instruments, lack of agreement among scales, indiscipline in observation and inconsistency of published collections—all these factors limited the achievements of early 18th-century climatology. A thoroughgoing quantitative treatment was not possible. Instead, meteorologists described the weather and summarized observations by calculating monthly and annual means of temperature and pressure and amounts of rainfall. They aimed, as

18. Feldman, "Applied mathematics," 137–8.

19. Jan van Swinden, *Dissertation sur la comparaison des thermomètres* (Amsterdam, 1778), ix. Van Swinden was one of the chief meteorologists of the latter part of the century.

20. Quoted in Heilbron, *Physics at the Royal Society*, 108.

21. Derham, "An abstract of the meteorological diaries," *PT* (1731–2) and (1733–4).

Derham put it, to give "a just notion of the state of every month...and that which was most observable in it."[22] Even when good quantitative data were available, they did not exploit it. From Jurin's observer at the Academy of Sciences in St. Petersburg Derham received records of the temperature, pressure, winds, and general state of the weather taken three times daily during 1724 and 1725. He felt, however, that the "observations (although very curious and useful), yet being too long, would be tedious to read at the Society's meetings."[23] What sort of quantitative treatment could there be when, as was the practice at scientific academies, Derham *read* his reports at the Society's weekly meetings?

Besides describing the weather, meteorologists also drew comparisons among the locations reporting to them. Derham, for example, compared the mean annual rainfall of half a dozen European towns.[24] Comparisons aimed particularly at coincidences in weather patterns at different locations, that is, at *Meteorologica parallela.*[25] Derham repeatedly pointed out agreements among prevailing winds and storms at the towns he was comparing, as well as parallel barometric motions, which were striking. When temperature observations were not comparable, as between Zurich and Upminster, he could still see that the maxima and minima of the two series—that is, warm and cold spells—coincided. "Yea, oftentimes any remarkable weather (especially if of somewhat long continuance) affecteth one as well as the other place."[26] During one month the weather at Zurich "constantly preceded ours here [at Upminster] by about five or more days." Pieter van Musschenbroek, who among his other services to natural philosophy sponsored a network of half a dozen Dutch observers, observed a similar parallelism: "when the south-east wind blows, it arises half a day sooner at Middelbourg than at Utrecht."[27]

22. Derham, "An abstract of the meteorological diaries" (1733–4), 105.

23. Derham, ibid., 101–2.

24. Derham, "A prospect of the weather," *PT, 24* (1704–5), 1877–81.

25. David Algöwer, *Meteorologica parallela* (Frankfurt and Leipzig, 1714), cited in Hellman, *Repertorium*, 880–1.

26. Derham, "Tables of the barometrical altitudes at Zurich in Switzerland in the year 1708," *PT, 26* (1708–9), 332–66, on 333.

27. Musschenbroek, *Essai de physique*, 2 vols. (Leyden: Chez S. Luchtmans, 1739), 895ff.; Derham, "An abstract of the meteorological diaries," *PT* (1731–2).

These coincidences resulted, of course, from the fact that single weather systems cover large parts of Europe. But only one or two meteorologists understood this before the 19th century. Derham saw this much: that "the weather in both places was influenced by the same causes, whether the Alpine hills and the cold, or the influx of the moon and other heavenly bodies, or any other cause."[28]

Meteorologica parallela were one type of correlation sought by early 18th-century meteorologists. The "weather rule" was another. Meteorologists hoped to discern patterns in their data that would allow them to predict the weather. Thus Kanold expected his collection to provide a "historical-theoretical attempt to predict one storm from another." Derham derived a number of rules from the observations of Jurin's network: "a cold summer is commonly a wet one"; "western clouds bring much wind"; "the falling of the quicksilver in dark and cloudy weather betokeneth rain; but the rain is always preceded by fair weather."[29] These and other "superstitious calendar-prognostications"[30] had long been common in almanacs and popular tradition. Meteorologists hoped to place them on a scientific footing.

These weather rules, meteorological parallela, and general descriptions and comparisons of the weather reflect an approach to the natural world that has been well characterized by Michel Foucault. In his study of 18th-century medical practice he wrote, "Disease is perceived fundamentally in a space of projection without depth, of coincidence without development." Disease appears in the "space" of the human body as in a space without character, flat; the different locations in this space do not affect the disease, so that for example dyspepsia in the lower abdomen, breathlessness in the chest, and epilepsy in the head represent the same disease.[31] Just so, meteorologists perceived the space of the earth's surface as characterless. We

28. Derham, ibid., *PT* (1733–4), 342.

29. Derham, ibid., *PT* (1731–2), 265, 267, and (1733–4), 101–2, 105. John Locke also hoped to extract weather rules from collections of observations. See Locke, "A register of the weather for the year 1692," *PT, 24* (1704–5), 1917–37; Kanold, *Sammlung von Natur- und Medicin- wie auch hierzu gehörigen Kunst- wie Literatur-Geschichten.*

30. Algöwer, *Meteorological parallela*, "so abergläubischen Calender-Prognosticis."

31. Michel Foucault, *Birth of the clinic* (New York: Vintage Books, 1975), 6, 10.

would see the space between Lund and St. Petersburg as a land (and water) mass of great extent, with a richly varied topography (and depth, embracing currents, varying temperatures, varying proximity to land masses, etc.), affecting the weather across its entire compass. For Derham and his colleagues this space might as well not exist. The weather is either the same or different at Lund and St. Petersburg; Derham could compare the weather at Upminster and Coventry, or at Upminster and Cambridge, New England, as easily. So characterless, or "flat," was his perception of space that Derham could call the Alps "hills" and say that the same causes influenced the weather at Zurich, in their midst, and at Upminster, hundreds of miles away in the plains of his island nation.

Just as space, of itself, did not affect the weather, so "there is no process of evolution in which duration [i.e., time] introduces new events of itself."[32] The weather rule, which correlates weather events at succeeding times with no sense of the creative role played by those times, exemplifies this approach. "Western clouds bring much wind," but we have no sense of the connecting skein of time, of the intervening process.

All this represents a lack of synthetic vision, which appears also in meteorologists' failure to describe the climates of places. They calculated mean temperatures and pressures, found days of monthly and annual extremes, and counted the number of days of rainfall. They did not, however, synthesize this information into a characterization of climate. To borrow once more from Foucault, their calculations named the "visible" aspects of the weather, but they did not penetrate to the hidden coherence among these aspects, that constitutes climate.[33]

Nor did they integrate the collection of places they studied into a the notion of a region—so that, a fortiori, they did not discuss regional climates.[34] (Two groups did discuss regions and their

32. Foucault, *Birth of the clinic*, 12.

33. Foucault, *The order of things* (New York: Vintage Books, 1973), 132.

34. Malcolm Nicholson has pointed out that 18th-century botany and plant geography similarly lacked the concept of regionality. See Malcolm Nicholson, "Alexander von Humboldt, Humboldtian science and the origins of the study of vegetation," *History of science, 25* (1987), 167–94.

climates: plant and human geographers, mentioned above, and mathematicians who calculated the effects of the sun's heat on different parts of the earth. They belonged to traditions distinct from meteorology; neither group drew on the weather observations considered here.) Climatology, then, existed in neither deed nor word in the early 18th century; what meteorologists gathered was a natural history—a collection of descriptions of the weather here and there, at this time and another. Given the piecemeal character of their data, it would have been difficult for them to proceed otherwise.

WEATHER OBSERVATION AND CLIMATOLOGY, 1770–1790

Between the time of Jurin's and Kanold's networks in the 1720s and 1730s and the last third of the century, interest in meteorology fell off.[35] The wars of the mid-18th century disrupted cooperative efforts and undoubtedly discouraged individual observations as well.

Around 1770 the situation changed abruptly. No meteorological treatise of Europe-wide reputation had appeared for more than a century.[36] Between 1770 and 1790, however, half a dozen authors in as many countries published treatises of international renown and substantial papers populated the journals;[37] meteorology was

35. Hellmann, *Repertorium*, 978, 986. Hellmann provided data on numbers of individual observers only for the Germans.

36. See Hellmann, "Entwicklungsgeschichte des meteorologischen Lehrbuches," in Hellmann, *Beiträge zur Geschichte der Meteorologie, 6*, Veröffentlichungen des Königlichen Preussischen Meteorologischen Instituts, 296 (Berlin, 1917), 1–133, on 52.

37. Important treatises of the period include Jean André Deluc, *Recherches sur les modifications de l'atmosphère*, 2 vols. (Geneva: J.A. De Luc, 1772) and *Idées sur la météorologie*, 2 vols. (London: T. Spilsbury, 1786); Louis Cotte, *Traité de météorologie* (Paris: Imprimerie Royale, 1774) and *Mémoires sur la météorologie*, 2 vols. (Paris: Imprimerie Royale, 1788); Horace Bénédict de Saussure, *Essais sur l'hygrométrie* (Neuchâtel: Chez S. Fauche père et fils, 1783); van Swinden, *Dissertation sur la comparaison des thermomètres*; Giuseppe Toaldo, *La meteorologica applicata all'agricultura* (Venice: G. Storti, 1775). Hellmann documented a sharp rise around 1770 in numbers of meteorological publications and of observing stations in Germany. The number of publications more than doubled from the 1760s to the 1770s and increased by a further 25 percent in the 1780s. The number of "station-years" of observations doubled twice in the two decades. Hellmann, *Repertorium*, 978, 986.

"zealously pursued throughout almost the whole of Europe."[38] Organized meteorology prospered. Throughout France, Great Britain, and the United States, economic, agricultural, and patriotic societies sponsored programs of meteorological observation and research. In France a national network of observers was established in 1778 under the auspices of the Société royale de médicine, while the Societas meteorologica palatina at Mannheim constructed an international network whose work was not surpassed for three-quarters of a century. Contemporaries spoke of meteorology as "a new science."[39] Ludwig Kämtz, the chief authority on meteorology during the first half of the 19th century, agreed. It was in the last half of the 18th century, he wrote, that "this part of physics first began to be treated scientifically."[40]

Interest in meteorology derived from a number of sources, of which two particularly affected weather observation and climatology. One was the application of meteorology to agriculture and public health. Prior to the advent of bacteriological theories of disease in the late 19th century, physicians followed the ancient Hippocratic doctrine that climate, topography, and living conditions—in short, the environment—are among the chief causes of disease. The particular content of the doctrine varied. In the original Hippocratic treatise "On airs, waters, and places" the seasons influence the balance of humors in the body by virtue of what was called the "constitution" of the air: the constitution of summer being hot and dry; of autumn, cold and dry; of winter, cold and wet; and of spring, a balanced mixture of all four qualities. These constitutions favor certain groups of diseases; an abnormal season or sudden changes in the weather also cause outbreaks of disease. Winds (airs) blowing from different directions, the orientation of towns (places) facing the winds, and the towns' water supplies (waters) similarly affect disease patterns. These

38. Jean Sénébier, "Sur les moyens qu'on pourrait employer pour perfectionner la météorologie," *Journal de physique, 27* (1785), 300–15, on 301.

39. "Sur le froid de 1776," Académie royale des sciences, Paris, *Histoire*, 1776, 1–14, on 9; quoted by David Cassidy, "Meteorology in Mannheim: The Mannheim Meteorological Society, 1780–1795," *Sudhoffs Archiv, 69* (1985), 8–25, on 8.

40. Ludwig Kämtz, *Lehrbuch der Meteorologie*, 3 vols. (Halle: Gebauer, 1831–6), *1*, viii.

groups or patterns of disease attacking a population were termed the epidemic constitution.[41]

While abandoning the theory of humors, the 17th and 18th centuries retained the notion that airs, waters, and places influence the epidemic constitution. The same four qualities—hot, cold, wet, and dry—were now held to act mechanically on the body; the air might also contain disease-causing effluvial exhalations from the interior or surface of the earth. Such theories led numerous physicians to keep weather observations in the expectation of correlating weather patterns with diseases. A typical mid18th-century product was Paul Malouin's annual "Histoire des maladies épidémiques, observées à Paris, en même temps que les différentes températures de l'air," which offered qualitative descriptions of each month's weather and epidemic diseases.[42] Agriculture enjoyed nothing like the corpus of theory relating the weather to disease, but agriculturalists hoped that regular observations would succeed in correlating the weather with the success of crops. The most prominent midcentury effort of this type was Duhamel du Monceau's "Observations botanico-météorologiques," which for forty years presented annual tables of weather observations and general remarks on crops and public health.[43] Duhamel's and Malouin's series became important models for later efforts.

This genre of medical and agricultural climatology became institutionalized in the last third of the century as European states increasingly intervened in matters of public welfare. Contemporaries referred to interventionist measures as "police"; "public administration" would be a modern synonym. Medical and agricultural police needed information on the environment, diet, hygiene, and living conditions of the populace, on agriculture, and on outbreaks of disease.[44] Dozens of state and private institutions arose to fill these

41. Genevieve Miller, "'Airs, waters and places' in history," *Journal of the history of medicine, 17* (1962), 129–40; Frederick Sargent II, *Hippocratic heritage: A history of ideas about weather and human health* (New York: Pergamon Press, 1982).

42. Académie royale des sciences, Paris, *Mémoires*, 1746–54.

43. Ibid., 1741–81.

44. George Rosen, "Cameralism and the concept of medical police," and "Mercantilism and health policy in eighteenth-century French thought," *From medical police to social medicine: Essays on the history of health care* (New York: Science His-

needs. Climatology became an essential component of their programs. So strong, in fact, was the link between climatology and medicine in France during the 18th century that the French term "température" retained its ancient medical significance: it meant not the degree of heat but the "temperament" or constitution of the atmosphere.[45]

The second stimulus to climatology was exact experimental physics. A call for reform of instruments was sounded by the Genevan meteorologist Jean-André Deluc in 1772, in his *Recherches sur les modifications de l'atmosphère*, which included extensive historical and critical surveys of barometers and thermometers and his own design for a portable barometer, accurate to between one-eighth and one-sixteenth of a line.[46] Deluc discussed problems of parallax in taking readings and the proper point of the meniscus from which to read the mercury column. He demonstrated the importance of boiling the mercury in the barometer tube, a procedure that multiplied the accuracy of the instrument by a factor of 10. He showed how to use the barometer and thermometer in extensive, systematic measurement, taking all possible precautions to avoid disturbing factors. In order to establish a barometric rule for heights (i.e., a formula relating barometric pressure to altitude), he took over 400 measurements of temperature and pressure at fifteen stations on a mountain near Geneva, using great care in the exposure of the instruments and correcting for the effect of heat on both the barometer and the

---

tory Publications, 1974), 120–41, 211–9, resp.; see André Bourde, *Agronomie et agronomes en France au XVIII siècle*, 3 vols. (Paris: S.E.V.P.E.N., 1967), *3*, 1533ff., for a discussion of agricultural police in late 18th-century France; L.J. Jordanova, "Earth science and environmental medicine: The synthesis of the late Enlightenment," in L.J. Jordanova and Roy Porter, eds., *Images of the earth* (Chalfont-St. Giles: British Society for the History of Science, 1979), 119–46, esp. 136; Caroline Hannaway, "Discussion," in Abraham Lilienfeld et al., eds., *Times, places, and persons, aspects of the history of epidemiology*, supplement to the *Bulletin of the history of medicine* (Baltimore: Johns Hopkins University Press, 1980), 39–42, on 40; Jean-Pierre Peter, "Disease and the sick at the end of the eighteenth century," in Robert Forster and Orest Ranum, eds., *Biology of man in history*, transl. Elborg Forster and Patricia M. Ranum (Baltimore: Johns Hopkins University Press, 1975), 103–5.

45. Note the title of Malouin's memoirs, "Histoire des maladies épidémiques...en même temps que les différentes températures de l'air." See also the usage of the term by Turgot and Cotte and in medical topographies, as discussed below.

46. Deluc, *Recherches*. A line is 1/12 inch.

column of air whose height he was measuring. Later he made eighty-seven observations atop Geneva's cathedral. Such methods were a great novelty, and the *Recherches* were hailed as "a revolution in this part of physics."[47]

The revolution spread quickly. Within a few years Jesse Ramsden, the great English instrument-maker, was constructing barometers accurate to 1/500 or 1/1,000 inch.[48] With them two surveyor-mathematicians, William Roy and George Shuckburgh, measured mountain heights to within 0.2 and 0.7 percent, adopting Deluc's methods and his rule for heights.[49] Hygrometry—the science of measuring humidity—presented greater difficulties than barometry and thermometry, but in the 1770s and 1780s Deluc and Horace Bénédict de Saussure designed reliable, reasonably accurate, and in de Saussure's case, comparable hygrometers. By the 1790s Alessandro Volta was measuring saturation quantities of water vapor and aqueous vapor pressures to within 4.5 percent.[50] Around the same time he and other electricians devised sensitive electrometers, although they were not sure just what their instruments measured.[51] Everyone knows about the adoption of exact instruments and methods in chemistry in France.[52]

These methods transformed experimentation from a descriptive art to a quantitative science. They provided the kind of numerical data that both inspired and confirmed mathematical laws. With their emphasis on discipline, rationalization, and standardization, they were closely related to the bureaucratic impulse of the late Enlightenment.[53]

47. Cotte, *Traité*, xii.

48. Liverpool Papers, British Library, Shuckburgh in Add. MS. 38,481, ff. 1v–4r, Roy Papers, Public Record Office, Kew, MS. WO30/119, f. 26v.

49. Feldman, "Applied mathematics," 178.

50. Feldman, *The history of meteorology*, 106, 71–2.

51. Heilbron, *Electricity,* 451.

52. French chemists relied on instruments made by compatriots—Alexis Megnié and Nicolas Fortin, among others. Their instruments were of a precision comparable to Ramsden's. Cf. Daumas, "Precision measurement."

53. Charles Gillispie has pointed out that quantification in the late Enlightenment often meant standardization, rationalization, and a "spirit of accountancy." *Science and polity in France at the end of the old regime* (Princeton: Princeton University Press, 1980), esp. chap. 1, sect. 6 and p. 65.

The two reform movements—exact experimental physics and medical and agricultural reform—offered solutions to the most pressing problems of observation and climatology. Instruments were generally more reliable and accurate. Exact experimenters stressed careful specification of their make and scales. Together with the general adoption of the Fahrenheit and Réaumur scales after midcentury, this meant that the readings of different instruments could be reliably compared. The emphasis on discipline in measurement meant that observers would now read their instruments several times daily instead of the single reading common earlier in the century. (Jurin had requested one daily reading by his observers.) They would keep their weather diaries over a period of years rather than submit an out-of-date register covering a single year or less. Bureaucrats in medical and agricultural police were anxious to recruit these observers into their programs and, especially in France, had the means to enforce a proper "labor discipline." A coherent collection of weather observations became possible, and with it a quantitative climatology. A survey of late 18th-century meteorological institutions will show just where and to what degree all these possibilities were realized.

The Smaller Societies

In the German states a number of agricultural, economic, and patriotic societies took up meteorology.[54] Their emphasis on a friendly, amateurish atmosphere and a preponderance of bureaucrats among their membership[55] did not favor rigorous observation. Typically they published occasional weather observations, made with or without instruments, and reports of unusual or hard weather and its effects on crops and public health. The Gesellschaft der Naturforschenden Freunde at Berlin was one of the more active groups. Over two decades (1775–95) the society published some twenty reports of unusual weather, accounts of fog, snow, northern lights,

54. Henry Lowood gives an excellent account of the more than 200 such societies in *Patriotism, profit, and the promotion of science in the German Enlightenment: the economic and scientific societies, 1760–1815* (Ph.D. dissertation, University of California, Berkeley, 1987; *DAI* 49/06-A, 1563).

55. Ibid., 39, 76.

and the like, as well as descriptions of improved meteorological instruments. These were brief, elementary discussions. A paper of 1787 used rainfall measurements made in Berlin in the 1730s to find monthly precipitation, numbers of rainy days, and the average precipitation on rainy days of each month.[56] Another contributor described a lightning rod that doubled as an electroscope for atmospheric electricity. He used the apparatus intermittently: "I had no opportunity to observe during the whole of 1791; there were few storms, and [they occurred] at inconvenient times."[57] The use of data half a century old and a casual attitude toward observation suggest that exact experimental physics had not penetrated the Gesellschaft der Naturforschenden Freunde.

Several German societies organized meteorological networks. In Silesia Ignaz Felbiger, abbot of the Monastery of Our Lady at Sagan, established a network under the auspices of the Patriotische Gesellschaft at Breslau. Felbiger's own enthusiasm for meteorology yielded papers on lightning rods, on the art of weather observation, and on the cold winters of 1783–5,[58] and prompted him to correspond frequently with Johann Heinrich Lambert, who advised him about organizing observers.[59] Felbiger saw the project as a second but better *Breslauer Sammlung*: "with the help of mathematics the necessary instruments and methods of observation have attained a far higher level of perfection since [Kanold's] time; without this exactness and precision it is impossible to compare observations."[60] Felbiger was ambitious: his observers were to record the temperature in Fahrenheit degrees and tenths; the barometer in Paris inches, lines, and tenths; weather conditions and cloud cover; quantity of rain; wind direction and strength; humidity, by means of Lambert's new hygrometer; optical phenomena such as rainbows and haloes; and the phase of the moon. These specifications, which

56. Rosenthal, "Bestimmung des Ganges des Niederschalges zu Berlin," Gesellschaft der Naturforschenden Freunde, Berlin, *Beobachtungen, 7* (1787), 484–9.

57. J.P. Pelisson, ibid., 398–407.

58. Hellmann, *Repertorium*, 126.

59. Lambert's considerable contributions to meteorology are discussed in Feldman, "Applied mathematics" and *The history of meteorology*.

60. Patriotische Gesellschaft, Breslau, *Ökonomische Nachrichten, 1* (1773), 2.

would have been impossible a decade earlier, reflect Felbiger's reading of Deluc's newly published *Recherches.*[61] Felbiger also stipulated standard times of observation—soon after dawn, midday (1–4 p.m.), and evening (around 10 p.m.)—and he provided scales for cloud cover, fog, rain, wind, and snow. He had set his sights too high, however. The few observers responding submitted only occasional contributions and several did not use instruments.[62] An invitation to cooperative weather observation from Prague's patriotic society to "all Bohemian patriots" met with even less success: the organizer found himself alone in supplying the society's journal with observations.[63]

Western European projects achieved better results than did those in central and eastern parts. In France the Société royale d'agriculture de Paris, founded in 1761, had lapsed into inactivity by the 1770s. It was revived in 1785 by a drought and by the growing urgency of agricultural reform. The Intendant of Paris, under whose jurisdiction the Society fell, appointed the energetic Auguste Broussonet as its permanent secretary, and the Academy began publishing *Mémoires*, holding public meetings, awarding prizes—in short, adopting the demeanor of a learned academy of the Enlightenment.[64] Between 1785 and its demise in 1793, the Society published in its *Mémoires* eight sets of "Observations géorgico-météorologiques," elicited by questionnaires distributed by the local authorities, and a paper on the cold winter of 1789. The observations, more georgical than meteorological, included detailed topographical descriptions, general monthly accounts of weather, descriptions of the effect of weather on crops and animals, harvest quantities, and grain prices.[65]

61. Ibid., 79.

62. For example, Kretzschmer, "Wetterbeobachtungen in Braunau, 1778," *Ökonomische Nachrichten, 6* (1778) and Freytag, "Einfluss der Witterung," ibid., 113–4, 180–1, 220–2.

63. Strnad, "Meteorologische Beobachtungen," 397–9.

64. Gillispie, *Science and polity*, 370–1, 375–6.

65. De Courset (le Baron), "Observations géorgico-météorologiques faites dans le Boulonnais," Société royale d'agriculture, Paris, *Mémoires,* 1786–8; Gallot, "Observations géorgico-météorologiques," ibid., 1787.

The mix of topics illustrates nicely the affinities between meteorology and agriculture.

Bern's Ökonomische Gesellschaft, which included among its members Albrecht von Haller, Daniel Bernoulli, and the important instrument-maker Michel du Crest, organized on a more ambitious scale. Since 1760 the Gesellschaft had published "Observations rurales" from seven towns and villages of the Canton. The observations included qualitative descriptions of the weather and its medical and agricultural effects; tables of monthly rainfall; degree summations of heat and cold at morning, noon, and night; and extremes of temperature and pressure. (Degree summations are sums of temperature readings taken over a given period. They indicate the total amount of heat available to plants and are a characteristic innovation of late 18th-century climatology, with its interest in agricultural applications.) In 1763 the Society resolved "to establish meteorological observers in at least six different places in the Canton, and to supply them with exact instruments." The network was part of a plan to collect information on topography, climate, and disease (i.e., on "airs, waters, and places") and on agricultural and industrial resources. It lasted a decade before submissions petered out.[66]

The Big Projects

These smaller societies lacked personnel and resources for significant meteorological research. The great projects of the late Enlightenment were carried out by institutions that enjoyed generous funding and the full support of the state: the Societas meteorologica palatina and the Société royale de médecine. The Royal Society of London, whose connection with the state was more tenuous, provides an interesting counterpoint.

The Royal Society had maintained its interest in meteorology after the end of Jurin's group, publishing about 150 meteorological papers between 1750 and 1770.[67] Most of these were brief notices of spells

66. Ökonomische Gesellschaft, Bern, *Mémoires*, 1763–72; quotation is in *Mémoires*, 1763, xvii; cf. Lowood, *Patriotism*, 202–3.

67. Feldman, *The history of meteorology*, 213–4.

of unusual or hard weather or of rainbows, haloes, and other optical phenomena, submitted by provincial correspondents. Typical were "Observations of the late severe cold weather" of the winter of 1753–4 and "Concerning a very cold day, and another very hot day" in 1748—observations of at most a few days' weather, reporting only its temperature and general character and using outdated instruments.[68] Between 1750 and 1770 brief reports like these outnumbered meteorological registers (weather diaries covering a longer period) by almost two to one.

The contrast with the two decades after 1770 is striking. Between 1770 and 1790 registers and theoretical and experimental papers on meteorology outnumber brief, descriptive reports by eight to five. Several registers covered five or six years' weather; one spanned ten times as much. Contributors used instruments from the best artisans—Nairne and Ramsden, for example—and took care to describe their construction, calibration, and exposure.[69] Several contributors were on assignment for the East India or Hudson's Bay Companies and observed the weather at their employers' and the Royal Society's request. The arrangement reflects the kind of relations the Society enjoyed with semigovernmental commercial and exploring ventures.[70] The Society published their registers and those of provincial English correspondents in full.

In 1774 the Royal Society began keeping its own register of observations. Henry Cavendish submitted the Society's instruments to an

68. William Arderon, in *PT, 48* (1753–4), 249–50, and Henry Miles, in *PT, 46* (1749–50), 208–13, resp.

69. The six-decade register is Thomas Barker, "Register of the barometer, thermometer, and rain at Lyndon," *PT, 61–88* (1771–98); the first year included a register of rainfall from 1736 to 1771. William Roxburgh, "A meteorological diary kept at Fort St. George in the East Indies," *PT, 68* (1778), and *80* (1790), observed at three fixed times daily with a Nairne thermometer and Ramsden barometer.

70. The best known example involving meteorology is reported in Thomas Hutchins, "An account of the success of some attempts to freeze quicksilver, at Albany Fort in Hudson's Bay," *PT*, 1776, 174–81, and Thomas Hutchins and Charles Blagden, "Experiments for ascertaining the point of mercurial congelation," ibid., 1783, 303–70; Roxburgh at Fort St. George provides another example. See also "Weather observations from Nain and Okak at Labrador," "Meteorological journal kept at Port Jackson in New South Wales," and others, Royal Society of London, Meteorological Archives, mss. 143, 146, etc.

exhaustive exact experimental investigation prior to the institution of twice-daily observations, which continued until 1843.[71] The registers, printed in full in the *Philosophical transactions*, included readings of exterior and interior thermometers, barometers, hygrometers, and instruments to measure wind, magnetic variation and dip, and rainfall. Together with the contributions of the Society's correspondents, they constituted a far more extensive, detailed, and reliable source of data than had been available earlier in the century. The Society's registers were analyzed by its secretary Samuel Horsley, who calculated monthly mean, extreme, and mean morning and evening temperatures, extreme and mean pressures, the number of days on which the wind blew from each of the sixteen points of the compass, and the proportional rainfall in each month and season of the year.

Horsley also tested for the influence of winds and of the moon on the barometer. The moon was supposed to affect the weather by causing atmospheric tides; because its motions were periodic, a long enough series of observations might reveal correlations that would enable meteorologists to predict the weather. The theory ("so improbable, so destitute of all foundation," wrote Horsley) was eventually discredited, although adherents could be found well into the 19th century. As for the influence of wind direction on the barometer, this hypothesis contributed in the 19th century to the theories of wind rotation of H.W. Dove and others. Counting mean monthly barometer heights and corresponding directions of monthly prevailing winds, Horsley found a correlation: the barometer stood higher in months with winds in the semicircle WSW-W-N-NE; W and NE winds accompanied the greatest mean heights, and in seven months out of twelve the highest barometer reading occurred during a NE wind. Horsley must have been surprised when his count of the number of changes in the weather occurring within three days of the moon's syzygies and quadratures supported the hypothesis of lunar influence. He remarked only that the observations had not continued long enough to draw a firm conclusion.[72]

71. Henry Cavendish, "An account of the meteorological instruments used at the Royal Society's house," *PT, 66* (1776), 375–401; Cavendish, *The scientific papers*, vol. 2, *Chemical and dynamical* (Cambridge: The University Press, 1921), 53.

72. Samuel Horsley, "An abridged state of the weather at London in the year

While Horsley was calculating, an epidemic—or, more properly, an epizootic—of cattle plague was raging through France. The quarantine and slaughter of thousands of infected cattle became necessary. Anne-Robert Turgot, Louis XVI's minister of finance, encountered bitter resistance from the peasantry; faulty communications between the central authorities and physicians in the provinces further inhibited effective action. In order to direct and enforce the necessary measures, Turgot called forth a Commission de médecine aux maladies epidémiques et epizootiques in 1776. It received letters patent in 1778 as the Société royale de médecine.[73]

The Society was charged with the medical police of France, formerly the responsibility of the intendants. Working through them, it centralized public health policy in a single state agency staffed by trained physicians. Its duties were varied: to regulate the distribution of patent medicines and mineral waters; deploy a network of provincial physicians to gather information on public health and the environment; establish sanitary measures and regulations; advise provincial physicians on the prevention and treatment of epidemics; and, in cases of epidemic outbreaks, coordinate and enforce appropriate measures.[74] The Society also took on all the attributes of a learned academy: titles of membership signed by its president and secretary, medals, prize questions, *éloges*, and a journal whose title, *Histoire de la Société royale de médecine, avec les mémoires de médecine et de physique médicale,* echoed that of its elder sister, the Paris Academy of Sciences.

Two aspects of the Society's operations concerned climatology. The first was the construction of "a medical and topographical map of all of France," toward which physicians "of all the cities of the realm" would contribute memoirs "on the nature of their climate (*de aere, acquis, et locis*), and on the temperament of those who live

1774," *PT, 65* (1775), 167–93.

73. Gillispie, *Science and polity*, 24–33; Jean-Paul Desaive et al., *Médecins, climat et épidémies à la fin du XVIII siècle* (Paris: Mouton, 1972), 9; Caroline Hannaway, "The Société Royale de Médecine and epidemics in the Ancien Régime," *Bulletin of the history of medecine, 46* (1972), 257–73, on 259ff.

74. Gillispie, *Science and polity*, 219ff. and 226ff.; Hannaway, "The Société Royale de Médecine," 258ff.; Desaive et al., *Médecins, climat et épidémies*, 11–2.

there."[75] The best of these medical topographies, as they were called, won annual prizes from the Society. They might cover nearly 100 pages of text with complete environmental and geographical descriptions of the author's region, including climate and topography, economy, hygiene, and endemic diseases.[76] By the outbreak of the Revolution, 226 topographies had been collected.[77]

The Society's network of provincial physicians also carried out a grand program of meteorological observation. Already in 1775, in the midst of the epidemic, Turgot had tested the waters by distributing questionnaires via the provincial intendants, asking physicians in their jursidictions for "the temperature and the [epidemic] constitution of the years 1772, 1773, 1774, and 1775."[78] The resulting medicometeorological correspondence was continued by the Commission for Epidemics. After 1778, the authors of this correspondence became the Society of Medicine's network of observers.[79]

The direction of the network was undertaken by Louis Cotte, Oratorian priest and France's foremost meteorologist. Just a few years earlier Cotte had published the first textbook of meteorology based on observations—that is, the first to include many observations at all and to derive its discussions of the weather from them.[80] Cotte was especially interested in agricultural applications of meteorology. He had originally planned his *Traité de météorologie* as an extension of Duhamel's "botanico-meteorological" investigations; it is agriculture, he wrote, "which I have had principally in view in this work." He included medicometeorology as well, excerpting Malouin's

75. Société royale de médecine, Paris (SRM), *Histoire,* 1776, xiv; ibid., 1782–3, 6–7.

76. For example, de Brieude, "Topographie médicale de la Haute-Auvergne," SRM, *Mémoires,* 1782–3, 257–340.

77. Hannaway, "The Société Royale de Médecine," 267.

78. Ibid., 3; "Questions à faire aux médecines," France. Académie nationale de médecine, Archives, SRM Carton 189, dossier 13, number 16.

79. France. Académie nationale de médecine, Archives, SRM Carton 172, dossiers 1–22; cf. Gillispie, *Science and polity,* 33, 202–3.

80. Cotte, *Traité de météorologie.* Cf. Gustav Hellmann, "Entwicklungsgeschichte des meteorologischen Lehrbuchs," in Hellmann, *Beiträge zur Geschichte der Meteorologie, 3* (Berlin, 1922), 1–14, on 1.

memoirs extensively. The goal of meteorology, he believed, is "the perfection of the sciences of agriculture and medicine."[81]

In the Royal Society of Medicine, Cotte had observers to pursue that goal. The requirements he imposed upon them were, in the words of one modern meteorologist, "very strict," and, in the judgment of another, comparable to mid19th-century standards.[82] "It is necessary first of all," Cotte warned, "to have good instruments." A Réaumur mercury thermometer was to be calibrated if possible against the Society's standard; the barometer, calibrated "with the greatest exactitude," should be equipped with a vernier and readable to an accuracy of tenths or twelfths of a line. The Society would recommend reliable artisans on request. Cotte specified the proper exposure for the instruments and set fixed times for observation.[83] "Great exactitude and a spirit of order—these are the principal requirements of the physicist who devotes himself to these sorts of observations."[84] These words might serve as a motto for late-Enlightenment physical science.

Cotte complained more than once that his observers' instruments were not comparable, that observers were not providing "an exact description" of their instruments, that the instruments were "defective...supplied by travelling barometer-peddlers." But he had to admire his observers' zeal.[85] By 1785, 150 provincial physicians were participating in the project; about 50 of them observed for more than a decade.[86] Modern climatologists agree that the Society's observers were the elite of the medical profession—otherwise they could not have followed Cotte's instructions at all. They generated a great mass of observations over the whole of France—the largest collection, most likely, prior to the foundation of national meteorological bureaus in the mid19th century. From these observations

81. Cotte to Macquer, 25 Aug 1768, France. Bibliothèque nationale, ms. Fr. 12305, ff. 209–11; Cotte, *Traité de météorologie*, ix, xvii.

82. Desaive et al., *Médecins, climat et épidémies*, 90; J.A. Kington, "A late eighteenth-century source of weather data," *Weather, 25* (1970), 169–75.

83. SRM, *Histoire*, 1776, xi–xiv.

84. Cotte, *Traité de météorologie*, 517.

85. SRM, *Histoire*, 1777–8, 104; ibid., 1782–3, 245–6; ibid., 1784–5, 204.

86. Ibid., 203; Peter, "Disease and the sick," 85, 87.

meteorologists have been able to reconstruct the climate of France in the last years of the Old Regime.[87]

Cotte did not publish these observations in full; "Tables are not pleasant for the reader," he said, echoing Derham across a half-century.[88] Instead he published monthly summaries of the weather at each reporting location, including its "temperature" in terms of the Hippocratic categories cold and wet, cold and dry, warm and wet, and warm and dry, along with the usual means, extremes, and descriptions of diseases. Although the summaries resembled those of Derham's time, they were more plentiful and more consistent. Cotte arranged them in great tables (tables were not to be dispensed with entirely) that amounted to the first descriptions of the climate of a nation based on detailed, regular observations.

Cotte's tables represented a collection of data dense enough over a sufficient area to reveal something of the importance of space for the weather. "These extreme temperatures take place," he said, "at the same time over a very great extent of country."[89] This remark, made early in the Society's career, sounds like one of Derham's coincidences, only it covers more ground. A few years later Cotte wrote that the tables showed clearly "the influence of great variations of the atmosphere over a very great extent of country on the thermometer, and principally on the barometer."[90] These isolated suggestions mark the limit of Cotte's insight. He did not draw from the tables a description of the climate of France. Moreover, the tables' monthly means and extremes could not reveal the density of the weather in time. Because Cotte did not publish the full record of observations, his observers' efforts were of as little use to other meteorologists as they seem to have been to himself. They lay hidden in the Society's archives until climatologists began to explore them around the middle of the present century.

87. Desaive et al., *Médecins, climat et épidémies*, 40–2, 97. O. Muller, ibid., has submitted the Society's observations to correlation analysis. Many, especially those of Paris and the Ile de France, show a correlation coefficient of 0.8 and higher.

88. Cotte, *Traité de météorologie*, 517.

89. "Temperature" here is taken in a broad sense to include barometric pressure, rains, storms, and drought. SRM, *Histoire*, 1777, 105.

90. SRM, *Histoire*, 1780, 261.

The Societas meteorologica palatina has enjoyed greater fame than its French sister, in part because it was an international organization whose members were the chief scientific institutions of Europe (more on this below). More significantly, the Palatine Society published the full record of its members' observations, which were excellent and became a valuable resource for 19th-century meteorologists.[91] The Society was a product of a late 18th-century enlightened state, the principality of the Palatinate or Kurpfalz. Cultural life flourished there under the patronage of the Elector Karl Theodor, who was keen on science and its contribution to his subjects' welfare.[92] Mannheim, Karl Theodor's capital, was one of Germany's centers of culture; the Elector lavished on it an Akademie der Wissenschaften, a Deutsche Gesellschaft, and a Deutsches Theater, and brought Gotthold Ephraim Lessing, Friedrich Gottlieb Klopstock, and Wolfgang Amadeus Mozart to write and play for him. He furnished an astronomical observatory with English instruments, then the best in the world, and directed his court chaplain and ecclesiastical councillor, Johann Jakob Hemmer, to assemble a *cabinet de physique* and offer lectures and demonstrations.[93] In the 1770s Hemmer investigated the electricity of flames, of dew, and of the atmosphere, and he was largely responsible for the erection of lightning rods throughout the Elector's extensive realm. His book on the subject went through two editions. Many German rulers bought it for distribution in their principalities.[94]

The idea for a Palatine meteorological society may have come from neighboring Baden. There, in 1778, the Karlsruhe professor of mathematics and physics Johann Böckmann founded the Badische

91. For example, H.W. Brandes derived the first synoptic analysis of European weather in 1823 from the Society's record for the year 1783. See Brandes, *Beiträge zur Witterungskunde* (Leipzig, 1820).

92. Hellmann, *Repertorium*, 896.

93. David Cassidy, "Meteorology in Mannheim," 12; Hellmann, *Repertorium*, 193.

94. J.J. Hemmer, *Anleitung, Wetterleiter an allen Gattungen von Gebäuden auf die sicherste Art anzulegen* (Mannheim, 1786, 1788). Cited by Adolf Kistner, *Die Pflege der Naturwissenschaften in Mannheim zur Zeit Karl Theodors* (Mannheim: Selbstverlag des mannheimer Altertumsverein, 1930), 87; Cassidy, "Meteorology in Mannheim," 14–5.

Witterungsanstalt, a network of sixteen observers within the margravate.[95] The project failed for lack of funds, but it probably inspired Hemmer, who hired Böckmann's instrument-maker and placed his own plans before the Elector.[96] Karl Theodor, who himself kept a weather diary, approved, and in 1780 the Societas meteorologica palatina received its charter as a three-member "Meteorologische Klasse" in the Mannheim Akademie der Wissenschaften. Hemmer, the Society's secretary, recruited observers, oversaw the construction of instruments for them, and edited the annual *Ephemerides meteorologicae*.

The Society's work was a model of exact experimental method. Hemmer followed Deluc closely in specifications for instruments. "We shall always search for ways to make observations more exact," he declared, "both for the sake of agriculture and for our health." Tubes of barometers and thermometers were to be exactly cylindrical, carefully cleaned, and filled with pure, boiled mercury, and the scales carefully prepared and exactly measured. Barometers were to be read to tenths of a line with the vernier against the cusp of the meniscus and thermometers to tenths of a degree. Care was to be taken to prevent disturbances from the observer's breath or candle. Observers could use their own instruments if they met these standards. Hemmer nonetheless dispatched to each of them a packet of instruments "whose comparability has been assured, sparing no expense": a barometer with correcting thermometer (the thermometer to correct for the effect of heat on the barometer), two thermometers (for exposure in sun and shade), and a hygrometer of Deluc's design (using a goose-feather quill that expanded with humidity). Selected observers received a magnetic declination compass as well. Hemmer also asked participants to observe wind direction and strength, cloud cover, precipitation and evaporation, river level, lunar phase, and medical and agricultural conditions. Hours of observation were set at 7 a.m., 2 p.m., and 9 p.m.; Hemmer supplied scales for wind

95. J.L. Böckmann, *Beiträge zur neuesten Geschichte der Witterungslehre* (Karlsruhe, 1781), cited in Hellmann, *Repertorium*, 889.

96. Kistner, *Die Pflege der Naturwissenschaften,* 96–8; Cassidy, "Meteorology in Mannheim," 17–8.

strength and cloud cover and symbols for precipitation and other "meteors."[97]

Along with these exacting specifications went a new strategy of recruitment. In order to secure continuity in observations, Hemmer invited scientific institutions rather than individuals to participate; each institution would appoint one of its members to observe. No fewer than thirty-seven academies, universities, and monasteries in Europe and America responded, and Hemmer soon had an international network staffed by the world's capable workers. Their registers—reliable, uniform, and, in a number of cases, extending over a decade—appeared in extenso in the Society's *Ephemerides.* This was a "princely plan of operations," in the words of the 19th-century meteorologist John Daniell,[98] carried out at princely expense: 1,500 gulden annually for the *Ephemerides* alone, or about a third more than the salary of a senior member of France's Académie royale des sciences.[99] The resulting collection, wrote Daniell, "contain[ed] more data for a correct history of the weather than all other works on the same subject taken together."[100] Such was the stature of the Palatine Society that its hours of observation—the so-called "Mannheim hours"—and the "Mannheim cloud cover and wind scales" were widely adopted and remained in use for more than a century.

Several of Hemmer's observers carried out extensive climatological analyses of their own weather records. Typical was the contribution of Nicholas de Béguelin of the Academy of Sciences in Berlin. He reported monthly maxima and minima for the thermometer and barometer, their monthly ranges, and annual and monthly means; the same for morning and evening thermometrical observations, together with maximum diurnal range; for the hygrometer and magnetic needle monthly mean morning, afternoon, and evening values, the mean of all three, and monthly extremes and extreme ranges; and monthly

97. Societas meteorological palatina, *Ephemerides meteorologicae, 1* (1781), 21, 7–14. See Hellmann, *Repertorium*, 898.

98. John Daniell, *Meteorological essays*, 3d ed., 2 vols. (1845), 305.

99. Cassidy, "Meteorology in Mannheim," 19; for salaries, see Heilbron, *Electricity*, 117, 122.

100. Daniell, *Meteorological essays*, 305.

frequencies for each direction and degree of wind strength and for each degree of cloud cover, all according to Hemmer's scales.[101] Participants contributed a wealth of other material, which Hemmer published indiscriminately: ten years' tidal observations from Padua, hourly barometric measurements over an entire month, hourly observations of magnetic declination from 6 a.m. to 10 p.m., Hemmer's own measurements of atmospheric electricity, and van Swinden's collection of temperature observations taken four and five times daily in nine Belgian towns during a cold spell in December 1783. From analyses of contributors' registers, Hemmer's assistant Karl König published tables like Cotte's but far more complete. A single table for each reporting location presented monthly summaries. A "general table" of all locations showed annual extremes, means, and ranges; extremes of monthly means for barometer, thermometer, and declination-needle; annual rainfall and prevailing winds; monthly sums of degrees of heat; and frequencies of wind strength and direction, "meteors," and cloud cover.

Although the collection was something of a miscellany, the observations and analyses were more consistent and rigorous than anything earlier in the century. In them we can see the first hints of a discipline of climatology: a standard and consistent practice in the gathering and analysis of weather data. This practice is reflected in the emergence of a standard set of climatological variables calculated by meteorologists such as Béguelin, König, and Horsley. Their work marks the beginning of a quantitative climatology.

The new climatology provided some definitive answers to old questions. Van Swinden, in an independent investigation, and two contributors to the Society's *Ephemerides* examined barometric observations taken five, six, and eighteen times daily over a year or more. They refuted the hypothesis of lunar influence but confirmed the diurnal variation, a daily fluctuation of the barometer that several earlier meteorologists thought they saw in their data.[102] These analyses

101. *Ephemerides*, 1784, appendix, 3–6.

102. Van Swinden, *Mémoire sur les observations météorologiques faites à Franeker pendant le courant de l'année MDCCLXXIX* (Amsterdam: M.M. Rey, 1780), 34; Joh. Jac. Planer, "Observatio oscillationis mercurii," *Ephemerides*, 1783, 250–7; Vincent Chiminiello, "De diurna nocturnaque oscillatione barometri," ibid., 1784, 230–4.

exploited the density of observations *in time*. The coverage of observations *in space* led meteorologists to a greater appreciation of its role in the weather. They could now trace the paths of weather events from one end of Europe to the other: the cold wave of the winter of 1775–6, for example, and the famous hailstorm of 13 July 1788, which by destroying crops across the most fertile regions of France contributed significantly to the trouble preceding the Revolution.[103] And they arrived at a clear statement of the spatial extent and of the motion of barometric variation. "Anyone who carefully examines and compares the barometric observations in volume I of the meteorological *Ephemerides*," proclaimed Coelestin Steiglehner, "cannot fail to conclude that oscillations longer than one day extend over many places of diverse longitude and latitude."

From observations at London, Regensburg, and St. Petersburg, Steiglehner determined that a barometric minimum occurring around Christmas 1775 had traveled from west to east, striking each town in succession. Further analysis of data from London, Regensburg, St. Gotthard, Buda, Mannheim, and Vienna confirmed the rule "early in the west, late in the east."[104] F.X. Epp had drawn a similar conclusion from the observations of the network of twenty-four Bavarian observers he directed. The Churbayrische Akademie der Wissenschaften, a sister society to the Mannheim Academy after Karl Theodor inherited Bavaria in 1777,[105] established the network in the same year that the Palatine Society was chartered. From parallel barometric motions recorded by his observers Epp concluded that the causes of barometric variation extend over wide regions, perhaps over

103. Van Swinden, *Observations sur le froid rigoreux du mois de janvier, MDCCLXXVI* (Amsterdam: M.M. Rey, 1779), 5–6; Cotte, in the "Feuille météorologique" of the *Journal de france*, cited in Buissart, "Mémoire sur l'orage du treize juillet, 1788," France. Académie nationale de médecine, SRM carton 141, dossier 40. On effects of the hailstorm, see Robert D. Harris, *Necker and the revolution of 1789* (Lanham, Maryland: University Press of America, 1986), 273; Michel Vovelle, *The fall of the French monarchy 1787–1782*, transl. Susan Burke (Cambridge: Cambridge University Press, 1984), 86.

104. Coelestin Steiglehner, "Excerpta ex dissertatione: atmosphaerae pressio varia," *Ephemerides*, 444–57.

105. See James E. McClellan III, *Science reorganized. Scientific societies in the eighteenth century* (New York: Columbia University Press, 1985), 118.

hemispheres. He confirmed this conclusion by examining one month's barometric variations in five European towns, the data for which he found in the Palatine Society's Ephemerides.[106]

Finally, a few meteorologists began to synthesize descriptions of regional climates from observations. The Society of Medicine's medical topography project aimed at a "medical and topographical map of all of France." Van Swinden planned to determine the climate of Frisia from his own observations, and from others' observations throughout Holland he began to construct a picture of its provincial climates.[107] Epp expected from his collection "a more exact knowledge of [Bavaria's] climate" and of the "physical character of the land."[108] If the stations were properly distributed the observations would yield information on local climates as well.[109] None of these projects, however, was completed.

CONCLUSION

The French revolutionary period brought a precipitous decline to meteorology. The correspondence of the Société royale de médecine ceased with the onset of the Revolution in 1789; along with all the Old Regime academies the Society was suppressed in 1793.[110] The Palatine Society, already in decline after the death of Hemmer in 1790, collapsed in 1795 when the French army crossed the Rhine, occupied Mannheim, and closed the Academy of Sciences.[111] In Germany the number of stations observing the weather dropped during the 1790s to a third of its peak value in the previous decade.[112]

106. Akademie der Wissenschaften, Munich, *Ephemerides, 1* (1781), 14; ibid., *2* (1782), 39–43.

107. Van Swinden, *Observations sur le froid rigoreux*, 134–6.

108. Akademie der Wissenschaften, Munich, *Ephemerides, 1* (1781), 4–5.

109. Akademie der Wissenschaften, Munich, *Anzeige an das Publikum, von den Gegenstände der Witterungslehre, und von der Art und Weise die Witterung zu beobachten* (Munich, 1781).

110. Busquet, "Les origines de l'Académie de médecine," Société française d'histoire de la médecine, *Bulletin, 20,* (1926), 391–6; Cotte to Thouret, Montmorenci, 9 thermidor an 4 (1796), France. Académie nationale de médecine, Archives, SRM carton 197, dossier 3, number 3.

111. Cassidy, "Meteorology in Mannheim," 19–20.

112. Hellmann, *Repertorium*, 985.

Meteorological contributions to the Royal Society of London's *Philosophical transactions* had already diminished in the second half of the 1780s.

No doubt other factors besides the physical disruption of war and revolution contributed to the decline. Politics distracted natural philosophers; van Swinden, for example, became involved in administrative and educational reform. When in 1800 Cotte requested some earlier observations from him, van Swinden wrote in regret of "the oblivion to which I have consigned them, regretting all the while that so much research produced so few useful results."[113] His remarks suggest a further reason for the loss of interest in meteorology. Meteorologists like Cotte had promised "to prove the usefulness of meteorological observation to those who deride [it]."[114] Nothing positive seemed to have come from the expenditure of so much effort on what many considered a tedious and trivial activity.

The program of late Enlightenment climatology had failed. Meteorology had not brought "the perfection of the sciences of agriculture and medicine." No useful correlations had been discovered between the weather and agriculture or disease, and correlations like the lunar influence that might lead to predictive rules remained in doubt. "All attempts at rules governing the weather have been in vain," remarked the astronomer Bode.[115] There were a number of causes for this failure. The meteorological projects of the late Enlightenment still confronted some of the limitations of the earlier decades of the century. Instruments had neared perfection but remained unavailable in the more remote French provinces and German states. Rigorous standards had been established for observers but were not always followed outside scientific centers. Institutional arrangements were weak in many areas; this was especially true in central and eastern Europe, where projects suffered from the small size and political fragmentation of the German states. And editors of

113. Van Swinden, "Lettres sur les grands hivers," *Journal de physique, 50* (1800), 277–96, on 277–8.

114. Cotte, *Traité de météorologie*, 248.

115. Rudolf Wolf, *Biographien zur Kulturgeschichte der Schweiz*, 4 vols. (Zurich: Orell, Füssli & Co., 1858–62), *1*, 451.

meteorological ephemerides retained the practice of publishing summaries rather than the full record of observations, to the frustration of generations of meteorologists.

Cotte and his colleagues no doubt believed that they might have reached some of their goals if the Revolution had not interrupted them.[116] They were mistaken. Their search for correlations between weather patterns and agriculture and disease belonged to the "classical" *episteme*, as Foucault called it, the *episteme* of natural history and nosology.[117] Meteorologists working within this *episteme* associated the weather with diseases and agriculture according to superficial correlations, without penetrating to the interior forces that govern their interactions. But climatology (parallel with the rest of Western knowledge, if Foucault was right) was moving in the last decades of the 18th century away from this approach toward the perception of these interior relations.

It was approaching this goal through quantification. Quantification here did not include mathematical theories of the weather. Other parts of meteorology did develop mathematical theories. In hygrometry, thanks to a new hygrometer and the techniques of exact experimentation, de Saussure and others announced Dalton's law of partial pressures for the special case of aqueous vapor: the total pressure of moist air at a given temperature is the sum of the pressures of its components, water vapor and dry air. From this elementary arithmetical relation de Saussure calculated the relative weights of dry and saturated air, using nothing more complicated than simple proportions.[118] More sophisticated mathematical laws were worked out in the 1790s governing the relation of vapor pressure to temperature.[119] In barometry Lambert, working in the tradition of mixed mathematics, had applied the integral calculus to the variation of air pressure with height in the 1760s.[120] Studies of the distribution of

116. This appears from Cotte's papers and his correspondence with van Swinden during the 1790s. See, for example, Cotte, ms. preface to *Nouveaux mémoires sur la météorologie* (1810), Laon, France. Bibliothèque municipale, ms. 568^2.

117. Foucault, *Order of things*.

118. See Feldman, *The history of meteorology*, sect. I, chap. 4.

119. Ibid., chap. 7.

120. Feldman, "Applied mathematics," 145–9.

heat over the earth employed trigonometry to calculate the effect of the sun's heat in different latitudes.[121] All these results lay in areas of contact of meteorology with the numerous disciplines discussed in the introduction to this chapter. Except for one or two isolated examples,[122] meteorology proper—the study of weather phenomena such as rain and snow, winds, and storms, as well as climatology—lacked any mathematical theory until well into the nineteenth century.

In climatology quantification meant precise instruments and a rigorous discipline of observation. The example of the Société royale de médecine shows how closely tied this discipline was with the bureaucratizing impulse of enlightened absolutism. Once observations were collected, meteorologists analyzed them using the most elementary statistical techniques: counting (the number of days of rainfall, the number of days on which the wind blew from different quarters), taking averages and ratios (means of temperature, pressure, etc., the proportion of rain falling in each season, the proportion of changes in the weather occurring within three days of the moon's syzygies and quadratures), and establishing correlations (between the weather and diseases, among the weather at different locations and times, the weather rule). It is no accident that a discipline so close to public health and administration should rely on statistical techniques.

Limited as it was, this type of quantification pushed climatology to the verge of a new existence. The "discipline" of regular observation and intensive calculation led toward the "discipline" of climatology: the practice of deriving a standard set of variables representing the typical weather of a location. By enforcing a finer coverage in time and space, this discipline led meteorologists' perception (or "gaze," as Foucault called it) toward the creative functions of time and space in weather and climate.

121. For example, Tobias Mayer, "A more accurate definition of the variations of a thermometer," in Eric G. Forbes, ed., transl., *Tobias Mayer's* Opera inedita (New York: American Elsevier Pub. Co., 1971), 53–61.

122. For example, James Hutton, "The theory of rain," Royal Society of Edinburgh, *Transactions, 1* (1788), 41–86. The mathematics involved here is elementary.

Even so, there is a tremendous gap between the climatology of the late Enlightenment and that of the early years of the 19th century. The most impressive result of late 18th-century climatology, the diurnal barometric variation, remained a correlation. The discovery of the great extent of barometric variation did not capture the full role of space in modifying weather and climate. Steiglehner's "early in the west, late in the east" still sounds like a weather rule.

The climatology of the first years of the 19th century looks vastly different. Alexander von Humboldt may stand as a representative of the new point of view. Humboldt's isotherms (lines of equal temperature), presented in an essay of 1817, are syntheses of mean annual temperatures across the globe, which present immediately to the eye the role of space in the development of climate.[123] In the same essay Humboldt speaks naturally and comfortably of the climates of the different regions of the earth; develops comparisons among the climates of eastern and western shores of continents and among coastal, continental, and island climates; and arrives at the notion of a "climatic system"—a region in which the different factors of climate vary in a continuous manner, so that in specifying for example the mean annual temperature one has fixed by implication the range of other climatic factors such as winter and summer temperature. Foucault argued that between 1775 and 1795 naturalists found the key to the new *episteme*—the notion of organic structure, in the case of biology—but still applied it to the old task of classification.[124] During the same decades exact experimental physics had provided the key to an appreciation of the role of space and time in climate and the weather, but meteorologists still applied that key to the search for correlations. A mature climatology awaited the age of Humboldt.

123. Alexander von Humboldt, "Sur les lignes isothermes et la distribution de la chaleur sur le globe," Société d'Arcueil, *Mémoires, 3* (1817), 462–602.

124. Foucault, *Order of things*, 226ff.

6 Accuracy, Rhetoric, and Technology: The Paris-Greenwich Triangulation, 1784–88

By Sven Widmalm

In the 18th century the technology and the scope of surveying were radically transformed. There emerged a national cartography based on comprehensive triangulation measurements and a scientific discipline, geodesy, which made use of these measurements. France was in the van: Jean-Dominique Cassini, his son, grandson, and great grandson—each of whom in turn became director of the Paris Observatory—had charge of a triangulation that, in the 1730s, developed into a national survey. The influence of the Cassini methods and style had started to spread to other nations by 1750. At first, large triangulations were employed mainly to measure arcs of meridians, providing points of reference for mapmaking and also empirical material for constructing theories of the shape of the earth. Full-fledged national surveys followed, first in Denmark, next in Great Britain.

The "Trigonometrical Survey" of Britain, known since the 1820s as the "Ordnance Survey," was founded in 1791, partly as a consequence of a geodetic measurement led by Major General William Roy and by Jacques Dominique Cassini in the mid1780s in order to connect the observatories of Greenwich and Paris. The work of Roy, which will be the main subject of this chapter, was at the time

I wish to thank the following for having commented on earlier versions of this chapter: J.A. Bennett, Matthew H. Edney, Robert M. Friedman, Mikael Hård, Derek Howse, and Simon Schaffer. Part of the research was carried out with financial support from the British Council, Stockholm. The following abbreviations are used in the notes: BOP, Bibliothèque de l'Observatoire (Paris); *DSB*, *Dictionary of scientific biography*; PRO, Public Record Office (Kew); RS, Archives of the Royal Society of London; *PT*, *Philosophical transactions* of the Royal Society of London.

considered unsurpassed in its technical refinement and accuracy, and it supplied the model for future British geodetical surveying. The chapter begins with an overview of the development of geodesy and cartography in the 18th century and ends with a discussion of the militarization of large-scale surveying all over Europe. The story is symptomatic of a general tendency around 1800 to amass quantified data for civilian as well as military planning and control.

GEODESY AND CARTOGRAPHY IN THE 18TH CENTURY

In Cassini de Thury's influential book on the measurement of the Paris meridian, published in 1744, questions relating to the shape of the earth and to the "geometrical description of the kingdom" were treated separately.[1] The division existed only on paper. These fields had developed side by side, ever since Jean Picard surveyed a degree of the Paris meridian in the years 1668–70. Picard's triangulation prepared for a national cartographic venture, desired by Jean Baptiste Colbert; it also produced data for calculating the size of the earth, if not its shape. Throughout the 18th century, most advanced geodetical measurements had to do with mapping projects.

Cassini de Thury and Europe

The French survey was begun in 1683 and, after frequent lengthy intermissions, produced in 1718 its first major accomplishment, the triangulation of over seven degrees of a meridian line through Paris. Operations resumed after political support for the project was regained in 1730. Prompted by Pierre Louis Moreau de Maupertuis' measurement of an arc of meridian in Scandinavia in 1736–7, members of the Paris Academy of Sciences resurveyed the Paris meridian.[2] The Cassinis had claimed that their measurements showed

1. César François Cassini de Thury, *La meridienne de l'Observatoire royale de Paris* (Paris: H.L. Guerin & J. Guerin, 1744), 26, quote, and *Relation des deux voyages faits en Allemagne par ordre du roi* (Paris, 1765), 1–2: "le seul moyen de perfectionner la géographie, étoit de suivre pour la description d'un pays la même méthode que l'on avoit employée pour la détermination de la figure de la terre."

2. Josef W. Konvitz, *Cartography in France 1660–1848* (Chicago: University of Chicago Press, 1987), 3–13.

the earth to have an oblong shape. But Maupertuis' result suggested otherwise; the remeasurement of the Paris meridian and a survey of an arc of meridian in South America made between 1735 and 1744 confirmed that the earth has an oblate shape. Thus measurements outside of France served to calibrate the geodetic work in France. The triangulation of the whole of France now went ahead, under the direction of César-François Cassini de Thury (third in the astronomer dynasty), although it was Nicolas-Louise Lacaille who did most of the fieldwork.[3] Between 1739 and 1744 almost 800 triangles, from Dunkerque to Perpignan, were measured, along with nineteen bases. The triangles, printed on eighteen sheets, provided a "geometrical skeleton" rather than a map. This was considered to be sufficient for their intended users, namely, engineers involved in public construction works.[4]

In the 1750s several projects were begun or completed in Europe to complement the measurements sponsored by the French in France, Lapland, and Peru. A leading promoter of these projects was the Jesuit Roger Boscovich, who obtained the support of the pope to measure an arc of meridian between Rome and Rimini. The survey, which ran from 1750 to 1752, had the dual purpose of adding to knowledge of the shape of the earth and updating maps of the papal state. A copy of the "toise of Peru," borrowed from Paris, allowed comparison of Boscovich's geodesy with French results. Boscovich persuaded the king of Sardinia, Carlo Emanuele III, to sponsor a measurement in Piedmont, which Giovanni Battista Beccaria, a Scolopian priest and professor of physics at the University of Turin, carried out between 1760 and 1764. Beccaria's result did not confirm the oblate shape deduced by the French and so contributed to debate about gravitational irregularities and the reliability of geodesy.[5] This

3. Charles C. Gillispie, *Science and polity in France at the end of the Old Regime* (Princeton: Princeton University Press, 1980), 113–5; Lloyd A. Brown, *The story of maps* ([1949] New York: Dover, 1977), 252.

4. Konvitz, *Cartography in France*, 15–28.

5. Antonio Marussi, "Italian pioneers in the physics of the universe. III: Geodesy," *Cahiers d'histoire mondiale, 7:2* (1963), 473–5; Elizabeth Hill, "Roger Boscovich: A biographical essay," in Lancelot Law Whyte, ed., *Roger Joseph Boscovich: Studies of his life and work on the 250th anniversary of his birth* (London: George Allen and Unwin, 1961), 42–6.

was true also of the first measurement of an arc of meridian in North America, for which Boscovich was also in some measure responsible. It was carried out between 1764 and 1768 by Jeremiah Dixon and Charles Mason.

Another Jesuit, Father Joseph Liesganig, director of the Observatory at the University of Vienna, began his survey of an arc of meridian between Vienna and Brün in 1759. Maria Theresa ordered this measurement following suggestions from Boscovich, and probably also from Cassini de Thury, who worked on a triangulation to connect the Paris meridian with Vienna, one purpose of which was to correct the military maps of France's ally, Austria. Cassini de Thury participated in Liesganig's work, which was calibrated with his own by means of a copy of the toise of Peru. In 1762 Maria Theresa authorized an extension of this arc and provided Liesganig with instruments and engineers from the military academy in Vienna.[6]

Boscovich and others, including Bouguer, had supposed that irregularities in the earth's mass distribution drew plumb bobs from the perpendicular and introduced serious error into geodetic measurements. His view was widely accepted by the 1770s. In 1775 Etienne Bonnot de Condillac made public a severe critique of existing geodetic investigations, and, in agreement with d'Alembert and later also Laplace, he judged that the different measurements of arcs of meridian were too contradictory to demonstrate that the earth has a regular ellipsoid shape. In particular, measurements in France and in Piedmont around the same degree of latitude did not agree. In Condillac's view measurements should conform to theory or else theory should be modified.[7]

6. Cassini de Thury, *Relation des deux voyages*, 3–4; Ernst Bernleithner, "Oesterreichs Kartographie zur Zeit des Grafen Ferraris," *La cartographie au XVIIIe siècle et l'oeuvre du comte de Ferraris (1726–1814)*, Collection histoire pro civitate, no. 54 (1978), 139–42; Johannes Dörflinger, *Österreichichsche Karten des 18. Jahrhunderts* (Vienna: Österreichische Akademie der Wissenschaften, 1984), 61–2.

7. Volker Bialas, *Erdgestalt, Kosmologie und Weltanschaung: Die Geschichte der Geodäsie als Teil der Kulturgeschichte der Menschheit* (Stuttgart: Konrad Wittwer, 1982), 158–9, 166–7, 181–7; Condillac, *Cours d'étude pour l'instruction du prince du Parme, III: De l'art de raisonner* (1775; Deux-Ponts: [Bodoni], 1782), 217, 275–8. Laplace later changed his mind and advocated the view that the earth is a true ellipsoid. Pierre Simon Laplace, *Oeuvres*, 7 vols. (Paris: Imprimerie Royale), *5* (1846), 14–6 (text of 1823).

The conflict that Condillac and others pointed out may be characterized as one of quantification versus geometric simplicity. Possessing only a few measurements of meridianal arcs and a theoretical model authorized by the *Principia*, scientists could maintain that the earth had a regular shape. By the 1780s, however, it had become possible to be a Newtonian and still entertain novel shapes for the earth. The Newtonian theory flowed from the hypothesis that the earth had once been in a fluid state. With the decreasing interest in hypotheses about first causes, which characterized the late 18th-century instrumentalist attitude to science, one could choose any curve that fit the data.[8] The work of William Roy offers an example.

Meanwhile, Cassini de Thury had created the first nationwide topographical map based on extensive triangulations. A team of draftsmen and engineers worked on the project between 1750 and its conclusion thirty-nine years later under Cassini de Thury's son, Jacques Dominique.[9] French geodesy had a lasting influence on European cartography. As early as 1736–7, when Maupertuis visited "Lapland," the Swedish Surveying Office made an inspired but, as it happened, premature attempt to appropriate the new technology for the benefit of Swedish cartography. In Denmark, the Scientific Society of Copenhagen was commissioned to carry out a national triangulation on the French model in 1762. The Austrian military surveys by Joseph Jean Ferraris in the 1770s drew on Cassini's example in the choice of scale (1:86,400) and in the triangulation methods. Other European states and Russia did not introduce large-scale triangulations until the early 19th century.[10]

In sum, French cartography and geodesy led European practices in both field and office. Meridianal arcs were triangulated as backbones for exact cartography; geodesists computed distances in terms of the

8. J.L. Heilbron, *Electricity in the 17th and 18th centuries: A study of early modern physics* (Berkeley: University of California Press, 1979), 71–3.

9. Gillispie, *Science and polity*, 480–1; Konvitz, *Cartography in France*, 16–21.

10. J. Svärdson, "Lantmäteriteknik," *Svenska lantmäteriet 1628–1928, 1* (Stockholm, 1928), 223; Asger Lomholt, *Det Kongelige Danske Videnskabernes Selskab, 1742–1942: Samlingar til Selskabets historie, 4* (Copenhagen: E. Munksgaard, 1961), 90–104; Bernleithner, "Oesterreichs Kartographie," 132–8; Brown, *Story of maps*, 270.

Peruvian toise, and draftsmen used the scale of the Cassini maps, sometimes even when cartographic methods remained traditional, as in Henri Mallet's maps of Switzerland.[11] This calibration and standardization owed much to Cassini de Thury. He envisaged a Europe geometricized to a uniform scale; to this end he extended the French triangulation to Flanders and to Austria, and tried unsuccessfully to win support for a similar extension into Italy.[12]

Britain kept outside the sphere of Cassini de Thury's influence until 1783. Then, about a month after the end of the American war of independence, he proposed to the British government that a triangular connection should be made between London and Paris. The resulting measurement became the starting point for the Ordnance Survey, a cartographic venture undertaken by military engineers.

British Military and Scientific Engineers

British military cartography before the Ordnance Survey was poorly developed. The only important surveys outside England had been conducted in response to emergencies, like the measurements in Scotland as part of the pacification of the Highlands after the 1745 rebellion. The young draftsman William Roy led the fieldwork of the Scottish survey and later became director of the whole operation.[13] After completing the work in 1755, he joined the corps of engineers under the Board of Ordnance and became a lieutenant in the army. Throughout his career Roy held double ranks in the engineers and in the army, the latter always being the higher.[14] This elevated him to a higher social as well as military position than that of mere engineers, who did not acquire military rank before 1757. The status of the mere engineers rose in the 1770s, partly as a result of a new policy of

11. M.-A. Borgeaud, "Cartographie genevoise de XVIe au XIXe siècle," *Archives internationales d'histoire des sciences*, no. 6 (1949), 363–74.

12. Marussi, "Italian pioneers," 475.

13. J.B. Harley and Yolande O'Donaghue, "Introductory notes," in *The old series Ordnance Survey maps of England and Wales, 1* (Lynpne Castle: Harry Margary, 1975), xi; R.A. Skelton, "The military survey of Scotland 1747–1755," *The Scottish geographical magazine, 83:1* (1967), 1–15.

14. R.A. Gardiner, "William Roy, surveyor and antiquary," *The geographical journal, 143* (1977), 441–50, on 443–4.

the Board of Ordnance that required cadets in the corps of engineers to have had some formal technical education; the requirement functioned as a social filter, favoring sons of army officers. At the same time the American war gave rise to a greater recognition of the importance of cartographic skills and to the acceptance of engineers as staff officers.[15] By the 1780s the engineers had achieved the same social status as army officers.

Because of their specialist abilities, members of the corps of engineers (and of the artillery) saw themselves as a part of the scientific community. Some achieved membership in the Royal Society. They figured among the opposition during the so-called dissensions of 1783–4, when Sir Joseph Banks, who had been president of the Society since 1778, was attacked by practically or mathematically oriented Fellows for favoring antiquarianism and natural history—that is, "gentlemen's science."[16] Banks' forces managed to defeat their opposition. It might therefore come as a surprise that the Banksians in the Royal Society, not the mathematical practitioners, effected the cooperation with Cassini de Thury.[17] The key figure in their mobilization was William Roy, who belonged

15. Douglas W. Marshall, *The British military engineers 1741–1783: A study of organization, social origin, and cartography* (Ph.D. dissertation, University of Michigan, Ann Arbor, 1976), 1–7, 119–24, 130, 317, 343–4.

16. David Philip Miller, *The Royal Society of London 1800–1855: A study in the cultural politics of scientific organization* (Ph.D. dissertation, University of Pennsylvania, 1981), 6–14, 36–104, 115, 120, 135; W.A. Seymour, ed., *A history of the Ordnance Survey* (Folkestone: William Dawson, 1980), 3, 29, 67; Charles Richard Weld, *A history of the Royal Society, 3* (London: J.W. Parker, 1848), 151–70; Hector C. Cameron, *Sir Joseph Banks* (Sydney: Angus and Robertson, 1966), 128–34; Henry Lyons, *The Royal Society 1660–1940* (Cambridge: Cambridge University Press, 1944), 202–5, 211–4, 342. There is no evidence that Banks suppressed mathematically oriented science in the *Philosophical transactions.*

17. These Banksians on the triangulation included Charles Blagden, Henry Cavendisch, John Lloyd, and John Smeaton, as well as Roy; all were members of the Royal Society Club, a center of Banksian sympathies. Letters of Blagden to Banks, 1783–4, as listed in W.R. Dawson, ed., *The Banks letters: A calendar of the manuscript correspondence of Sir Joseph Banks* (London: British Museum, 1958), 61–2, 64–5; Archibald Geike, *Annals of the Royal Society Club* (London: Macmillan, 1917), 174–5; Seymour, *A history of the Ordnance Survey*, 134.

socially and intellectually to *both* the Banksian and the mathematical camps.[18]

In the 1760s Roy had settled in London and had become a Fellow of the Royal Society and a firm friend of Joseph Banks. Roy's strong interest in antiquities resulted in a major work on Roman Britain, published after his death; thus he qualified as a member of the "Banksian Learned Empire." By the early 1780s, Roy had also established a scientific reputation through painstaking work in the field of barometric hypsometry. As a major-general in the army and a lieutenant-colonel in the corps of engineers, he was an influential proponent of a national military survey, the realization of which was his main objective in the Paris-Greenwich triangulation.[19]

THE PARIS-GREENWICH TRIANGULATION

In his *mémoire* submitted to the British government in October 1783, Cassini de Thury argued that a triangulation between Greenwich and Paris would give the relative position of the two observatories with more certainty than celestial observations at Greenwich had done.[20] The *mémoire* made its way to the Royal Society, whose president arranged for his friend William Roy to lead the British part of the project. George III granted the Royal Society money to order a theodolite from Jesse Ramsden and the Board of Ordnance supported the project with manpower, both soldiers to undertake the manual work and officers to oversee it. The measurement required cooperation not only between the French and the

18. Seymour, *A history of the Ordnance Survey*, 5–6; Yolande O'Donaghue, *William Roy, 1726–1790: Pioneer of the Ordnance Survey* (London: British Museum, 1977); Gardiner, "William Roy," 441–2; Theodore S. Feldman, "Applied mathematics and the quantification of experimental physics: The example of barometric hypsometry," *Historical studies in the physical sciences, 15:2* (1985), 127–97, esp. 156–77.

19. Gardiner, "William Roy," 446–7. Cf. H.B. Carter, *Sir Joseph Banks* (London: British Museum [Natural History], 1988).

20. Nevil Maskelyne, "Concerning the latitude and longitude of the Royal Observatory at Greenwich; with remarks on a memorial of the late M. Cassini de Thury," *PT, 77* (1787), 151–87, on 151–2. The *mémoire* is also printed in Charles Arden-Close, *The early years of the Ordnance Survey* ([1926] Newton Abbot: David and Charles, 1969), 13.

British but also between military officers and "scientists."[21] In 1784 Roy supervised the measurement of a baseline on the outskirts of London, and three years later he led the survey from London to Dover and the cross-channel triangulation. The French part was carried out by Jacques Dominique Cassini (whose father had died in 1784), Pierre François Méchain, and Adrien-Marie Legendre.[22] In 1788 Roy completed the survey between London and Dover. He published two elaborate reports in the *Philosophical transactions*, which included no fewer than sixteen finely engraved plates.[23] In 1791 the duke of Richmond, Master General of Ordnance, decided that Roy's work should be continued under military direction. He bought a copy of the Royal Society's theodolite from Ramsden and engaged the civilian mathematician Isaac Dalby as the first employee of the Trigonometrical Survey. Dalby had participated in the triangulation of 1787–8 and was highly praised by Roy, who died in 1790.[24] The employment of Dalby ensured continuity between Roy's work and its military sequel. The connection between the survey and the Royal Society continued until 1803, when the last report appeared in the *Philosophical transactions*.

It may be that Banks responded to Cassini de Thury's overture because it carried an implicit criticism of the astronomers of Greenwich. The Astronomer Royal Nevil Maskelyne belonged to the rebel camp within the Royal Society. One charge leveled against Banks during the "dissensions" was that he cared more for social status than for scientific ability; but by promoting the triangulation he showed himself to be a "very worthy President of the Royal Society,

21. Although the term "scientist" was not in use in the 1780s, its functional equivalent, "scientific person," was. See Edward Williams, William Mudge, and Isaac Dalby, "An account of the Trigonometrical Survey carried on in the years 1791, 1792, 1793, and 1794," *PT, 85* (1795), 432.

22. Seymour, *History of the Ordnance Survey*, 15–7.

23. William Roy, "An account of the measurement of a base on Hounslow Heath," *PT, 75* (1785), 365–480; "An account of the trigonometrical operation whereby the distance between the meridians of the Royal Observatories of Greenwich and Paris has been determined," *PT, 80* (1790), 111–270.

24. Seymour, *History of the Ordnance Survey*, 21–2; Roy, "An account of the trigonometrical operation," 118–9. Dalby did practical as well as mathematical work. Ibid., 155.

ever zealous in the cause of science."[25] Further, by turning the triangulation into an event of social significance, by bringing George III and other members of the highest stratum of society to visit the site of the baseline measurement, Banks also proved that patronage did pay off. Unlike his predecessor John Pringle, Banks was on good terms with the king, who financed a large part of the operation.

The Longitude of Greenwich and the Shape of the Earth

Jacques Dominique Cassini wrote in 1791 that the "sole aim" of the triangulation had been to ascertain the longitudinal difference between the observatories of Paris and Greenwich.[26] On the other side, Roy had asserted in 1787 that the "chief and ultimate goal has always been considered of a still more important nature, namely, the laying the foundation of a general survey of the British Islands."[27] But there is nothing to indicate that Roy had official support for such a scheme at this stage. Although he wished to create a national survey, the measurement he was actually conducting had a specific purpose within a *scientific* context.[28] This purpose he had to defend. Consequently Roy accepted the French criticism depicting positional astronomy—or anyway British astronomy—as less accurate than geodesy. The fact that the expensive and time-consuming triangulation went forward without the cooperation of the astronomer at Greenwich suggests the strength of Roy's position within the Royal Society.

The disagreement between Roy and Maskelyne was ventilated in two papers published in 1787. Maskelyne wrote a belated reply to Cassini de Thury's *mémoire*; Roy, annoyed at the delay, threatened

25. Maskelyne, "Concerning the latitude and longitude," 151–2; Roy, "An account of the measurement of a base," 425, quote.

26. Jacques Dominique Cassini, Pierre François Méchain, and Adrien-Marie Legendre, *Exposé des opérations faites en France en 1787 pour la jonction des observatoires de Paris et de Greenwich* (Paris, [1791]), 65.

27. William Roy, "An account of the mode proposed to be followed in determining the relative situations of the royal observatories of Greenwich and Paris," *PT, 77* (1787), 188–226, on 188.

28. Seymour, *History of the Ordnance Survey*, 21–2. George III insisted that the financial support he gave go for "the science of astronomy"; Harley and O'Donoghue, "Introductory notes," xxi.

to publicize Maskelyne's idleness in the *Philosophical transactions*—thereby demonstrating that he had the official support of the Royal Society, whereas Maskelyne did not.[29] Roy accentuated the criticism of British astronomy by adding to the uncertainty of eleven seconds of time in longitude between Greenwich and Paris, which Cassini de Thury had pointed out, an uncertainty of three seconds between Oxford and Greenwich. Geodetic methods would narrow the margin of error; Roy considered them "infallible," since they could always be controlled by verifying the bases. The planned triangulation would give a value for the longitudinal difference "sufficiently near the truth, and. . .probably considerably nearer than it will be brought for many years to come, by a mean of the best observations of the heavenly bodies."[30]

Roy probably did not know that in 1785 Maskelyne had equipped his assistant Joseph Lindley with a number of watches and sent him on a secret "chronometer run" to Paris, to determine the time difference between the capitals. Lindley's result (9 minutes 20 seconds) verified Maskelyne's astronomically deduced value, published in 1787, which was later found to agree with the result of Roy's triangulation.[31] Roy avoided admitting this embarrassing consistency, which he had earlier denied, by misquoting Maskelyne's data. In his final report he simply plucked from Maskelyne's paper a number about 10 seconds larger than the figure on which the astronomer finally settled, and claimed it as the astronomically deduced value.[32] He thus demonstrated the superior accuracy of the geodetic method.

29. William Roy to Nevil Maskelyne, 11 Dec 1786 (RS, DM.4.14). A copy of this letter is in PRO, O.S.3/4. Cf. Roy, "An account of the mode proposed," 213; Eric G. Forbes, "The geodetic link between the Greenwich and Paris observatories in 1787," *Vistas in astronomy, 28* (1985), 173–81, on 174. Maskelyne complained that he had not been shown the *mémoire* until a year after preparations for the triangulation had begun. See Maskelyne, "Concerning the latitude and longitude," 153.

30. Roy, "An account of the mode proposed," 213–4.

31. Eric G. Forbes, *Greenwich Observatory*. Vol. 1: *Origins and early history (1635–1835)* (London: Taylor and Francis, 1975), 149–50, and "The geodetic link," 174; Maskelyne, "Concerning the latitude and longitude," 183–6; Roy, "An account of the trigonometrical operation," 231.

32. Roy, ibid., 231. Roy's maneuvers to protect himself from the suspicion of deliberately misquoting Maskelyne are sufficiently transparent. Cf. Roy to Maskelyne, 11 Dec 1786 (PRO, O.S.3/4).

The question of the earth's shape came in as an important scientific side issue in the determination of longitude differences. Roy had developed a new technique for geodetic investigation, which involved astronomical observations of a kind different from those usually associated with longitude determinations. Nevertheless, as Isaac Dalby was to point out, it constituted perhaps the weakest link in Roy's geodetic work. By taking the angles between three mutually remote stations and at the same time observing the angles between the stations and the polestar, Roy calculated the longitudinal differences between pairs of stations by spherical trigonometry. The relationship between these differences and the distance on the ground gave a value for the length of a degree of longitude at a particular latitude. Polestar observations made at only a few stations furnished the basis for calculation of the longitude difference of the whole chain of triangles. Roy calculated the latitudes of the stations in relation to that of Greenwich from a spheroidal model of the earth devised by Pierre Bouguer.[33]

To justify use of this spheroid, Roy computed the lengths of the arc between Greenwich and Perpignan (the southern extremity of the Paris meridian) on ten different hypotheses about the shape of the earth. In the model Roy favored, the lengths of degrees of latitude increased with the fourth power of the sine of the latitude. One of the models he rejected was an ellipsoid based on data from the six earlier arc measurements Roy thought most consistent. To achieve consistency, however, he had had to take a mean between the arcs of Cassini in France and Liesganig in Austria, since comparison between them gave an "absurd" result—that is, an oblong earth.[34] Roy combined the six arcs into fifteen pairs and calculated the flattening for each. ("Flattening" is defined as the ratio of the equatorial axis to the difference between the equatorial and polar axes.) Values ranged from 100 to 850, but Roy did not present the extremes: he exhibited only the mean flattening of 190, well within

33. Roy, "An account of the mode proposed," 216–20, and "An account of the trigonometrical operation," 200, 225–7.

34. Roy, "An account of the mode proposed," 206–8.

the limits of what was considered reasonable.[35] The same was true of the other six ellipsoids in Roy's table, resulting from other combinations of measurements or from hypothetical premises, and of the two spheroids as well.

Roy nonetheless proffered Bouguer's spheroid as the most probable alternative because, unlike the other hypotheses, it gave values sometimes above and sometimes below the lengths of the measured arcs: "a never failing proof" that it was "exceedingly near the truth."[36] Roy hid the wide discrepancies that actually existed between different measurements behind the reassuring surface of averages. Mathematical analysis of error was then only in its infancy. Roy employed another, more visual technique for comparing the different solutions to the problem of the shape of the earth. He presented the results on the different hypotheses in tabular form so that the reader could judge, "by simple inspection only, which of the theories agrees best with actual measurement." He also gave the lengths of degrees of meridians, parallels, and oblique great circles according to Bouguer's spheroid—not only for the portion of the earth covered by his own triangulation, but for the whole earth, so that others could use these figures until, in the distant future, the shape of the earth would "ultimately" become known. Meanwhile, Roy thought, the spheroid would furnish data of "general utility."[37]

Dalby disagreed with the use of Bouguer's spheroid and with Roy's method of finding differences in longitude. He criticized the method as too sensitive to observational errors. An error of one second of arc in the angles between the stations and the meridian would cause an error of six seconds in the longitude difference between Greenwich and Dunkerque.[38] Roy had said that he

35. The usual magnitude of the flattening ranged between 170 and 540—the values suggested by Maupertuis and Christian Huygens, respectively.

36. Roy, "An account of the mode proposed," 210–1. In early mathematical analysis of observational error by Boscovich and Laplace, this condition usually came with another: that the sum of the absolute values of the errors should be minimized. See Stephen M. Stigler, *The history of statistics: Measurement of uncertainty before 1900* (Cambridge, Mass.: Belknap Press, 1986), 11–61, esp. 47, 51.

37. Roy, "An account of the mode proposed," 201, 222.

38. Isaac Dalby, "The longitudes of Dunkirk and Paris from Greenwich, deduced from the triangular measurement in 1787, 1788, supposing the earth to be an ellipsoid," *PT, 81* (1791), 236–45, on 237, and "Remarks on Major-General Roy's ac-

determined the longitude difference "by the instrument itself," meaning that the extreme accuracy of the theodolite guaranteed the precision of the result. Dalby challenged this assertion and rejected Roy's spheroid in favor of his own ellipsoid shape, with the flattening of 229 predicted by Newton.[39]

As we know, the ellipsoid shape depended only on theories of gravity and mechanics and on the assumption that the earth had once been fluid. The theoretical implications of Bouguer's spheroid, on the other hand, are unclear, and probably they did not matter very much to Roy. He adopted it because it gave a good fit to existing measurements. Ramsden, not Newton, was the arbiter of exact geodesy; the theodolite could make errors "totally vanish." Dalby had no special interest in defending the elaborate technology adopted by Roy, and consequently he was happy to accept Maskelyne's value for the longitude difference between Paris and Greenwich as *support* for the result of the measurement in which he himself had participated.[40] In modern terms, Roy's attitude was instrumentalist, whereas Dalby's might be called realist. Condillac, who advocated a strict empiricism, criticized both. Condillac's view eventually won out. When the concept later to be christened the geoid was developed in Germany in the early 19th century, irregularities in the earth's mass distribution ceased to be regarded as anomalies and became instead constitutive of the "real" shape of the earth.

Instruments of Competition

The technology for making accurate angular measurements under field conditions was highly refined in the second half of the 18th century, when the quality and quantity of instrument-making rose dramatically, especially in Britain. An expanding market for navigational equipment, intensified surveying, and escalating scientific

---

count of the trigonometrical operation," *PT, 80* (1790), 593–614, on 607–8. Roy's method for obtaining longitudes is described in Roy, "An account of the trigonometrical operation," 206–25.

39. Roy, "An account of the mode proposed," 219; Dalby, "The longitudes of Dunkirk and Paris," 327–8.

40. Roy, "An account of the mode proposed," 219, quote; Dalby, "The longitudes of Dunkirk and Paris," 245.

demands for accuracy all pushed development. The accuracy of angular measurements increased from about 2 seconds to about 0.5 seconds, verniers and microscopes for reading off the scales became common equipment, and the achromatic lens and the dividing engine set new standards for precision.[41]

The level of accuracy of surveying technology was raised significantly through the Paris-Greenwich triangulation, which for a while turned the coast of the English Channel into an arena for technological rivalry between Britain and France. The competitive spirit no doubt helped both the French and the British parties to gain financial support from their respective monarchs. George III financed the Ramsden theodolite, and Louis XVI paid for a repeating circle by Etienne Lenoir. William Roy wrote that "The honour of the nation is concerned in having at least as good a map of this as there is of any other country."[42] Governments and the military knew that geodesy yielded accurate maps that could facilitate the exercise of political power and the waging of war. The French astronomers were directed not from the Académie des sciences but from Versailles. Cassini was to report on everything concerning the operations to the minister De Breteuil in Paris, who acted as intermediary with the court. Cassini and Méchain were instructed to undertake a little industrial espionage while in London for the triangulation, paying special attention to the telescopes of Herschel and the instruments of Dollond and Ramsden.[43] They were also to try to talk Ramsden into

41. Maurice Daumas, *Scientific instruments of the seventeenth and eighteenth centuries and their makers* (London: Batsford, 1972), 53–5, 153–6, 176, 189–204; Allan Chapman, "The accuracy of angular measuring instruments used in astronomy between 1500 and 1800," *Journal for the history of astronomy, 14* (1983), 133–7; Willem D. Hackman, "Instrumentation in the theory and practice of science: Scientific instruments as evidence and as an aid to discovery," Istituto e museo di storia della scienza di Firenze, *Annali, 10* (1985), 103–6.

42. Breteuil to J.D. Cassini, 9 Jun 1787 (BOP, D5–7); Roy, "An account of the trigonometrical operation," 162–3. Cf. Daumas, *Scientific instruments*, 184.

43. Breteuil to J.D. Cassini, 9 Jun and 30 Aug 1787 (BOP, D5–7). Cf. J.D. Cassini, "De la jonction des observatoires de Paris & de Greenwich, & précis des travaux géographiques exécutés en France qui y ont donné lieu," Académie des Sciences, Paris, *Mémoires*, 1788 (1791), 706–54, esp. 714–7.

joining the staff at the Paris observatory or taking French apprentices.[44]

The French repeating circle (fig. 6.1) and the British theodolite (figs. 6.2, 6.3) were vastly improved versions of older instruments. They soon replaced the quadrant as the preferred instrument in large surveys. They also represented two attitudes to precision measurement that henceforth would prevail. The theodolite represented the very best in British instrument-making.[45] The circle was also well constructed, and in that sense signified a breakthrough for French instrument-making. Equally important, however, its accuracy depended on a new principle—a method of averaging errors mechanically—which paralleled the theoretical notions of error under way in the 1780s and fully developed after 1800.[46]

Everybody involved in the Paris-Greenwich triangulation recognized that it set new standards for surveying. Elaborate descriptions of the new instruments were published, although, in the case of the theodolite, not all the details, which (as Roy indelicately put it) would have been "a disgusting labour." Cassini and his coworkers expatiated on the merits of their repeating circle in their book on the triangulation.[47] The circle made possible high precision by repeating the angular observations an arbitrary number of times over the whole of its limb, so that irregularities in its construction would eventually even out. The circle offered advantages of cost and size over the theodolite: it weighed only about 20 pounds; the British instrument, over 200 pounds. The cross-channel triangulation (fig. 6.4) served as an important check on the accuracy of the French method; Cassini wrote that the information about the relative merits of the circle and the theodolite would perhaps be its most interesting result. Triangles

44. Konvitz, *Cartography in France*, 27.

45. Cassini called the theodolite the "chef-d'oeuvre du plus habile Artiste qu'il y ait en Europe." Cassini, Méchain, and Legendre, *Exposé des opérations faites en France*, 58. Cf. Daumas, *Scientific instruments*, 129, 174, 186–7.

46. Gillispie suggests a direct connection between Legendre's work on the theory of the repeating circle and his development of the method of least squares. Gillispie, *Science and polity*, 127.

47. Roy, "An account of the trigonometrical operation," 135–6; Cassini, Méchain, and Legendre, *Exposé des opérations faites en France*, 23–7.

were closed to within a few seconds using either the circle or the theodolite.[48] The measurement therefore served as a kind of *calibration* as well as demonstration of the new instruments. The size and cost of the theodolite had at least one advantage. Roy argued that it would be a great waste not to use it for a national survey toward a good map of Britain, on which, as he put it, the country's honor depended.

The Rhetoric of Accuracy

The Paris-Greenwich triangulation increased the efficiency and reliability of surveying. At the same time the praises and promises of accuracy that surrounded it served as rhetorical devices, upholding the close relationship between scientific and technological applications of precision measurement. William Roy's accomplishment depended on his command of both the rhetorical and technological resources of accuracy.

The eminent metrologist Jean André Deluc had stated: "We are obliged to take up with probability in Nature in so many respects, that it is perhaps of more importance to us to investigate the physical rules of probability than to attend to its mathematical rules upon hypotheses." Deluc advocated that research be directed toward precision measurement rather than mathematical analysis of error, and he predicted that "we shall be led to seek for exactness in every thing."[49] Roy, who knew Deluc's work well, acted on this metrological precept. He treated every measurement with extreme care, but as an isolated event, and he did not take the accumulation of possible errors into account even when estimates of the precision of individual measurements might have been made. Like his contemporaries, Roy did not think in terms of significant figures; for example, he might add figures with five and two decimals, and give the sum to four.[50]

48. Cassini, Méchain, and Legendre, *Exposé des opérations faites en France*, 57–65. Cf. J.D. Cassini, "De la jonction des observatoires," 713.

49. Jean André Deluc, "An essay on pyrometry and aerometry, and on physical measures in general," *PT, 68* (1778), 493, 545–6. Cf. Feldman, "Applied mathematics and the quantification of experimental physics," 150–6.

50. Roy, "An account of the measurement of a base," 401–2, 441–61, 476–8. Roy wrote: "The hypotenuse length of the base, as measured by 1369.925521 glass rods of twenty feet each + 4.31 feet...has been shewn to be 27402.8204." This

In fact, he loved decimals, which abound in his writings far beyond practical, but perhaps within rhetorical, efficacy.[51]

In Britain the rhetoric of accuracy extended to the prestige of the instrument-maker. Unlike their French colleagues, British instrument-makers could be highly respected members of the scientific community. Jesse Ramsden became a Fellow of the Royal Society (FRS) in 1786 and won the Copley medal in 1795. Ramsden's name was synonymous with accurate measurement, and his instruments assured the quality of the British triangulation.[52] The implicit reasoning went like this: because Ramsden was such an "ingenious artist" (although "dilatory"), his instruments were "rendered extremely perfect," hence the measurements showed a "wonderful degree of accuracy."[53] Most of the devices manufactured for Roy existed in one copy only; not even specialist readers were likely to get their hands on a Ramsden theodolite, but had to be convinced of its excellence verbally. Besides the theodolite, Ramsden constructed a surveying chain "which would measure distances much more accurately than anything of that kind had ever done before," a pyrometer "of such accurate construction that it seems not easy to improve it," and other smaller instruments. Furthermore, Ramsden himself sometimes took part in especially important measurements; he was called upon for advice during the Paris-Greenwich triangulation and later in the Ordnance Survey.[54]

The commendation of Ramsden's achievements, like the long rows of decimals and the frequent references in the work of Maskelyne,

---

figure was rounded off to one decimal, because "the most accurate measurement imaginable is still more liable to err in excess than in defect." Ibid., 478. Cf. Seymour, *A history of the Ordnance Survey*, 35.

51. Cf. Heilbron, *Electricity in the 17th and 18th centuries*, 76, 83; see Theodore S. Feldman, chap. 5 in this volume.

52. Daumas, *Scientific instruments*, 102–6, 241–3; E.G.R. Taylor, *The mathematical practitioners of Hanoverian England, 1714–1840* (Cambridge: Cambridge University Press, 1966), 43; Gillispie, *Science and polity*, 113; R.S. Webster, s.v. "Ramsden, Jesse," *DSB, 11*, 285.

53. Roy, "An account of the mode proposed," 189, and "An account of the trigonometrical operation," 136, 149.

54. Roy, "An account of the measurement of a base," 394, 402, 416–7, 435, 462; Williams, Mudge, and Dalby, "An account of the Trigonometrical Survey," 432, 435, 438.

Cassini, and Roy to measurements made to "the last exactness" or to "mathematical exactness," may be considered the rhetoric of exact science.[55] This rhetoric papered over a serious rift between Roy and Ramsden, which is worth uncovering for its illumination of underlying social realities. Roy criticized Ramsden severely for his dilatoriness in the draft of the last report on the project. Most of the complaints did not reach print because Ramsden filed countercomplaints with the Royal Society and Roy's death allowed some freedom with his text. Ramsden's complaint gives an unusual glimpse of the working relationship between a scientist and an instrument-maker during the late 18th century. He claimed the credit for the high precision achieved in the triangulation, and was sorry that Roy—a *gentleman*—treated him with such disrespect: it was not "consistent with common sense, that a Tradesman or Mechanic, should suffer his *professional character* in particular to be publicly traduced in so respectable a place as at [a] meeting of the Royal Society."[56] Ramsden could get no satisfaction from Banks (who threatened to defend Roy's position "with every drop of my blood") and therefore appealed directly to the Council.[57]

Ramsden asserted that Roy was not a competent judge of the technological aspect of the work; that he, Ramsden, had decided what kind of instrument should be made for the triangulation; that he, Ramsden, had written the description of the theodolite for the *Philosophical transactions*; that he had constructed every single piece of apparatus used in the measurements; and that Roy had not given him full credit. He claimed further that Banks and Roy had granted him a free hand to construct a theodolite that was "superior in point of accuracy to any thing of whatever radius yet made,"[58] and

55. Maskelyne, "Concerning the latitude and longitude," 186, "the last exactness"; Roy, "An account of the trigonometrical operation," 165, "mathematical exactness." Cf. Roy, ibid., 149, 154, 166, 244, 161; and Cassini, Méchain, and Legendre, *Exposé des opérations faites en France*, 23, 34, 58–9.

56. Seymour, *A history of the Ordnance Survey*, 17; Roy to Banks, 3 May 1790, *The Banks letters*, 721; Jesse Ramsden to the Council of the Royal Society, 30 May 1790 (RS, MM.3.30).

57. Banks to Roy, 30 Jan 1790, *The Banks letters*, 720–1.

58. Jesse Ramsden to the Council of the Royal Society, 30 May 1790 (RS, MM 3.30); cf. RS, MM.DM.4.44, an unsigned statement by Ramsden about his technical work for Roy. Roy did indeed fail to give Ramsden credit for the technical descrip-

complained when the innovations took time to perfect. These complaints showed that Roy did not understand the character of precision technology.

If we credit any of this, Roy's dependence on Ramsden's abilities was even closer than the official reports suggested. It does seem clear that Roy controlled the successful achievement of his goal—to make a measurement and create a technology of unprecedented accuracy—not so much by mastery of the necessary technique or technology, but rather by command of the social situation through his influence in the Royal Society and in the Ordnance. Roy could purchase Ramsden's and Dalby's know-how and employ them to write the difficult technological and mathematical passages in his reports. Like Ramsden, Dalby complained that Roy could not judge his work competently, and he had to append a long list of corrections to Roy's final paper in 1790.[59]

The rhetoric of accuracy that helped to cover up those disputes was meant to inspire *faith* in a vast and costly undertaking like a national survey.[60] When Roy invited the citizens of London to confirm the accuracy of his measurements by stepping out on their rooftops and sighting the angles between buildings that had been used as triangulation stations, the cartographic entrepreneur was speaking.[61] When the state or the public supported the geodetic and cartographic sciences, they were offered accuracy in return for their money. When they were promised "absolute" or "mathematical" accuracy, they no doubt expected new and superior technology.

---

tion of the theodolite; he mentioned Ramsden's name only a few times, one of them libelously, in his last report. See Roy, "An account of the trigonometrical operation," 135–60.

59. Roy, "An account of the trigonometrical operation," 118–9, 193; Blagden to Banks, 31 Aug, and 23 and 26 Sep 1790, *The Banks letters*, 76–7; Dalby, "Remarks on Major-General Roy's account."

60. In September 1783 Roy wrote to Banks that the government would probably grant a sum of £1000 toward the triangulation, "if they are assured that it shall be frugally and faithfully applyed to the above, or any other scientific operation." Roy to Banks, 28 Sep 1783 (RS, DM.4.4).

61. Roy, "An account of the trigonometrical operation," 258–9.

Accuracy and Technology

Another of Roy's rhetorical devices was his repeated use of the word "truth" in describing the intended goal of exact measurement.[62] The quest for "truth" meant a quest for accuracy; and in surveying it covered all the steps of the operation, not just the final result. Since everything had to be superior to previous surveys, everything had to be *new*. Because of their novelty, the methods and instruments had to be meticulously described so that the public could affirm their value.[63] Hence the many pages of technical description and numerous plates that made Roy's reports a résumé of the state of the art of trigonometrical surveying. Delambre's account of the methods used in the metric survey would play the same part ten years later.[64]

The high precision of Roy's work made it possible to calibrate simpler and more convenient methods. His measurement of the baseline on Hounslow Heath is a case in point. At first Roy used the traditional wooden rods as measuring sticks. He rejected these because their contraction and expansion varied erratically with humidity, and replaced them with glass tubes, whose lengths depended only on the temperature in a way accurately determinable by Ramsden's pyrometer.[65] Roy knew about the properties of glass from his work with the barometer: the use of glass tubes for the baseline measurement was an imaginative piece of technology transfer from barometry to geodesy.[66] Roy used the glass tubes to control

62. Roy, "An account of the mode proposed," 214, "sufficiently near the truth," 219, "differ very little from the truth"; cf. Roy, "An account of the trigonometrical operation," 128–9, 186, 230–1, 247.

63. Roy, "An account of the trigonometrical operation," 247; Roy, "An account of the measurement of a base," 406, 461. Cf. A.W. Richeson, *English land measuring to 1800* (London: Cambridge, Mass.: MIT Press, 1966), 180–1.

64. J.-B.-J. Delambre, *Rapport historique sur les progrès des sciences mathématiques depuis 1789, et sur leur état actuel* (Paris: Imprimerie Press, 1810), 66.

65. Roy, "An account of the measurement of a base," 434–5, 461–6. Cf. Seymour, *A history of the Ordnance Survey*, 34.

66. Roy, "An account of the measurement of a base," 435. Roy's papers at the Public Record Office contain several documents relating to investigations of the behavior of rods and tubes of glass at different temperatures (PRO, O.S.3/3). Cf. Feldman, "Applied mathematics and the quantification of experimental physics," 172–4.

measurements with a surveyor's chain, also from Ramsden's workshop. The chain did well, and since it was easier to handle than glass tubes or metal bars, Roy recommended it for surveying. The chain would be good enough for the national survey, for which, as Roy put it, "there would not be any necessity for that wonderful exactness" requisite in the Paris-Greenwich triangulation.[67] Thus "scientific" exactness became a benchmark in the creation of a solid and efficient surveying technology.

Simplifications of surveying practices were not an accidental spin-off from scientific work: Roy had identified the development of more efficient surveying techniques as an explicit goal of the triangulation.[68] Cassini de Thury had likewise pointed out that instruments had to be simplified and made portable in order to make practical the technology of surveying.[69] The repeating circle accomplished both objectives. The British theodolite did not meet the needs of the ordinary surveyor, but was practical in an organization that could maintain a work force to transport it, erect solid stations, operate cranes, and so forth. In short, British technology was serviceable in a military organization.

THE MILITARIZATION OF CARTOGRAPHY

In 1762, in the course of extending the French triangulation into Austria, Cassini de Thury and the Heidelberg professor Christian Mayer measured a baseline in an unusual location. They used the alley, almost three leagues long, leading up to the new observatory in the main building of the Palatine Elector Karl Theodor's summer residence, situated between Mannheim and Heidelberg. Karl Theodor was delighted: "the position of the new base and [that] of the

67. Roy, "An account of the measurement of a base," 394, 449–50; Roy, "An account of the trigonometrical operation," 268.

68. Roy, "On the advantages that are likely to rise from the operations on Hounslow Heath" (RS, DM.4.6).

69. Cassini de Thury, *Relation des deux voyages*, 1–2.

observatory were two monuments that should coincide."[70] The base was a monument to the scientific patronage of the creator of the Mannheim Academy of Sciences and to Cassini's knack for fraternizing with nobility and royalty. It symbolized the union of political power and science, so important in the fields of geodesy and cartography at this time and in the next century when national cartography in Europe was militarized. With the technology of triangulation, the geometrical grid—formerly present only in the mind, on maps, or in architecture—was imposed on the landscape itself. Geodetic surveying quantified geographical information, which became atomized and readily presentable in tabular form. This kind of extensive, yet compressed, knowledge was an attractive resource for centralized governmental and military planning. In the terminology of Michel Foucault, the Ordnance Survey and similar organizations brought an expansion of military surveillance from the limited disciplinary units of the garrison and the fortress to the full physical extent of a nation.[71]

The social ascent of engineers and artillery men was a major ingredient in the genesis of national triangulation as a military operation in Britain, where it first occurred. The new strategy of the Revolutionary and Napoleonic wars also promoted the fusion of military and national cartography. Larger and more mobile armies now appeared, and the battle conquered the siege as the decisive element in war. That made every place a potential theater of war and created a demand for knowledge of the whole geography, making comprehensive and accurate cartography an essential preparation for warfare.[72]

70. Ibid., xxiii, xxxiii, 94–5; Adolf Kistner, *Die Pflege der Naturwissenschaften in Mannheim zur Zeit Karl Theodors* (Mannheim: Selbstverlag des mannheimer Altertumsverein, 1930), 52–3.

71. Michel Foucault, *Discipline and punish: The birth of the prison* (Harmondsworth: Penguin Books, 1977), 184–9, 195–203.

72. William H. McNeill, *The pursuit of power: Technology, armed forces, and society since A.D. 1000* (Oxford: Basil Blackwell, 1982), 161–3, 170–1; Richard A. Preston and Sydney F. Wise, *Men in arms: A history of warfare and its interrelationships with Western society*, 4th ed. (New York and London: Holt, Rinehart and Winston, 1979), 133–48, 173–80; Alfred Vagts, *A history of militarism: Civilian and military*, 2d rev. ed. (London: Hollis & Carter, 1959; 1st. edn. 1938), 58, 75–91, 114; N.H. Gibbs, "Armed forces and the art of war," *The new Cambridge modern*

In Britain, fortifications were not a major part of the defense system, which relied mainly on the navy. When William Roy proposed in 1766 that a "General Military Map of England" be drawn on the French model, he pointed out that should an enemy reach the British Isles, engagement would necessarily take place in the field, about which detailed geographical knowledge would be required.[73] In the 1780s the question as to whether Britain should embark on a program of extensive fortification brought heated debate.[74] The Duke of Richmond advocated defense works for the docks of Plymouth and Portsmouth; he was supported by Pitt, and vehemently opposed by others, who carried Parliament in 1786 by one vote. Richmond's opponents argued that large fortifications would lead to an "unconstitutional" militarization of British society.[75] Thousands of men living in barracks would foster and spread a militarism, which might become a stronghold for royalist and other anti-Parliament sympathies. The agitation included some jibes at engineers:[76]

> If this Military projector [Richmond] was not checked in his career, none could know what consequences might ensue. A Master General, with his Committee of Engineers, like the Laputan philosophers in their flying island might hover over the kingdom in an Ordnance balloon, descend in a moment, and seize on any man's house and domain..., draw out their scales and compasses, or sketch out their works. The country Gentlemen would find their terraces converted into bastions, their slopes into glacis,

history (Cambridge: Cambridge University Press, 1965), *10*, 62–9, 74–5; Henning Eichberg, "Geometrie als Barocke Verhaltensnorm: Fortification und Exerzitien," *Zeitschrift für Historische Forschung, 1* (1977), 40–50.

73. George III, *The correspondence of King George the Third, from 1760 to December 1784*, John Fortescue, ed. (London: Macmillan and Co., 1927), *1*, 328–9.

74. [James Glenie], *A short essay on the modes of defence best adopted to the situation and circumstances of this island*, 2d ed. (London, 1785); [Charles Lennox, Duke of Richmond] *An answer to "A short essay..."* (London, 1785); [James Glenie], *A reply to the answer to a short essay...* (London, 1785), and *Observations on the Duke of Richmond's extensive plans of fortification* (London, 1794). Cf. Alison Gilbert Olson, *The radical Duke: Career and correspondence of Charles Lennox third Duke of Richmond* (London: Oxford University Press, 1961), 81–6.

75. *An authentic account of the Debates in the House of Commons, on Monday, February 27, and Tuesday, February 28, 1786, on the proposed Plan for Fortifications, by His Grace the Duke of Richmond* (London, 1786), 1–2, 20, 34, 56–7.

76. *An authentic account*, 48.

> their pleasure grounds into horn works and crown works to which they have hitherto borne an irreconcilable aversion.

Richmond returned from his defeat five years later with a new, subtler, and more successful scheme for militarization: the Ordnance Survey. In the words of one of Richmond's critics, the survey would provide essential information to "a resistance, which is not confined to particular spots, but is capable of operating every where."[77] It also provided the British public with detailed information about their land: in 1801 the first one-inch map, of Kent, was published.

The work of Roy and of the Ordnance Survey won acclaim outside of Britain as a model of scientific accuracy, for example in India, where military surveying using Roy's methods began in 1802.[78] It was above all the French who exported geodetically founded military cartography to the rest of Europe, by force as well as by example. Napoleonic military cartography was founded on the methods developed by Delambre and Méchain during the metric survey of 1792–8, and therefore also on the French experiences during the Paris-Greenwich triangulation. Delambre's formulas and Lenoir's repeating circle came to dominate French military cartography, and by 1810 "an army of astronomers armed with chronometers, telescopes, and sextants" had invaded Europe.[79]

77. [Glenie], *A short essay*, 30–1.

78. Clements R. Markham, *A Memoir on the Indian Surveys* (London, 1878; reprinted Amsterdam: Meridian, 1968), 60–6; [Franz Xaver von Zach], "Ost-Indische Gradmessung, der Länge und Breite," *Monatliche Correspondenz zur Beförderung der Erd- und Himmelskunde, 12* (1805), 485–94. Roy's measurement was praised by Delambre, for example, although at the *Dépôt de la Guerre* it appeared "dégénère en science purement spéculative." J.-B.-J. Delambre, *Grandeur et figure de la terre* (Paris: Gauthier-Villars, 1912), 334–9; Delambre, *Rapport historique*, 74–5; "Des opérations géodesiques de détail," *Mémorial topographique et militaire*, an XI (1803), 1–56, 126.

79. Delambre, *Rapport historique*, 77–8 ("une armée d'astronomes munis de chronomètres, de lunettes et de sextans"). General Nicolas Sanson, director of the *Dépôt de la Guerre*, prescribed the use of the repeating circle as well as the use of Delambre's mathematical methods. See Berthaud, *Les ingénieurs géographes*, 307; review of Sanson's "Instruction sur la disposition et la tenue des régistres de calculs géodesiques," *Monatliche Correspondenz zur Beförderung der Erd- und Himmelskunde, 11* (1805), 49–66; Delambre, *Rapport historique*, 66. Cf. J.L. Heilbron, chap. 7 in this volume.

The French government had taken control of cartography in 1793 by expropriating the Cassini map; a quarter century later, it initiated a new survey, supervised by the general staff. The map of the *État-major* (published 1833–81) deployed information and expertise assembled in the French army after more than a decade of military cartographic expansion.[80] In 1801 Napoleon had started local Topographical Bureaus to extend the *Carte de Cassini* to the areas conquered by the French army; soon bureaus existed in Hanover, the Rhineland, Bavaria, Switzerland, Savoy, and Italy.[81]

Austria set up similar services in 1806, and the Netherlands followed suit in 1816.[82] In the same year, in response to its experience of French warfare, Russia founded a military cartographic organization that was to become the largest in Europe.[83] The leader of the Franco-Italian Topographical Corps in Milano, the Swede Gustaf Wilhelm Tibell, founded a similar corps in Sweden in 1805, which made use of the technology imported from France for the remeasurement of Maupertuis' arc of meridian in Lapland in 1801–3.[84] We see in Sweden the same interplay between military cartography and science-related geodesy that occurred earlier in Britain, and which recurred in Hanover in 1821–5 when a military survey took over Carl Friedrich Gauss' measurements.[85]

Major Johann Jacob Baeyer, of the Prussian Army, was co-leader on Friedrich Wilhelm Bessel's famous survey of East Prussia in 1831–6; and Friedrich Georg Wilhelm Struve and general C. de

80. Konvitz, *Cartography in France*, 59–61.

81. Henri Marie Auguste Berthaut, *Les ingénieurs géographes militaires 1624–1831: Étude historique* (Paris: Service Géographique de l'Armée, 1902), *1*, 239–40, 305–431.

82. See the collection *La cartographie au XVIIIe siècle*; Frans Depuydt, "The large scale mapping of Belgium, 1800–1850," *Imago mundi, 27* (1975), 21–4; Brown, *The story of maps*, 274–5.

83. George M. Wheeler, *Report upon the Third International Geographic Congress and Exhibition at Venice, Italy, 1881, accompanied by data concerning the principal government land and marine surveys of the world* (Washington, D.C.: Government Printing Office, 1885), 365–79.

84. Ulla Ehrensvärd, "Fortifikationsofficeren som kartograf," in Bertil Runnberg, ed., *Fortifikationen 350 år, 1635–1985* (Stockholm: Fortifikationskåren, 1986), 115–9.

85. W. Grossmann, "Gauss' geodätische Tätigkeit im Rahmen zeitgenössischer Arbeiten," *Zeitschrift für Vermessungswesen, 80* (1955), 371–84.

Tenner found what they called "the problem of their lives" in their measurement of a meridian of more than 25° (1816–55).[86] Baeyer subsequently founded the *Europäische Gradmessung*, which met seventeen times between 1864 and 1912; it brought together astronomers, military cartographers, and civilian administrators to coordinate and promote geodetic measurements all over continental Europe.[87] The twenty national cartographic organizations that existed in Europe and its colonies in 1885 were all under military direction.[88] The cartographic situation paralleled the use of military manpower, technology, and administrative expertise in the construction of railroads and canals and in the development of mechanized industrial production.[89]

As the store of geodetic data increased, the notion of a regular shape of the earth came into general doubt. The long-disputed ellipsoid had been put out of court by Delambre and Méchain, whose demonstration that meridians are irregular was corroborated by the Ordnance Survey in 1803.[90] As more surveys comparable in

86. Friedrich Wilhelm Bessel, *Gradmessung in Ostpreussen und ihre Verbindung mit Preussischen und Russischen Dreieckskketten* (Berlin, 1838), iv, ix–x; Friedrich Georg Wilhelm Struve, *Arc du méridien de 25 × 20′ entre le Danube et la Mer Glaciale* (St. Petersburg, 1860), *1*, xv ("problème de leur vie").

87. Of those participating in the meeting in 1867, twenty-four were directors of observatories or academics, eleven were military officers, and five were civilian administrators. See C. Bruhns, W. Foerster, and A. Hirsch, eds., *Bericht über die Verhandlung der vom 30. September bis 7. October 1867 zu Berlin abgehaltenen allgemeinen Conferenz der Europäischen Gradmessung* (Berlin, 1868), 3–4. Bialas, *Erdgestalt*, 242–6.

88. Brown, *The story of maps*, 280–1.

89. See, e.g., Merritt Roe Smith, "Introduction," in Merritt Roe Smith, ed., *Military enterprise and technological change: Perspectives on the American experience* (Cambridge, Mass.: The MIT Press, 1985), 1–37; and Barton C. Hacker and Sally L. Hacker, "Military institutions and the labor process: Noneconomic sources of technological change, women's subordination, and the organization of work," *Technology and culture, 24:4* (1987), 743–75. Cf. Svante Lindqvist's and Robin Rider's contributions in this volume (chaps. 10 and 4, resp.).

90. For example, see the report of the international commission that investigated the metric survey. Jan Hendrik van Swinden, "Rapport sur la mesure de la méridienne de France, et les résultats qui en ont été déduits pour déterminer les bases du nouveau systéme métrique," Institut national des sciences et arts, *Mémoires*, Sciences mathématiques et physiques, *2* (an VII), 47, 49–52. Several contributions from the heated debate concerning the British meridian are reprinted in Olinthus Gregory, *Dissertations and letters...tending either to impugn or defend the Trigonometrical Survey of England and Wales* (London, 1815). Delambre defended the

magnitude to Delambre's and Méchain's were carried out and interconnected, the ellipsoid came to be viewed as a convenient fiction. The notion of the geoid, developed by Friedrich Wilhelm Bessel and Gauss around 1830 but not christened thus until 1873, became the preferred theoretical tool to describe the physical shape of the earth (or rather of the geopotential surface).[91] The shape of the geoid could not be predicted; it had to be measured, over and over again, whenever instrumental improvements promised refinements. Increased data caused geometrical simplicity to give way to dynamic complexity, which could be managed only by the systematic work of well organized institutions like the military.

British results; see Delambre, *Grandeur et figure*, 363–72.

91. Irene Fischer, "The figure of the earth—changes in concepts," *Geophysical surveys, 2:1* (1975), 3–54, esp. 20–8; Bialas, *Erdgestalt*, 234–5.

7 The Measure of Enlightenment

By J.L. Heilbron

The quantifying spirit united with the purposes of the Enlightenment and the Revolution to produce the metric system of weights and measures. This exemplar of 18th-century rationalism was directed against the foggy and feudal metrology of the Ancien Régime; supported by the rhetoric of the *philosophes*; realized through the politics of scientific societies and revolutionary tribunals; and resisted by the common man in whose interest the reformers, *philosophes*, scientists, and revolutionaries claimed to act.

THE PLIGHT OF THE PEOPLE

Cain's Legacy

God may have made the world according to weight and measure, but it was Cain who invented weights and measures, and thus—we have this from Flavius Josephus—"converted the innocent simplicity in which man had lived into a miserable existence dominated by fraud and deceit."[1] The existence of French men and women around 1790 was made miserable by, among other things, 700 or 800 differently named measures and untold units of the same name but different sizes. A "pinte" in Paris came to 0.93 liter; in Saint-Denis, to 1.46; in Seine-en-Montagne, to 1.99; in Précy-sous-Thil, to 3.33. The aune, a unit of length, was still more prolific: Paris had three,

The following abbreviations are used in the notes: *HAS* and *MAS*, Académie des sciences (Paris), *Histoire* and *Mémoires*, resp.

1. Josephus, *Jewish antiquities*, cited by Witold Kula, *Measures and men* (Princeton: Princeton University Press, 1986), 3.

each for a different sort of cloth; Rouen had two; and France as a whole no fewer than seventeen, all in common use and all different, the smallest amounting to just under 300 lignes, royal measure, the largest to almost 600.[2]

France possessed nonuniform measures in law as well as by custom. Their multiplicity went with other relics of the feudal system, which maintained arbitrary rents and duties usually to the disadvantage of the peasant. A landlord wanted his bushels of grain or hogsheads of beer in the biggest measures in use in the neighborhood, and he preferred to sell according to the smallest. Nor were all seigneurs above enlarging the vessel in which they collected their rents; and since in many cases they possessed the only exemplars of their patrimonial bushel, no one could be certain that it did not grow in time. But one suspected. A frequent complaint in the *cahiers*, or notebooks of desiderata brought by representatives of the people to the meeting of the Estates General in 1789, was that "the nobles' measure waxes larger year by year." These same representatives castigated the oppressive confusion of customary measures as barbaric, ridiculous, obscurantist, gothic, and revolting, and demanded an end to them, and the establishment of a system of unchanging and verifiable weights and measures throughout the country, or at least throughout their region. Many urged that the King's measure, the royal foot, be made the law of the land.[3]

Sharpers and crooks whose practices were not sanctioned by ancient rights and wrongs and middlemen acting in analogy to money changers opposed the rationalization that menaced their livelihood. In 1747, shortly after returning to Paris with the vision acquired while measuring a piece of a meridian in Peru, Charles-

2. Georges Bouchard, *Un Organisateur de la victoire. Prieur de la Côte-d'Or membre du Comité de salut publique* (Paris: R. Clavreuil, 1946), 286; Charles Maurice de Talleyrand-Périgord, *Proposition faite à l'Assemblée nationale, sur les poids et les mesures* (Paris: Impr. nationale, 1790), in John Riggs Miller, *Speeches in the House of Commons upon the equalization of the weights and measures of Great Britain...together with two letters from the Bishop of Autun* (London: J. Debrett, 1790) 60–3.

3. Kula, *Measures and men*, 164–226, esp. 191–6, 230, 236; Beatrice Fry Hyslop, *French nationalism in 1789, according to the general cahiers* (New York: Octagon, 1968), vii, 56.

Marie de La Condamine identified and condemned this special interest, which he proposed to abolish with the confusions that engendered it. His accusation echoed in France for decades and eventually bounced across the Channel. We read in the *cahiers* from Orléans that multitudinous measures "expose[d] people daily to swindlers" and in the records of Parliament that John Riggs Miller, an obscure and verbose M.P., declared that they had but one purpose, "the perplexing of all dealings, and the benefitting knaves and cheats."[4]

Reformers laid down several requirements for a new system of weights and measures. It should not rest on an arbitrary unit, especially not on a king's foot; it must not offer enticements to cheaters; and it had to be easily reproduceable were its exemplars lost. Further, it had to be rational, so as to recommend itself to all nations, and become universal. The measuring stick used by La Condamine and his colleagues, the "toise de Pérou," had attained some currency in France and in a few other countries, and the units used in Paris also had more than local authority. It would not do to impose them, however, as Talleyrand wrote Miller, since they had not been derived from nature or constructed "with the ceremony necessary to settle once and for all the opinion of all enlightened nations." Last and also first, the reformed system had to be simple, or, as Miller preferred to say, "on a level with the lowest and humblest capacity." It must not require "skill in calculation beyond what...the inferior orders of men commonly possess"; everyone should be able to confirm for himself the correctness of all transactions of interest to him, "the meanest intellect...on a par with the most dexterous."[5]

This last paragraph contains many buzz-words of the Enlightenment.[6] The replacement of the arbitrary and the capricious, of the

4. Charles-Marie de La Condamine, "Nouveau projet d'une mesure invariable, propre à servir de mesure commune à toutes les nations," *MAS*, 1747, 489–514, on 492–5; Kula, *Measures and men*, 209; Miller, in Thomas C. Hansard, *Parliamentary history*, *28*, col. 322, and Miller, *Speeches*, 18.

5. Talleyrand, *Proposition*, in Miller, *Speeches*, 68–9; Miller, ibid., 16, 35, and in Hansard, *28*, col. 321.

6. Cf. Maurice Crosland, "'Nature' and measurement in eighteenth century France," *Studies on Voltaire and the eighteenth century*, *87* (1972), 277–309, on 297.

feudal and historical, by the natural is the message of all the *philosophes* from Montesquieu to Condorcet. The natural coincides with the rational and the universal: when people cast aside customary belief and established abuse, they can reach agreements that all others, guided by their own reason, will accept. This reason is not the property of a few great intellects; the truth is accessible to all, or at least to every man. Everyone has the right to know, and to recognize, the truth. Any system that claims universal assent must be universally intelligible.

Mathematics and the Rights of Man

From the most remote times philosophers have taken number as the exemplar of the intelligible. Once one has grasped a principle in geometry, it was said, not even God Himself could understand it better, although, to be sure, He might know more theorems. Calculating people think for themselves; they despise the unintelligible, capricious, unfounded, authoritarian, and feudal as infringements on their thoughts and actions. In brief, mathematics is a science for free people. Or, to say the truth as the 18th century saw it, for free men. Newton's doctor, Sir John Arbuthnot, praised mathematics for giving "a manly vigour to the mind"; a tonic Newton took to such good effect that he mastered all "the noble and manly sciences" and became "the greatest man that ever liv'd."[7] All this will help to construe the remark made by the representatives of the revolutionary government in 1799, on accepting the prototypes of the meter, liter, and kilogram. The metric measurers, they said, reaching for their highest compliment, had carried through their work "with the confidence of a male and republican spirit."[8]

If mathematics is male, Europe grew increasingly manly during the 18th century. When Samuel Pepys became clerk of the king's ships in

7. Quoted from, respectively, John Arbuthnot, *Miscellaneous works*, 2 vols. (Glasgow: J. Carlisle, 1751), *1*, 9, 36, and James Jurin, in Royal Society of London, *Philosophical transactions, 34* (1727), dedicatory epistle to Martin Folkes.

8. P.L.C. Baudin, 4 messidor an VII (19 June 1799), in P.F.A. Méchain and J.B.J. Delambre, *Base du système métrique decimal, ou Mesure de l'arc du méridien compris entre les parallèles de Dunkerque et Barcelone*, 3 vols. (Paris: Baudoin, 1806–10), *3*, 650.

1660, he had to recruit his strength (he was then 27) and repair his education (which was excellent) by learning the multiplication table. In the resulting fit of *machismo*, he forced his wife to learn arithmetic.[9] Most of learned Europe in the 17th century was "innumerate," if by analogy to "literacy" we take "numeracy" to mean familiarity with numbers. During the 18th century the requirements of government bureaucracies, commerce, colonization, mining, rational agriculture, forest management, the military, and so on, brought dramatic increases in numeracy. Before 1750 *Poor Richard* enjoins us in rhymes to go early to bed to secure health, wealth, and wisdom; just after 1800 a book called *The young man's guide* delivers the same message by multiplication. If you stay up to eleven o'clock every night, the *Guide* admonishes, in fifty years you will uselessly expend $182.50 on candles.[10]

Perhaps the most significant sign of this burgeoning rudimentary numeracy was the multiplication of tables of numerical equivalence. A good survey of the tables in use in commerce has yet to be made; but there is no doubt that their number increased dramatically after the Seven Years' War. They came in several sorts: conversions of weights, measures, and moneys; total cost of goods, tabulated by size and unit cost; tables of interest and annuities; agricultural yields; and so on. It did not require great prowess at mathematics to use these compilations; rather, a degree of numeracy comparable to the literacy of one who could read but not write. The spread of the metric system depended on this widespread rudimentary numeracy, and raised its level.

These tables were required not only because of the number and uncertain equivalence of feudal weights and measures, but also, and

9. Samuel Pepys, *Diary*, ed. Robert Latham and William Matthews, 11 vols. (Berkeley: University of California Press, 1970–83), *3*, 134, quoted by Patricia Cline Cohen, *A calculating people. The spread of numeracy in early America* (Chicago: University of Chicago Press, 1982), 26; Helen M. Wallis, "Geographie is better than Divinitie: Maps, globes, and geography in the days of Samuel Pepys," in Norman J.W. Thrower, ed., *The compleat platt-maker. Essays on chart, map, and globemaking in England in the seventeenth and eighteenth centuries* (Berkeley: University of California Press, 1978), 1–43, on 2, 6. Cf. Pepys, *Diary, 3*, 131, 135, 137.

10. Cohen, *Calculating people*, 110–3.

perhaps primarily, because the arbitrary multiples and submultiples of the various units made computation burdensome and complex. Calculation of the price of a piece of cloth 2 yards 1 foot 4 inches square at 3 pence 2 farthings the square foot was a sufficient challenge. To change it into aunes, pieds, livres, and deniers, and to proceed to a problem in bushels and cubic king's feet, would have puzzled Archimedes. According to the Paris Academy, referring to the situation in 1790, people at ease with money computations could not handle weights and measures. "In the present state of affairs, a man who can calculate with sous and deniers cannot calculate with toises, pieds, pouces, and lignes, with livres, onces, gros and grains."[11]

The Paris Academy and many other scientific reformers supposed that by dividing the new standards and the revised coinage decimally they would eliminate the need for specialist computers. The decimal was not free from arbitrariness; but its simplicity and convenience could not be gainsaid, at least by practiced calculators, and, as the Academy observed, although not universal it is as natural as the human hand.[12] Only the hand of the learned had so far employed decimal arithmetic, and by no means universally, as Lavoisier pointed out in his *Elements of chemistry*, when urging his colleagues to state weights in decimal parts of whatever units they used. This natural arithmetic, "previously locked up in the domain of the sciences," was precisely what the reformers thought they sought. "Those who knew little will know everything; others will hurry to forget what they no longer need to know; all will accept as a true benefit a method of calculation that will save them time, study, and chances for error."[13]

Prieur de la Côte d'Or, a former military engineer who served on the all-powerful Committee of Public Safety (Comité du salut public), expected the decimal calculus to be the technical language of Utopia.

11. Charles Borda, J.L. Lagrange, A.L. Lavoisier, Matthieu Tillet, and M.J.A.N. Caritat, marquis de Condorcet, "Rapport" (27 Oct 1790), *HAS*, 1788, 1–6, on 5.

12. Ibid., 6; cf. Crosland, "'Nature' and measurement," 299.

13. France, Commission temporaire des poids et mesures republicaines, *Instruction abrégée sur les mesures deduites de la grandeur de la terre, uniformes pour toute le République, et sur les calculs relatifs à leur division décimale* (Paris: Impr. nationale exécutive du Louvre, an II), xvii, 171a; Antoine Laurent Lavoisier, *Elements of chemistry* (Edinburgh: William Creech, 1790), 295–6.

"How happy we will be not to be forced to consult anyone about our prosperity, property, expenses, and drink, and to have nothing to do any more with people who often seek only to profit from our ignorance."[14] The connection between democracy and the decimal was made plain and explicit by Condorcet during the first year of the Revolution. Decimalization, he said, fit perfectly with the political program and mandate of the National Assembly. "It [the Assembly] wants to insure that in the future all citizens can be self-sufficient in all calculations related to their interests; without which they can be neither really equal in rights...nor really free."[15] Long after the promulgation of the metric system, Laplace advised Napoleon that its chief advantage as understood by its creators was not the destruction of feudal metrology but the division by tens.[16]

Republican zeal is not easy to curb, and the decimalization of everything measured or metered figured among the excesses of the French Revolution. The Paris academicians demonstrated solidarity with the regime by dividing a right angle into 100 revolutionary degrees, and each such degree into 100 minutes; and they found much pleasant recreation in recomputing the trigonometrical functions in what Jean-Baptiste Joseph Delambre, the most assiduous of the metric measurers, later extolled as "the vastest [calculation] that had ever been done, or even conceived." The innovation neither saved the Academy nor suppressed the reckoning of the Babylonians, with which it still coexists in France. A less enduring initiative, the revolutionary calendar, which came into operation retrospectively on the day of the autumnal equinox of 1792, divided the year into twelve parts of thirty days each, grouped in ten-day blocks. The five

14. C.A. Prieur, *Instruction sur le calcul décimal, appliqué principalement au nouveau système des poids et mesures* (Paris: Impr. de la République, germinal an III [1795]), 4–5; cf. Miller, *Speeches*, 39.

15. Condorcet, *Mémoires sur les monnoies* (Paris, 1790), 3–4, quoted in Ruth Inez Champagne, *The role of five eighteenth-century French mathematicians in the development of the metric system* (Ph.D. dissertation, Columbia University, New York, 1979), 60.

16. Text of 7 May 1811, in G. Bigourdan, *Le système métrique des poids et mesures. Son établissment et sa propagation graduelle* (Paris: Gauthier-Villars, 1901), 193.

or six additional days required to make up the year were intercalated as necessary.[17]

The unsystematic concession to lunar motions represented by the numbers 12 and 3 in the divisions of the revolutionary year was offset by a revolutionary day of ten "hours," each containing 100 "minutes" and 10,000 "seconds." Several clocks ticking 10,000 "seconds" a day were made, but most clock makers and watchers preferred to divide their time in the manner of the servile peoples of Europe. On 18 germinal an III (7 April 1795), the revolutionary government suspended republican time sine die on the official ground that it was of interest only to scientists. The calendar survived longer, until 1 January 1806, when Napoleon put an end to it.[18]

The cost of conversion to decimal units was borne by the people whose lives the system was intended to ease. In computing prices of old goods in the new currency, sellers naturally rounded up to their advantage. The common man as naturally opposed the change. So did the common woman. According to a squib of 1791, the prostitutes of the Palais Royal, whose rates had been recomputed, complained that the 100 sols they now received for their services devalued their charms, "which opinion had previously reckoned at an écu and six livres," by a sixth.[19] It is not easy to anticipate the effects of reform.

Another conflict between geometry and system arose in the naming of the decimal divisions of the meter. In 1792 the Academy discussed the relative advantages of system (in which divisions would be

17. *Exposition abrégée du nouveau systême des poids et mesures, d'après le mètre définitif* (Avignon: Seguin, an X), 13–4; J.B.J. Delambre, *Rapport historique sur les progrès des sciences mathématiques depuis 1789, et sur leur état actuel* (Paris: Impr. impériale, 1810), 8; L. Rondonneau, *Concordance des calendriers républicain et grégorien*, 6th ed. (Paris: Rondonneau and Decle, 1812), v–vii; Georges Villain, "Etude sur le calendrier républicain," *La Révolution française, 7* (1884), 451–9, 535–53 and *8* (1885), 623–56, 740–58, 830–54, 883–8.

18. France, Commission temporaire des poids et mesures républicains, *Instruction*, 33–4, and *Avis* (Paris: Impr. de la République, frimaire an IV [1795]), 8; Bouchard, *Prieur*, 308–10; Thomas Bugge, *Science in France in the revolutionary era,* ed. M.P. Crosland (Cambridge, Mass.: MIT Press, 1969), 203–4; Rondonneau, *Concordance*, xii–xiv; Villain, "Calendrier républicain," 885–8.

19. Bouchard, *Prieur*, 310–1.

designated by the prefixes "deci-," "centi-," "milli-," and the only multiple would be "milliaire") and familiarity (in which submultiples of the meter would have common names, like "palme" and "doigt"), and plumped for the familiar. They erred in calculating revolutionary zeal. The Committee on Public Instruction (Comité d'instruction publique) preferred a clean sweep and adopted the systematic names. This occurred on 1 August 1793, on the eve of the suppression of the Academy.[20] One of the Academy's successors in metric matters explained that "it is almost impossible to reason correctly without a language aptly made."[21]

The language did not please the people. Classicists objected that the prefixes (enriched by "déca-," "hecta-," and "kilo-") violated the grammar of ancient languages, while the uneducated could make no sense of them at all. "These names," declared a delegate to the Convention, "novel and unintelligible to the large majority of our citizens, are not necessary for the maintenance of the Republic."[22] He did not know what he had escaped. Prieur de la Côte d'Or, who had taken an active part in the reform of weights and measures from the onset of the Revolution, had names of his own, derived, he said, from ancient languages and Low Breton. The irrationality of the Convention did not extend to adopting "kilicymbe," "myriadore," "ladedix," "pèzeprime," "centicadil," or "decidol," and Prieur, bending to the political wind in 1795, drew up what became the definitive metric names. The many centicadils of bitterness he then swallowed and decidols of crow he then ate were to damage the work of the metric reformers.[23] All concerned might have spared themselves the trouble. For decades the people stubbornly opposed decimalized units and their jabberwocky names.

20. *HAS*, 1789, 1–18 (text of 1792); Bigourdan, *Système métrique*, 78–82.

21. France, Agence temporaire des poids et mesures, *Aux citoyens redacteurs de la feuille du Cultivateur, en réponse à des objections contre la nomenclature nouvelle, inserées dans le No. 38 de ce journal* (Paris: Impr. de la République, thermidor an III [1795]), 10.

22. Bigourdan, *Système métrique*, 82 (text of 11 Aug 1795).

23. Bouchard, *Prieur*, 297–9; J.B.J. Delambre, *Grandeur et figure de la terre,* ed. G. Bigourdan (Paris: Gauthier-Villars, 1912), 212.

THE PROGRAM OF THE ACADEMY

Professional Imperatives

By 1789 the French had made three measurements of arcs along the meridian passing through Paris. The earliest, made by Jean Picard between 1668 and 1670, on an arc of about 1°20′, gave a value for the length of a degree of latitude near Paris that differed by less than three parts in ten thousand from what the metric measurers later obtained with much greater labor and much better instruments.[24] The second arc, extending over 8°30′ from Dunkirk to Perpignan and completed under Jacques Cassini in 1718, authorized a new Atlantic coastline that brought some French towns a hundred or more kilometers East of their previous positions. Louis XV lost more land to his cartographers than his successors have to the Germans. The measurements were not accurate enough, however, to settle the much agitated question of the shape of the earth. In the 1730s, the Paris Academy sent out its expeditions to Peru and to Lapland to measure arcs far enough apart to reveal the departure of the earth's profile from perfect sphericity. Their results, which confirmed Newton's conclusion that mechanics required the equatorial axis to exceed the polar, agreed with the third measurement along the Paris meridian, made during 1739 and 1740 by the Abbé Nicolas-Louis de Lacaille.[25]

As the general shape of the earth came into view, its finer features loomed ever larger in academic minds. Lacaille contributed another data point by measuring an arc at the Cape of Good Hope. The Pope, the Austrian-Hungarian Empress, and the King of Sardinia commissioned surveys of meridians traversing their domains, with the consequence, as one of the surveyors put it, that the more the earth

24. Jean Picard, *Mesure de la terre* (Paris: Impr. royale, 1671); republished in *MAS*, 1666–99, *7*, 133–90, and in P.L. Moreau de Maupertuis et al., *Degré du méridien entre Paris et Amiens déterminé par la mesure de M. Picard* (Paris: Martin, Coignar, and Guerin, 1740), 1–106. Picard gives 57,060 toises (fathoms) per degree, which was nominally 15 toises off the definitive value. In fact, the difference came to about 30 toises when Picard's measurements were corrected for refraction and a small error in his baseline; see Méchain and Delambre, *Base, 1*, 7–8.

25. Delambre, *Grandeur*, 63–72.

was measured, the more uncertain its shape became.[26] The British then ran an arc through Greenwich, to serve as the backbone for a military map of Scotland. The Paris academicians looked upon it rather as an opportunity to improve geodesy, and in 1783 they proposed a linking of the Paris and Greenwich observatories by triangles based on the meridians already determined. The proposal had a sporting side. The linking would pit the small repeating circle then recently invented by Charles Borda, a prominent academician and a former naval officer, against the large theodolite built by the world's leading instrument-maker, Jesse Ramsden, for the triangulations in England.[27]

Jean Dominique Cassini (Cassini IV), who took nominal charge of the French party, retrospectively threw down the challenge: "we dared to flatter ourselves that we had on our side an instrument that yielded nothing to the English in point of precision." The showdown took place in the late summer of 1787. The British turned their great theodolite toward Calais; the French aimed their delicate circle at Dover. The British took their single observations quickly; the French multiplied their repetitions slowly. The season grew late. Fog descended. The weather did not remit long enough to permit the full circle of French measurements, and the contest between the repeater and the theodolite ended indecisively.[28] As Delambre wrote much later, however, "a more important occasion soon presented itself to demonstrate the advantages of the new instrument."[29]

26. R.J. Boscovich and Christoph Maire, *Voyage astronomique et géographique dans l'Etat de l'Eglise: Entrepris par l'ordre at sous les auspices du pape Benôit XIV* (Paris: Tilliard, 1770), 492, cited in Djordje Nikolic, "Roger Bošković et la géodésie moderne," *Archives internationales d'histoire des sciences, 14* (1961), 315–35, on 315.

27. Jean Mascart, *La vie et les travaux du chevalier Jean-Charles de Borda (1733–1799)* (Lyon: A. Rey; Paris: A. Picard, 1919), 370–1, 381–3, 488–9, 501–2. Cf. S. Widmalm, chap. 6 in this volume.

28. J.D. Cassini, "De la jonction des observatoires de Paris et de Greenwich, et précis des travaux géographiques exécutés en France, qui y ont donné lieu," *MAS*, 1788, 706–17, on 710–3.

29. J.B.J. Delambre, *Histoire de l'astronomie au dix-huitième siècle* (Paris: Bachelier, 1827), 758.

Institutional Considerations

In exchange for their salaries, status, and grants for special projects like geodetic surveys, the Paris academicians acted as technical advisers to the Crown. Their advice tended to be conservative and elitist, especially in respect of unsolicited proposals from inventors asking for state subventions or monopolies for their novelties. Experience confirmed what arrogance had suspected: most of the proposals were worthless and most of the proposers ignorant. The Academy suggested that artisans might be licensed, but only after passing an examination in geometry and other useful arts. During the Revolution, the Academy faced the hostility directed at all enclaves of privilege and the special anger of frustrated inventors, manufacturers, and would-be scientists who had suffered at its hands.[30]

Well into 1790 the Academy held its meetings, read papers, and planned projects as usual. Some academicians favored the new regime, but few wished it to go further; almost all lamented the disturbances that kept them from their work. Many recognized, however, that they would have to reorganize their company in keeping with the new style, especially after the establishment in 1791 of a patent law that took from them their chief public service.[31] As a partial offset, academicians became increasingly involved in government technical projects. Of these, the most important, "repeatedly paraded as a prime example of science's potential value to the nation and a concrete instance of the Academy's proper function in society," was the reform of weights and measures.[32]

30. C.C. Gillispie, *Science and polity in France at the end of the ancien régime* (Princeton: Princeton University Press, 1980), 97–9, 461–3; Roger Hahn, *The anatomy of a scientific institution. The Paris Academy of sciences, 1666–1803* (Berkeley: University of California Press, 1971), 118–21; Gillispie, "The *Encyclopédie* and the Jacobin philosophy of science: A study in ideas and consequences," in Marshall Clagett, ed., *Critical problems in history of science* (Madison: University of Wisconsin Press, 1959), 255–89, on 257, 268–74.

31. C.P. Molard, *Description des machines et procédés spécifiés dans les brevets d'invention, de perfectionnement et d'importation, dont la dureé est expiré* (Paris: Huzard, 1811), 7–27; Hahn, *Anatomy*, 186–9; Henry Guerlac, *Essays and papers in the history of modern science* (Baltimore: Johns Hopkins University Press, 1977), 467–8.

32. Hahn, *Anatomy*, 162–3.

A month after the storming of the Bastille, academician Jean Baptiste Le Roy, physicist, mathematician, and one-time clockmaker, suggested that the Academy propose to the National Assembly a dissolution of the over-rich metrological heritage of the Republic. The proposal, probably drawn up by the Academy's secretary, the marquis de Condorcet, and certainly presented to the Assembly by Talleyrand, set aside all existing units in favor of the length of a pendulum that beat seconds at the 45th parallel of latitude, which passes just north of Bordeaux. Talleyrand proposed further that the British be invited to join in the determination and in promoting the result, "so that all nations might adopt it."[33]

On 8 May 1790 the Assembly considered the Talleyrand or Academy proposal together with several others to the same effect, notably one by Prieur, who expressly opposed using an arc of the meridian as a basis. "Besides the magnitude of the fundamental operation required, the difficulty of verifying it, and the impossibility of doing so daily, it is not easy to decide how exact the method might be." Here Prieur spoke as a military engineer familiar with surveying practice and with the lingering uncertainties over the earlier measurements of arcs. The same considerations caused Thomas Jefferson to give up his project of taking a decimal part of Cassini's degree as the basis of a new American unit: "the various trials to measure various portions of [meridians], have been of such various result, as to show there is no dependence on that operation for certainty."[34]

The seconds pendulum had been proposed as a standard-setter for over a century. The pioneer geodecists preferred it: Picard, in 1671, proposed defining the pied as a third, and the toise as twice, the length of the seconds pendulum at a convenient place; Jacques Cassini observed that all domestic measures could usefully be referred to

33. Hahn, *Anatomy*, 163–4; Bigourdan, *Système métrique*, 14–5; Talleyrand in Miller, *Speeches*, 59.

34. Bouchard, *Prieur*, 287–90; William David Pattison, *Beginnings of the American rectangular land survey system, 1784–1800* (Chicago: University of Chicago Press, 1957), 48–9, after Jefferson, *Papers*, ed. Julian P. Boyd et al. (Princeton: Princeton University Press, 1950+), 7, 25, 150–60; Bigourdan, *Système métrique*, 10.

such a unit, and even all European measures, since the length of a seconds pendulum is about the same throughout the Continent; La Condamine proposed international cooperation and a pendulum regulated at the equator, as determined by himself and "the hands of nature"; and Turgot, as controller of finance under Louis XVI, almost initiated a nationwide reform based on the seconds pendulum at 45°.[35] Just before the Revolution, Gaspar de Prony, who would play an important part in the metrication of France, declared the pendulum to be the ultimate arbiter of length; and after the promulgation of the meter, a compiler of dictionaries, no doubt lifting from his predecessors, held that "the length of the simple pendulum, an invariable quantity always easy to recover, seemed given by nature to serve as a measure in all countries."[36] The British had also considered the advantages of Cassini's suggestion, and of calibrating their yard by the pendulum, in order that "all future generations [may] obtain similar measures of length, capacity, and weight, and thereby render it altogether needless to cut them on stone, or to engrave them on brass, to perpetuate their existence."[37]

Miller was about to introduce proposals for the reform of English weights and measures based on the pendulum when a letter from Talleyrand inviting Britain to join France in finding the length of a seconds pendulum came to hand. The plan was simple, and a natural successor to the linking of the observatories of Paris and Greenwich, "which all Europe would take as a guarantor of rigorous exactitude."[38] While Parliament pondered the opportunity, the Revolution became too hot for Anglo-Saxon reformers, and the Paris Academy removed the rationale for collaboration by deciding not to take the

35. Bigourdan, *Système métrique*, 6–11; Jacques Cassini, *De la grandeur et de la figure de la terre* (Paris: Impr. royale, 1720), 250; La Condamine, "Nouveau project," 501–5, 511; Charles Henry, *Correspondance inédite de Turgot et Condorcet* (Paris: Charavay, 1883), xxv, 234–5.

36. Gaspar de Prony, "Discours préliminaire," in William Roy, *Description des moyens employés pour mesurer la base de Hounslow-Heath* (Paris: Didot, 1787), xvi–xvii; Lunier, *Dictionnaire des sciences et des arts*, 3 vols. (Paris: Normant and Nicolle, 1806), 2, 615, s.v. "mesure."

37. John Whitehurst, "An attempt toward obtaining invariable measures," in *The works* (London: W. Bent, 1792), iii (text of 1787); Miller, *Speeches*, 44.

38. Talleyrand in Miller, *Speeches*, 71.

pendulum as primary. Its decision seemed odd to many.[39] The rationale must be sought, not in the requirements of measurement, but in the circumstances into which the Revolution propelled the Academy.

The National Assembly accepted Talleyrand's proposal and sent it and a question about the most useful division of weights, measures, and monies to the Academy, which referred both matters to a committee composed of Borda, Lagrange, Lavoisier, Tillet, and Condorcet. The committee reported on 27 October 1790 that everything should be decimal.[40] It then handed what remained of its charge to a committee consisting of Borda, Lagrange, Laplace, Monge, and Condorcet. On 19 March 1791 these geometers reported that the pendulum was the poorer of the two universal and natural units they could imagine. They plumped instead for a piece of the Paris meridian and yet another measurement of it.

Their objections to the pendulum suggest a hidden agenda. In obtaining a length from pendulum beats, they wrote, a unit of time, which has nothing to do with distance, must be invoked; and this unit, the 86,400th part of a day, had the additional blemishes of being both arbitrary and nondecimal. "It is much more natural, in fact, to refer distances from one place to another [the academicians were thinking of a standard for cartography, not for commerce] to a quarter of a terrestrial circle rather than to the length of a pendulum." The committee did appreciate the greater convenience of the pendulum standard, for which they provided: the Academy would undertake to determine the length of a seconds pendulum at the Paris Observatory, to serve as a secondary reference, the primary to be one ten-millionth of the distance from pole to equator.[41]

39. Hansard, *28*, cols. 315–17; Miller, *Speeches*, 43–4; George Evelyn Shuckburgh, "An account of some endeavors to ascertain a standard of weight and measure," Royal Society of London, *Philosophical transactions, 88* (1798), 133–82, on 165–6.

40. Borda, Lagrange, Lavoisier, et al., "Rapport," 1–6; Méchain and Delambre, *Base, 1*, 14; Bigourdan, *Système métrique*, 16.

41. Charles Borda, J.L. Lagrange, and Gaspard Monge, "Rapport. . .sur la systême général des poids et mesures," *HAS*, 1789, 1–18, on 1–6 (decimalization), 7–16 (arc over pendulum). Cf. Bigourdan, *Système métrique*, 17–8; Méchain and Delambre, *Base, 1*, 14–9.

Although the ambition of the academicians did not extend to measuring the full quarter meridian, it opened a task sufficiently large. Borda's committee proposed to redo the arc from Dunkirk to Perpignan and to extend it to Barcelona, to obtain latitudes astronomically, to lay out new baselines, to observe the pendulum, to determine the weight of an exactly measured volume of distilled water at the temperature of melting ice, and to compare all the old units in use in France with the new standards. They saw no advantage in British cooperation: "we have excluded from our advice every arbitrary determination, we have used only the common property of all nations....In a word, if the memory of all our work disappeared and only the results remained, they would disclose nothing to show what nation conceived the idea and carried it through."[42] The Academy as a whole did not readily accept the recommendations of its committee of interested mathematicians: some objected that the arc had received enough attention; others, that the pendulum was easier. In the end, however, the Academy endorsed the recommendations and sent them to the National Assembly.[43]

The speciousness of the argument that the choice of Borda's committee was the most satisfactory and least arbitrary stands forth from their anticipation of the charge that enlightened people everywhere might not regard a section of the meridian through Paris and lying almost entirely within France as a unit dictated by nature. They argued: the section should extend equal distances on either side of the 45th parallel because there the seconds pendulum and the size of a degree have their mean values; it is a mere coincidence that the 45th parallel runs through France. But all meridians are bisected at 45°. Why take one through Paris? Because only there do meridians have arcs bisected at 45° that terminate at either end at sea level and that are short enough to measure. "There is nothing here that can give the slightest pretext for the reproach that we wished to assert any sort of dominance." Or, as Laplace put the point in a lecture in 1795, "had savants from all countries come together to fix the

42. Borda, Lagrange, and Monge, "Rapport," 13–6, 19.
43. Bigourdan, *Système métrique*, 19–21.

universal measure, they would not have made a different choice."[44] This flim-flam was perfectly clear to Jefferson: "The element of measure adopted by the National Assembly excludes, *ipso facto*, every nation on earth from a communion of measurement with them; for they acknowledge themselves, that a due portion for admeasurement for a meridian crossing the forty-fifth degree of latitude, and terminating at both ends at the same level, can be found in no other country on earth but theirs."[45]

Jean-Baptiste Biot wrote in 1803, when surveying the progress of science since the Revolution: "if the reasons that the Academy presented to the Constituent Assembly were not altogether the true ones, that is because the sciences also have their politics: sometimes to serve men one must resolve to deceive them." Biot gave as the hidden agenda the Academy's wish to settle the shape of the earth once and for all.[46] According to Delambre, Borda convinced his committee to opt for the arc because his circle, as suggested by the Paris-Greenwich measurement of 1787, made possible a determination of the meridian far more accurate than Lacaille's. This consideration left a trace in the committee's report to the Academy in March 1791. Today's instruments, it said, are so good that future improvements would not sensibly change the length of the meter that they determine; "or at least the length of time separating us from an age when everyday transactions would require and could attain such a precision is so great in comparison with the life of a man as to amount to infinity itself."[47]

44. Borda, Lagrange, and Monge, "Rapport," 15–6; Bigourdan, *Système métrique*, 56; Laplace, *Oeuvres complètes, 14* (Paris: Gauthier-Villars, 1912), 141, 145 (quote).

45. Jefferson to William Short, 28 July 1791, quoted by C. Doris Hellmann, "Jefferson's efforts towards decimalization of United States weights and measures," *Isis, 16* (1931), 266–314, on 286. The academicians' grantsmanship has succeeded with some historians: Léon Bassot, "Notice historique sur la fondation du système métrique," in France, Bureau des Longitudes, *Annuaire*, 1901, D. 1–43, on 2, 16; C.C. Gillispie, "Laplace," in *Dictionary of scientific biography, 15* (New York: Scribners, 1978), 273–403, on 334–5.

46. Jean Baptiste Biot, *Essai sur l'histoire générale des sciences pendant la révolution française* (Paris: Duprat and Fuchs, 1803), 355–6. Cf. Adrien-Marie Legendre, "Suite du calcul des triangles qui servent à déterminer la différence de longitude entre l'Observatoire de Paris et celui de Greenwich," *MAS*, 1788, 747–54, on 753.

47. Delambre, *Rapport* 5; Crosland, in Bugge, *Science*, 20; Borda, Lagrange, and

To these objectives—the old scientific imperative and the desire to vindicate and promote French instrumentation—should be added the social and strategic concerns of forging bonds with the new state. The metric project had high priority since the abolition of feudal rights had raised the gathering of rents and taxes in kind to a new level of confusion and litigation.[48] During the several years the project would last, the Academy could expect to enjoy strong government support and a useful flow of cash. On 8 August 1791 some 100,000 livres, more than the Academy's annual state subvention, was placed at its disposal; estimates of the entire cost of the project ran from 300,000 livres into the millions; and one can scarcely criticize the Academy if it saw in this commitment a pledge on the part of the government to see it through troublous political times.[49]

THE LABOR OF ACADEMICIANS

Pit and Pendulum

The National Assembly approved the Academy's revised proposal on 30 March 1791. The Academy immediately divided its work among five commissions: triangulations and latitude determinations (Cassini IV, Méchain, Legendre), baselines (Meusnier, Monge), pendulum of Paris (Borda, Coulomb), weight of water (Lavoisier, Haüy), and comparison of old and new measures (Tillet, Brisson, Vandermonde).[50] The whole enterprise was to be directed by Borda, Condorcet, Lagrange, and Lavoisier. Soon death and disinclination reduced this extraordinary mobilization of the brains of France. All geodetic work fell to Delambre and Pierre-Francçis-André Méchain, an excellent compulsive astronomer; Borda and Cassini took on the pendulum, and Lavoisier the water. The first year went by making instruments. Only one Borda circle, that of 1787, existed in 1791;

Monge, "Rapport," 15.

48. Kula, *Measures and men*, 236–7, 243.

49. Hahn, *Anatomy*, 163–4, 253; Adrien Favre, *Les orgines du système métrique* (Paris: PUF, 1931), 121–30; Bigourdan, *Système métrique*, 32.

50. Bigourdan, *Système métrique*, 22–6.

two more, readable to 3 or 4 seconds of arc, were ready in the summer of 1792, when Méchain set out for Spain and Delambre started north. The instrument answered its advertisements: although difficult to use in a cramped church tower, its precision with enough repetitions was, in Delambre's words, "nearly incredible."[51]

Great accuracy was also achieved in the pendulum experiments, which took place during the summer of 1792 in a pit in the Paris observatory. The general technique, by no means original with the metric project, was to observe coincidences between a swing of a simple pendulum of length L and that of a clock that accurately beat seconds. Let t_1 be the clock time when both pendulums move through the midpoint of their swing together, and let t_2 be the time of the next coincidence. In the interval $\Delta t = t_2 - t_1$, therefore, the number of swings of the two pendulums differs by one and the half period of the simple pendulum is $T = \Delta t/(\Delta t \pm 2)$. Thence the length of the seconds pendulum L_s can be deduced from the formula $T^2{:}1 = L{:}L_s$. If Δt is large, L_s can be obtained very accurately.

Borda designed the apparatus, which was made by Etienne Lenoir, who also supplied the repeating circles. The pendulum bob hung from a 12-foot steel cord suspended from a knife-edge mounted on a subterranean wall in the Paris Observatory. Borda and Cassini observed coincidences through a telescope pointed perpendicularly to the wall. The largest departure from the mean of the twenty coincidences they recorded was one part in 100,000. They corrected their computed L_s for the dependence of the period on the amplitude of the swing, on temperature, on air pressure, on the flexure of the support and the steel cord, on the moment of inertia of the bob, and on much else. The result: $L_s = 440.5593$ lines of the toise used in the expedition to Peru.[52] "Thus," wrote the Commissioners who presided over the metric project in 1794, "the pendulum can be considered the depository of the unit of measure, or even a method of

51. Méchain and Delambre, *Base, 1,* 20, 23, 43–4, 97–9; Delambre, *Rapport,* 6.

52. Borda and J.D. Cassini, in Méchain and Delambre, *Base, 3,* 337–401, and in C.J.E. Wolf, ed., *Mémoires sur la pendule, précédés d'une bibliographie,* 2 vols. (Paris: Gauthier-Villars, 1899–91), *1,* 17–64.

measuring the earth; and nothing is more suited to instill admiration of physics and geometry than seeing an undertaking that would appear to require travel from one end of the world to the other with great machines reduced to a very simple experiment done in one place with an instrument of modest dimensions."[53] That would have made an inexpensive as well as a reasonable standard. Lenoir's bill for the circles, the rules, the water apparatus, and the pendulum was just over 34,000 livres. The Academy then estimated—this is from a progress report of 2 May 1792—that the total project would cost 300,000 livres.[54]

Roughing It

When Méchain and Delambre set out in the summer of 1792, the Revolution had become, in Delambre's words, "truly frightening." Méchain's instruments perplexed the southern peasants, who jailed him as a counter-revolutionary. Officials who understood numbers eventually procured his release. He decided that he would be safer in Spain, and so began his mapping there, with great efficiency until an injury compelled him to break off early in 1793. By the time he had recovered and completed the arc in Spain, with an immaculate determination of the latitude of the southern terminus at Montjouy, the Spanish government had closed the border to Frenchmen. Méchain passed his enforced leisure by observing stars in Barcelona, also with exquisite accuracy; but, to complete his misfortunes, the difference in latitude between Montjouy and Barcelona, as deduced from the stars, came out 3.24 seconds greater than the difference Méchain had calculated from measurements made on the ground.[55] These miserable three seconds tormented Méchain for the rest of his life: he thought that Borda's circle permitted perfection and he blamed himself for blundering. Méchain had measured his angles as exactly as Borda's circle allowed. The discrepancy of 3 seconds—about the angular width of a penny observed at a distance of 450 miles—arose from

53. France, Comm. temp., *Instruction*, 29–30.
54. Bigourdan, *Système métrique*, 22.
55. Méchain and Delambre, *Base, 1*, 52–6; Delambre, *Histoire*, 759–60.

unevenness in the terrain, which dragged Méchain's plumb bobs off the vertical.[56]

While Méchain stagnated in Barcelona, Delambre had one unpleasant adventure after another in the north of France. Anticlerical peasants had knocked down many of the church towers that had served his predecessors; when he hung lanterns as substitute sights, he was suspected of signaling to the enemy. (France had been at war with most of Europe since April 1792.) Once, having been detained by local patriots, he tried to explain his mission in an impromptu lecture on geodesy. To no avail. He was packed off to a higher jurisdiction, in Saint Denis. The square teemed with volunteers waiting to go to the front, who did not have the prerequisites for the crash course in triangulation they forced him to give. "Evening came on. . . .The audience was very large: the front rows heard without understanding; those behind heard less and saw nothing. They grew impatient, and grumbled; some proposed one of those quick means [of dealing with suspect people] then so much in vogue." Were it not for a quick-witted official, who rushed Delambre into protective custody, the arc might have ended in Saint-Denis.

The National Assembly found time while busy transforming France into a Republic, which it declared on 22 September 1792, to issue an order for Delambre's release.[57] That did not signify a regard for academicians. On 8 August 1793 the National Assembly closed the Academy along with other unrepublican corporations. The suppression came only a week after the Assembly had given cause for reassurance to those who expected that the Academy's labors on measures in the national interest would protect it. That first week in August the Assembly had affirmed the decimal system and the meridianal definition of the meter, ordered the continuation of the work, and decreed that the Academy provide for the manufacture,

56. Méchain and Delambre, *Base, 2*, vii–ix, 617–20, 624–6, 630; Delambre, *Histoire*, 760–7.

57. Méchain and Delambre, *Base, 1*, 23–34, 41–4; Vulfran Warmé, *Eloge historique de M. Delambre* (Amiens: Caron-Duquenne, 1824), 15–16; Henri Marie Auguste Berthaut, *La carte de France, 1750–1898,* 2 vols. (Paris: Service géographique, 1898–99), *1*, 121–2; Méchain and Delambre, *Base, 1*, 32–4.

distribution, and explanation of provisional meters for general use while it prosecuted its measurements. This provisional meter was defined as a ten-millionth of ninety times the average degree in France as determined by Lacaille, or 443.444 lines of the Peruvian toise. It differed from the definitive meter by about a quarter of a millimeter.

It is not advisable to assign an important job to a specialized government agency one week and to abolish the agency the next. The Assembly therefore immediately accepted the recommendation of the Committee of Public Safety that—because its work would erase the last vestiges of feudal divisions—the Academy's commission on weights and measures should be reestablished, with the same personnel, as an independent temporary commission (Commission temporaire des poids et mesures républicains). It consisted of Borda, Brisson, Cassini, Coulomb, Delambre, Haüy, Lagrange, Laplace, Lavoisier, Méchain, Monge, and Vandermonde.

The Academy's confidence in metrology as an anchor against political storms thus received partial justification. The job of overseeing the making and distribution of the provisional standards brought another large and pressing public responsibility to the academic rump and an important cash flow. The minimum cost of the standards, as estimated by Lenoir and others, would be some 200,000 livres. This minimum provided for copper standards for the geographical departments into which the Revolution had divided the old provinces and for iron standards for the prefectures. The Assembly preferred to avoid this whiff of discrimination and voted an additional 60,000 livres or so to make all the standards of copper except for the platinum prototypes.[58] The temporary commission expected the copper standards to be accurate to around one part in 100,000 for the meter and the grave (kilogram) and one part in 10,000 for the pinte (liter).[59]

58. Bigourdan, *Système métrique*, 30–5; Champagne, *Role*, 102–3; Bouchard, *Prieur*, 293; Bugge, *Science*, 206–7; Charles Cappelli, *Rapports métriques* (Turin: Felix Buzan, an X), 7; C. Doris Hellman, "Legendre and the French reform of weights and measures," *Osiris*, *1* (1936), 314–39, on 322–3.

59. Bigourdan, *Système métrique*, 36–40.

The arrangements of August and September proved provisional in more ways than one. Lavoisier was arrested on 28 November 1793; the temporary commission requested his release as necessary to its work; the Committee on Public Safety of the Convention (into which the Assembly had transformed itself on declaring a republic) refused, and, for good measure, purged the temporary commission of Borda, Brisson, Coulomb, Delambre, and Laplace, who had not, in its opinion, shown a proper hatred for kings. The rejection of the appeal for Lavoisier and the order for the purge were drawn up by Prieur, who thus revenged himself on academicians who had opposed his political views and metrological proposals. When news of the purge, which was made public on 23 December 1793, reached Delambre, he was shivering from cold on the top of a signal tower 64 feet tall that he had erected near Orléans in the service of the people. On his return to Paris, a revolutionary committee studied his manuscripts without making much of them and pounced on a diploma from the Royal Society, which bore the arms of George III. Delambre found it prudent to disappear.[60]

What remained of the temporary commission did little more than print explanations of the new system and revolutionary rhetoric. "Soon [our] vision will no longer be affronted by those old weights and measures that still tell of the odious remains of times and things sullied by tyrants....Every child will know [the system]; and it will help diminish inequality among men."[61] There is no reason to put off enjoyment of the great benefit; the provisional meter will be perfectly satisfactory for commerce; the definitive will only serve to make the whole operation "more worthy of the powerful and enlightened nation that has undertaken it." Neither the rhetoric nor the practical business of furnishing provisional metersticks long protected the pitiful rump commission. In the spring of 1794, it requested 50,000 livres to pay its artisans; Prieur arranged to give 10,000 and to terminate its existence.[62]

60. Méchain and Delambre, *Base, 1*, 46–52.

61. Bigourdan, *Système métrique*, 55–6 (address to the Convention, 19 Jan 1794).

62. France, Comm. temp., *Instruction*, xv–xvi; Bouchard, *Prieur*, 295–6, 299–301, 461–4; George Sarton et al., "Documents nouveaux concernant Lagrange," *Revue d'histoire des sciences, 3* (1950), 110–32, on 127.

A Military Connection

Delambre believed, probably correctly, that the Committee on Public Safety and many members of the Convention wished to kill the meridian measurement and declare the provisional meter definitive.[63] At this point a powerful advocate of geodesy came to the rescue: General E.N. de Calon, formerly of the Royal Geographical Engineers (Ingénieurs géographes), which had been suppressed in 1791. Meanwhile Calon had become a deputy and a Jacobin; his rampant republicanism perhaps may be traced to unfulfilled ambition and a sense of injury suffered in the late 1770s, when, though an officer of twenty years' standing, he had been ordered to do survey work usually assigned to his juniors.[64] He was a numerary as well as a passionate man. His few recorded statements in the Convention and the Jacobin Club of Paris were precise and numerical, as in the following complete discourse: "A vehicle drawn by six horses has just been stopped at Neuf-Brisach; it contains a ton of gold and of white and black uniforms."[65]

The suppression of his former corps presented Calon with a vacuum in which to fulfill his ambition. With the corps went the map department of the Dépôt de la Guerre, the headquarters and storehouse of military cartography. The temporary organization put in its place could not handle the flood of maps, and Calon, who became head of the Dépôt in April 1793, managed to recapture most of its archives. He then purged the place of suspect employees installed by his aristocratic predecessor. To repopulate the Dépôt, Calon set up a school there in astronomy, geography, engineering, history, languages, and engraving; and to staff the school and get on with his major mission, providing maps of the front and of captured territories to the Army, he tried to attach to his service all the unoccupied savants he could find. At its height in the autumn of 1796, Calon's

63. Bigourdan, *Système métrique*, 50–81.

64. Henri Marie Auguste Berthaut, *Les ingénieurs géographes militaires, 1624–1831*, 2 vols. (Paris: Service géographique, 1902), 1, 43–6, 66, 102–3, 122–3.

65. F.-A. Aulard, *La Société des Jacobins. Recueil des documents pour l'histoire du Club des Jacobins de Paris,* 6 vols. (Paris: Jouaust and Noblet, 1889–97), *3*, 161, 522, 679.

cartographical program, printing establishment, and school ran at an annual cost of 300,000 livres. His expansiveness caused inefficiencies and inspired jealousies, and by the spring of 1797 he and his empire had fallen.[66]

Among the academicians Calon recalled to government service were Méchain, who had managed to get himself to Genoa in September 1794, after a narrow escape from corsairs, and Delambre. It was agreed, probably early in 1795, that they would triangulate the new departments and that, to prepare as firm a base as possible for the work, they would resume the measurement of the meridian. Then at the zenith of his influence, Calon persuaded the Committee on Public Safety to grant 220,000 livres for map-making, half for extending the Cassinis' Carte de France into the Rhineland and half for finishing the meridian between the Loire and the Pyrenees.[67] This largesse, given on 13 May, fit well with proposals Prieur had made in March for the prompt completion of the metric project.[68] Prieur recommended setting up a temporary agency (Agence temporaire des poids et mesures) to push through the manufacture of provisional standards and to oversee their deployment. He estimated that much could be done for 500,000 livres. He had to request another 500,000 livres the following September. Meanwhile the Army lent a hand by collecting old metal, chiefly church bells and discarded measures, to serve materially and symbolically as raw ingredients for metric measures.[69]

Prieur's program became law on 7 April 1795. It provided for using the "provisional" meter in everyday transactions even after the definitive platinum prototype came into existence; confirmed the basic names, meter, liter, and gram; and ordered resumption of the measurement of the arc. It established a triumvirate under the Commission of Public Instruction to oversee the manufacture, distribution, and explanation of the provisional standards, using machines

66. Berthaut, *Ingénieurs, 1*, 126–7, 130–2, 135–50.

67. Delambre, *Histoire*, 756–62; Méchain and Delambre, *Base, 1*, 57–8; Berthaut, *Ingénieurs, 1*, 162–4.

68. Bigourdan, *Système métrique*, 61.

69. Kula, *Measures and men*, 244.

wherever possible, "so as to combine facility and swiftness with precision, and consequently to allow citizens to buy the new measures at a reasonable price"; and it specifically ordered the new agency to provide graphic representations of conversions requiring no calculations by users.[70] Later that month the scientific side of operations was entrusted to academic survivors of earlier commissions: Borda and Brisson were to bring in the exemplar of the copper provisional meter within a "décade" (ten days); Méchain and Delambre, to do their thing; Delambre, Laplace, and Prony, to fix a baseline near Paris; Borda, Haüy, and Prony, to determine the standard of weight; and Berthollet, Monge, and Vandermonde, to oversee the preparation of the platinum to be used in the definitive primary standards.[71]

Matters being thus composed, Delambre left Paris on 28 June 1795 in the capacity of "astronomer of the Dépôt de la Guerre" to resume his observations at Orléans. He had the usual trouble finding intact steeples, and the cost of erecting signals quickly depleted his cash. No one wished to accept the assignats (government IOUs) he carried; and once again Calon tided him over. Three years of republican rule had not made peasants less suspicious and superstitious. They now tore down Delambre's signals on the theory that they attracted storms.

Méchain had his hardships, too, as he crawled north across the mountains toward their rendezvous in Rodez. In one place, on a high outcrop only twelve feet wide, surrounded by precipices, he sat for days waiting for the clouds to lift long enough to work the Borda circle. He arrived within signaling distance of Rodez around 1 September 1797, shortly after Delambre had completed his part of the work. There they stopped. Méchain would not accept Delambre's help and could not go on. "In this cruel situation," he wrote his collaborator, "I prefer to stay in this terrible exile, far from what I most cherish in the world; I will sacrifice everything, give up everything,

70. France, Convention nationale, *Loi relative aux poids et mesures [18 Germinal an III]* (Paris: Impr. de la République, [1795]), 4–7; Bigourdan, *Système métrique*, 67–70. Cf. Hellman, "Legendre," 324–7; Bouchard, *Prieur*, 306–7. The agency consisted of François Gattey, A.M. Legendre, and C.E. Cocquebert.

71. Méchain and Delambre, *Base, 1*, 58–64; Bigourdan, *Système métrique*, 71–7.

rather than return without finishing my part of the job." Delambre felt that he could not insist and Méchain finished the arc the following September.[72]

After concluding his triangles at Rodez, Delambre joined Laplace at Melun to fix the ends of the baseline near Paris. Borda had devised special rulers for the purpose, which were not ready until the spring of 1798. Their employment was extremely tedious. Delambre could make only about ninety lengths (about 360 m) a day; it took thirty-three days to cover the entire distance of 6075.90 toise. In the summer he redid the Perpignan baseline. That took six or seven weeks. Earlier surveyors had done it in twelve hours. The length of the Perpignan base as calculated from the length of the Melun base and the triangles differed from the measurement on the ground by less than a foot. On 1 brumaire 1798 an VI (22 October 1798), everything required to calculate the length of the arc from Dunkirk to Barcelona was in hand. Méchain and Delambre returned to Paris, where an international committee had assembled to examine, approve, advertise, and propagate their results.[73]

THE RESPONSE OF THE PEOPLE

The Academy of Sciences resumed its existence late in 1795 as the principal part of the new Institut de France, a government monopoly of the most learned new republicans. On 4 April 1796, the Institut took over the metric project through a commission composed of the reborn academicians Berthollet, Borda, Brisson, Coulomb, Delambre, Haüy, Laplace, Legendre, Méchain, Monge, and Prony. They had developed skill in the political as well as the technical aspects of their charge. Laplace proposed the convening of an international body of savants to check and accept the prototypes of the new units. "You realize [Laplace wrote Delambre] that all this is only a formality, to

72. Berthaut, *Ingénieurs, 1*, 165; Bigourdan, *Système métrique*, 135–45; Méchain and Delambre, *Base, 1*, 66–80; Méchain to Delambre, 20 brumaire an VI (22 October 1797), ibid., 83.

73. Méchain and Delambre, *Base, 1*, 84–90; Delambre, *Rapport*, 6, 53–4; Berthaut, *Carte, 1*, 123–4.

enable them to consider the system on their own, to do away with national envy, and to make them adopt the measures." Borda opposed the meeting; Napoleon supported it; the Directory ordered it.[74] The irrepressible Talleyrand, now minister of foreign affairs, issued the invitations, and had the satisfaction this time of omitting Britain altogether.

The international committee of experts convened appropriately in a military setting, in the Dépôt de la Marine, on 28 November 1798, having waited in Paris for a few months for Delambre and Méchain to remeasure a baseline. These patient experts represented client states and friendly powers, to wit: the Bavarian, Cisalpine, Helvetian, Ligurian, and Roman Republics, and Denmark, Sardinia, Spain, and Tuscany. The meeting divided into three working committees, one to control the weight of the kilogram, a second to compare the scales used in the work with the old toise, and the third—composed of four members calculating independently—to deduce the length of the meter from the cornucopia of measurements made by Méchain and Delambre. Their calculations fixed the definitive meter over-precisely at 443.295936 lignes of the toise de Peru taken at 17.6°C, some 0.144 lignes shorter than the provisional meter by then distributed throughout France. (The difference was meaningless for commerce, since it amounted to one foot in a kilometer.) Independent observers had nothing but praise for Borda's circle and for the care with which the units of volume and weight were deduced from the meter and the density of water.[75]

The Danish delegate retained enough independence to observe that although Méchain and Delambre had been as precise as was humanly possible, "it cannot be assumed that the determination of the meter is so absolutely and completely borrowed from nature that there is no doubt at all about its final accuracy." No echo of this

74. Hellman, "Legendre," 333–5; Crosland, in Bugge, *Science*, 197; Laplace to Delambre, 29 Jan 1798, in Yves Laissus, "Deux lettres de Laplace," *Revue d'histoire des sciences, 14* (1961), 285–96, on 287–8.

75. Crosland, in Bugge, *Science*, 197–200; Bugge, ibid., 205–11; Méchain and Delambre, *Base, 1*, 92–4. Praise for the circle: Bugge, *Science*, 204; Delambre, *Rapport*, 232–3, quoting F.X. von Zach.

discordant and irrelevant observation was heard in the speeches on 22 June 1799, when the joint committees formally presented the prototype standards to the people. In accepting them, the president of the upper legislative body, the Conseil des anciens, observed that it was a work of genius to have found an inviolable base of measurement in nature; "that, citizens, is the immortal service that the National Institute has rendered to the French Republic, or rather, the benefit that it offers to the entire human race."[76] It remained only to convince the race to accept the benefit.

Resistance

The experts returned to their countries encouraged to propagate the metric system at home for the improvement of commerce and the tightening of ties to France. The sort of pressure they were under appears from a hint earlier given by the Convention's Committee of Public Instruction to a newly liberated territory. The Committee pointed to the great "advantage it would be to the union just established between the French and Batavian republics to spread the system of uniform and decimal measures beyond French territory." The Cisalpine Republic, consisting of Lombardy and Emilia-Romagna, also felt the pressure early.[77] Annexed territory, such as the Département du Mont-Tonnerre carved out of the Rhineland, felt a heavier hand. By a decree of 18 June 1801, the Département would be obliged to use the metric system exclusively; all old weights and measures found in shops and all meter sticks on which former units of length were indicated would be confiscated.[78]

In friendly but not subject territory the system was recommended for its combination of abstract science and practical utility. "We [in the Kingdom of Sardinia] are obliged to the nation that has

76. Bugge, *Science*, 204, and in Crosland, "'Nature' and measurement," 305–8; P.C.L. Baudin, "Réponse," in Méchain and Delambre, *Base, 3*, 652. The president of the lower legislative body, the Conseil des cinq cents, spoke similarly; ibid., *3*, 649.

77. Quoted by Laissus, "Deux lettres," 288.

78. *Tables de comparaison entre les mesures anciennes en usage dans le Département du Mont-Tonnerre et les nouvelles mesures républicaines* (Mayence: Pfeipfer, an X), xxxii.

sustained with unexampled courage the most terrible political upheavals for having put forth, in the very midst of a frightful tumult, the most abstract speculations of mathematicians as a rule for every-day transactions." "Nature and not France brings [the new measures] to us [Spaniards]. Let us accept them following our natural ally with whom we have so many commercial relations....It behooves Spain [because it helped Méchain] to set other nations the example of adopting these units."[79]

None of this cajoling or commanding had much immediate effect. Even in France, where the new units were to oust the old in land transactions on 1 vendémiaire an X (23 September 1801), the people declined to benefit from the republican measures created ostensibly in their interest.[80] One of the orators at the presentation of the prototypes in July 1799 had imagined the satisfaction a peasant would feel in computing his holdings in square meters. "The field that supports my children is such and such a fraction of the globe. In that proportion I am a co-owner of the world."[81] In practice the peasant and his surveyor, builder, and draper refused to give up their toise sticks. So brutish a rejection of the natural and the rational perplexed the savants. "There is no respect in which the reform of weights and measures has not been advantageous," declared the president of the Agricultural Society of Paris (Société d'agriculture de Paris) in 1809, "and, consequently, if reason always was listened to, the success of the reform would have been complete."[82]

Alas, the meter had only complicated ordinary life. Artisans brought home their measures in familiar units and sat up all night converting them into meters, and into a crowd of irrelevant decimals, in order to write estimates and contracts in the obligatory figures.

79. Capelli, *Rapports*, 5; Gabriel Ciscár, *Memoria elemental sobre les nuevos pesos y medidas decimales fundados en la naturaleza* (Madrid: Imprenta real, 1800), on 33–4.

80. François Gattey, *Eléments du nouveau systême métrique, suivis des Tables de rapports des anciennes mesures agraires avec les nouvelles* (Paris: Bailey and Rondonneau, an X), 65. Cf. Bugge, *Science*, 205, and Maurice Crosland, "The congress on definitive metric standards, 1798–1799: The first international scientific congress?" *Isis, 60* (1969), 226–31, on 229–31.

81. Quoted by Crosland, "'Nature' and measurement," 285.

82. L. C[hessiron], "Mesures," in *Nouveau cours complet d'agriculture théorique et pratique, 8* (Paris: Delerville, 1809), 291–314, on 299–300.

Why did they not measure in meters to begin with and enjoy "the happy effects of the most useful of the gifts that scientists have been able to make to society"?[83] Because they were creatures of habit, vain, traditional, pious, ignorant, and ungrateful.[84] And also canny and suspicious. According to a British metrologist, who had it from a French merchant, consumers opposed the new units on the theory that they would be an advantage to shopkeepers, and shopkeepers opposed them for fear that consumers would be able to figure out true costs. The government hoped that a play with words would overcome suspicion. On 4 November 1800, it allowed the use of old words for multiples of the new measures, for example, "toise" for "two meters," "pinte métrique" for "litre," and so on. The reformers thought the concession damaging and absurd: if the people could use Greek words like "chirurgien" and "apothécaire," why not "kilogramme"?[85]

Because the people did not like change and cared nothing for system. So stubborn and irrational were they that in 1812 the imperial government sacrificed the jewel of the reform, the key to democracy, that is, reckoning by tens. Two laws, of 12 February and 28 March of that year, indicate the state of affairs. The first ordered that the official system be taught in all schools throughout the empire, that it alone be used in public administration, markets, and commercial transactions. The second destroyed the official system by introducing "common measures" ("mesures usuelles") divided by twos and threes: a toise of 2 meters divided into 6 pieds; an aune of 1.2 meters divisible into halves, quarters, eighths, and sixteenths, and also into thirds, sixths, and twelfths; and similarly for weights and coins. Our British metrologist enjoyed the spectacle: "Thus, after twenty years of

83. Ibid., 292.

84. Concocted from compliments in ibid.; Méchain and Delambre, *Base*, *3*, 651 (text of 1799); Ciscár, *Memoria*, 38–9; Jean Guillaume Garnier, *Traité d'arithmétique*, 4th edn. (Paris: Houdin, 1818), 281.

85. P. Kelly, *Metrology; or, an exposition of weights and measures, chiefly those of Great Britain and France* (London: Lockington et al., 1816), 25–6; Gattey, *Eléments*, 9–10; Chessiron, "Mesures," 296. The convenience of halves and doubles of the fundamental units was widely urged as an advantage of the reform, e.g., in *Journal de physique, 47* (messidor an VI [1798]), 59, 60.

trouble, mystery, and litigation, no advances are made, except that of having one common standard."[86] Otherwise confusion had only been compounded: the puzzled citizen had the provisional meter of 1795, the definitive of 1799, the compromised names of 1801, the common system of 1812, and, unofficially and omnipresent, the ancient units of the Kings of France. It is easy to change governments, not hard to reform currency, but almost impossible to alter the most common weights and measures. "Nothing has a greater tendency to grow worse, or more obstinately resists improvement." One took the long view. "This new metric system, all of whose parts fit together so simply, will no doubt triumph in the end over the obstacles that habit still opposes to its general adoption," so reads a text of 1820, "especially if it continues to be obligatory."[87]

Spread

A gauge of the rate of penetration of the new units is the pattern of their use in books requiring specification of units, like atlases, compendia of architectural drawings, and manuals of surveying, building, and engineering. Although such works employing old units exclusively were published into the 19th century,[88] the common pattern up to around 1815 was to present measurements and calculations in both meters and toises and their multiples. A good indication for the early years is the collection of architectural designs awarded prizes by the Institut de France and other government bodies between 1795 and 1803. Forty of these designs have units: three use toises only, two use meters only, and the balance uses both. Before 1800 toises have priority when both scales are present; after

86. Kelly, *Metrology*, xiii; ibid., 27–30, for text of the laws of 1812.

87. Respectively, Kelly, *Metrology*, xvi, and Louis Puissant, *Traité de topographie, d'arpentage et de nivellement*, 2nd edn. (Paris: Courcier, 1820), 245 (1820).

88. See, for example, Charles-François Viel, *De la construction des édifices publics sans l'emploi du fer* (Paris: Perronneau, 1803); L. Ducrest, *Vues nouvelles sur les courans d'eau, la navigation intérieure et la marine* (Paris: Perronneau, 1803); Molard, *Description*, 1; J.Ch. Krafft, *Recueil d'architecture civile* (Paris: Bance ainé et al., 1812); Bernard Forest de Belidor, *La science des ingénieurs, dans la conduite des travaux de fortification et d'architecture civile*, ed. C.L.M.H. Navier (Paris: Firmin Didot, 1813).

1800, meters.[89] A collection of plans for large structures published in 1823 shows a very considerable change: of seventy-six plates with scales, two have toises only, eleven have both, and sixty-three meters only; and the old measures occur exclusively in one section of the collection, on stone bridges.[90] The obvious extrapolation does not hold, however; a set of 105 house plans published in 1843 under the promising title *Paris moderne* suggests that builders of private dwellings then still preferred to express themselves in the measures of the old regime. About 30 percent of the plans use toises only; another 30 percent, meters only; and 40 percent, both. All the plans are dated, the earliest to 1815; the strongest showing of the toise against the meter occurs not at the beginning, but in the period 1835–9. The meter takes over after 1840, the year in which the metric system at last became obligatory in France.[91]

Up-to-date technical manuals show the same equivocation as architectural drawings. An authoritative indicator is the huge treatise on the building arts (four volumes in five of text, three of plates) published between 1805 and 1816 by Jean-Baptiste Rondelet, chief architect of the church of Sainte-Geneviève and architectural consultant to the government. Rondelet's plates strongly favor the old measures; his text appears to be the work of a schizophrenic. The earliest volume (1805) uses pieds, pouces, and lignes and also metric units, sometimes converted but often not, in tables and in calculations; at one point it gives dimensions in meters and computes in toises.[92] In the volumes published in 1812, Rondelet usually proffers all dimensions in both the old and new style, but without the advantage of decimal notation: for example, "31 pieds 3/7 (10 mètres 209 millimètres)."

89. *Projets d'architecture et autres productions de cet art qui ont merité les grands prix accordés par l'Académie, par l'Institut national de France, et par des Jury du choix des artistes ou de gouvernment* (Paris: Detourelle, 1806).

90. Louis Bruyère, *Etudes relatives à l'art des constructions* (Paris: Bance ainé et al., 1823).

91. Louis Marie Normand and G.E. Lemmonier and *Paris moderne, ou choix de maisons de campagne et constructions rurales des environs de Paris*, 3 vols. (Liège: d'Avanzo, 1843), and several other editions; Bouchard, *Prieur*, 313. The obligation was laid down in a law of 4 July 1837; it took effect on 1 January 1840.

92. Jean Baptiste Rondelet, *Traité théorique et pratique de l'art de bâtir* (Paris: chez l'auteur, 1805–16), *3*, 57, 66–7, 83 ff., 102, 164, 179–83, 269 ff., 388–96.

His preference for the old appears from his way of obtaining specific gravity, which he understands as the weight in grams of a cubic centimeter of a substance. He measures the weight of a sample, in air and in water, in onces and gros, deduces the weight of a cubic pied, and converts to the metric system. The preference for the old units persists into the last volumes, published in 1816. Then—after 1,839 pages of schizophrenic computations, the last of which provides the cost of old bricks in old money—Rondelet introduces some "notions about metric measures."[93] It is instructive that this instruction precedes a recomputation of costs in metric measures and decimalized currency. The relative ease of such computations was the great benefit that the reformers had advertised.

From about 1810 many technical manuals by professors and government officials used metric measures exclusively. For example, the professor of mathematics at the Ecole impériale militaire, Louis Puissant, and Pierre Pommiés, professor at the Lycée Napoléon, both influential teachers of surveying and geodesy, used meters exclusively and without explanation in many textbooks beginning in 1807.[94] Jean-Nicholas Hachette, professor at the Ecole polytechnique, brought the metric system into the preface of his textbook on machines, without comment, as the only satisfactory way "to express their effect in numbers." The chemist Jean-Antoine Chaptal, one-time minister of the interior and long-time educational reformer, naturally used nothing but modern measures in his survey of the state of French industry after the fall of Napoleon. Still there are hints of backsliding where least expected. The official surveyors of France might have worked exclusively in meters and understood the higher geodesy; but the local land measurer often enough made do with old units and little geometry into the reign of Louis Philippe. A manual for such measurers, "especially people who have not studied geometry," was published in 1833 by an inspector of the cadastral survey. It contains an explanation of the metric system, an injunction

93. Ibid., *1*, 99, 101–2, 158–60; *4*, 583–604, 609–44.

94. Puissant, *Traité de topographie*, 139, 154, and Michel Pommiés, *Manuel de l'ingénieur du cadastre* (Paris: Impr. impériale, 1808), 166–9, illustrate the type.

to use it, and tables for the easy conversion of old local measures into meters, "and vice versa."[95]

As it spread, the metric system did help to realize the reformers' ambition to improve the numeracy of Europeans. In the Ancien Régime, despite its baroque abundance of metrological units, most people had need only for a single system; during the Empire, because of the government's policy of driving the meter home, more and more citizens had to learn how to convert one set of measures into another. If the first advance of European "Rechenhaftigkeit"—"the inclination, habit, and ability to resolve the world into numbers and to bring these numbers together into an artificial system of income and outgo"—may be likened to literacy in one language, the domestication of the metric system made reckoners bilingual.[96]

The early treatises on metric calculations propagated not only decimal arithmetic, including the concept of place, but also the idea of significant figures and the operation of rounding off. An *Exposition abrégée du nouveau système*, published in the provinces in 1802 or 1803, may stand for the genre. It explains that the many decimals generated during conversion are artifacts of multiplication and division, and demands that they be dropped from the final answer.[97] Rounding off, says Rondelet, in his belated account of metric computations, is essential to the new system. Rounding off might in practice mean only rounding up, as readers of an *Arithmétique pratique* of 1800, which recommends changing 0.7411440 to 0.75, learned to do; a practice that would favor shopkeepers, who could round up each of their many small transactions.[98] Even arithmetic may have its social bias. The utility of metric conversions in teaching numeracy

95. J.A. Chaptal, *De l'industrie Françoise*, 2 vols. (Paris: Renouard, 1819); A. Lefevre, *Guide practique et mémoratif de l'arpenteur, particulièrement destiné aux personnes qui n'ont point étudié la géométrie* (Paris: Bachelier, 1833), i–vi, 15–21.

96. Quote from Werner Sombart, *Der Bourgeois. Zur Geistesgeschichte des modernen Wirtschaftsmenschen* (Munich: Duncker and Humblot, 1913), 164.

97. *Exposition abrégée*, 23–4, 29–30. See also the contemporary *Tables de comparaison*, xiv–xxix; Capelli, *Rapports*, 49–57.

98. Rondelet, *Traité, 4:2* (1816), 610–3; Adrien Poittier, *Arithmétique pratique et démontrée, pour réduire les anciennes mesures en nouvelles* (Paris: Bernard and Moutardier, nivôse an VIII), 146–52. Gattey, *Eléments*, 32–3, rounds up only if the part dropped is ≥ 0.5 of the last figure kept.

appears further from popular textbooks of arithmetic that use such computations as exercises. This practice persisted long after the universal adoption of the metric system in France.[99]

According to the president of the Paris Agricultural Society, the great majority of people who could read and write in 1810 knew nothing more of arithmetic than addition and subtraction. He had a prescription for correcting this innumeracy and a vision of the happy consequences. "If decimal calculations could be introduced into primary schools along with the use of the new measures, not only would the housewife be able to make all the computations she requires, but also the worker could measure without difficulty and, by adding the use of the rule and compass for tracing geometrical figures, he would be able to draw all his plans himself, and the farmer would have no problem with surveying."[100] This grand project has been realized in large measure. The male academic and quantitative spirit of the late Enlightenment found a fertile if capricious partner in *La belle Marianne*, the spirit of revolutionary France.

99. See, for example S.F. Lacroix, *Traité élémentaire d'arithmétique, à l'usage de l'Ecole centrale des quatres nations*, 2d edn. (Paris: Courcier, 1805), 115–46; Garnier, *Traité, 1*, 263–75, 281–93; and, for the later period, A. Thinon, *Leçons sur le système métrique et sur les applications usuelles de l'arithmétique* (Paris: Dezobry and Magdelleine, 1860), 131.

100. Chessiron, "Mesures," *8*, 293.

FIG. 1.1 Besides the strict mathematical method, which was intended only for the learned reader, Wolff also used the physicotheological method for a popular audience. According to the latter method, everything in nature must be both perfect and useful. Demonstrating or proving this thesis became a rationale for 18th-century natural history. The scientist should look for "greatness in small things" (*maxima in minimis*), which phrase served as the motto for Friedrich Christian Lesser's *Insecto-theology* (Frankfurt, 1740). The study of nature reveals the infinite diligence of the wasps and the ants, the beauty and strange development of the butterfly, the artfulness of the spider, the utility of the silkworm—these and other features of nature provided inspiration for 18th-century research and speculation.

FIG. 1 Albrecht von Haller called Linnæus "the second Adam." He appears in this role in a vignette for the eleventh edition of *Systema naturae* (ed. Lange, Haller, 1760). As Adam, Linnæus is sitting naked, naming all the animals (cf. Genesis 2:19). Paradise is crowded with animals and plants, monkeys climbing the trees, whales swimming in the sea; everything is at peace. The fall is yet to come. The figure in the middle is Diana polymammae, symbol of Nature and its generative force. Linnæus—with "his" flower *Linnea borealis* at his feet—is urging numbers and names, *Numeros et nomina*, from Diana. He is Nature's bookkeeper.

FIG. 2 This instructive vignette, from Charles Bonnet, *Oeuvres d'histoire naturelle, 4* (Neufchâtel, 1781), depicts man at the top of the scale of nature. As man belongs to both the material and the spiritual world, he keeps his head in the clouds. On lower steps of the scale follow the monkey, the lion and the dog, the eagle, fish, lizards, corals, and so on. Everything in nature is thus to be understood both in hierarchy and in continuity.

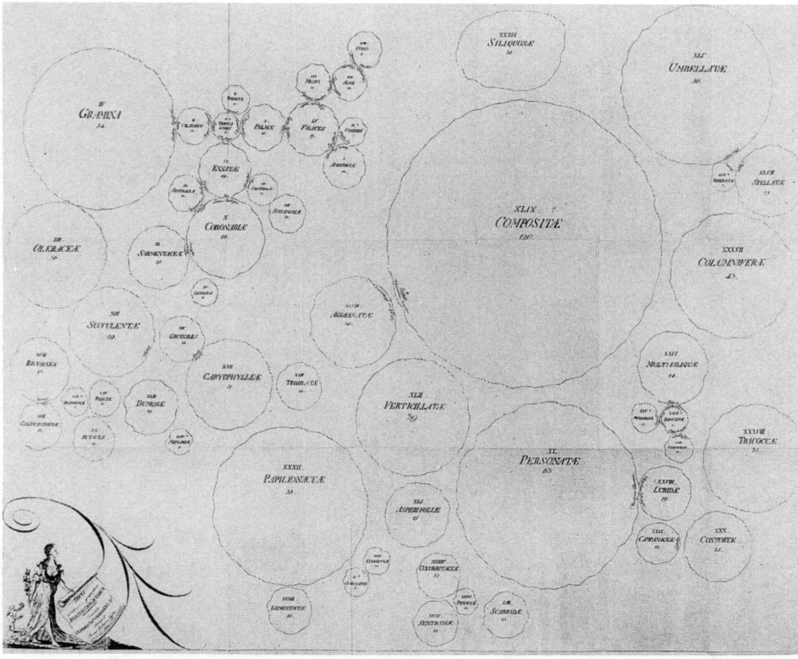

FIG. 3 In private lectures Linnæus discussed the possibility of arranging natural orders linearly. With so many uncertainties in establishing these orders and with so many more plants still to discover, however, such an arrangement must wait. This map, believed to be based on a drawing by Linnæus, shows continuity between some groups, as well as gaps requiring more knowledge: Linnæus, *Praelectiones in ordines naturales plantarum*, ed. P.D. Gieseke (Hamburg, 1792), facing p. 623.

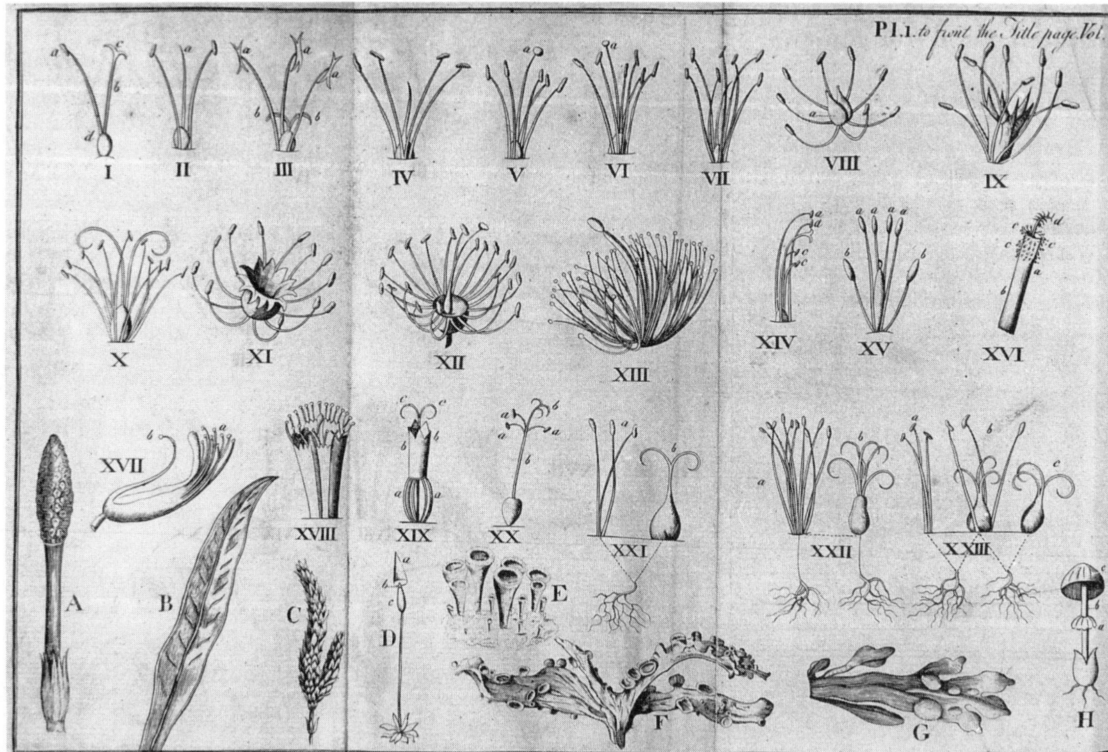

FIG. 3.1 Illustration of the Linnæan sexual system, from William Withering, *A botanical arrangement of all the vegetables growing in Great Britain* (1776). See chapter 3, page 77.

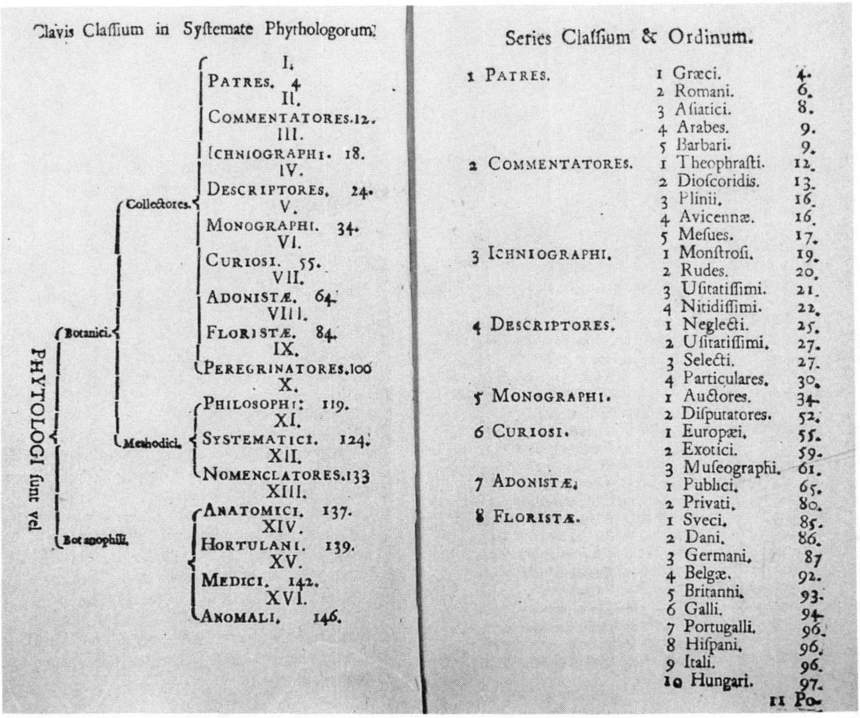

FIG. 3.2 Linnæus' classification of botanists, from his *Bibliotheca botanica* (1736). See chapter 3, page 79.

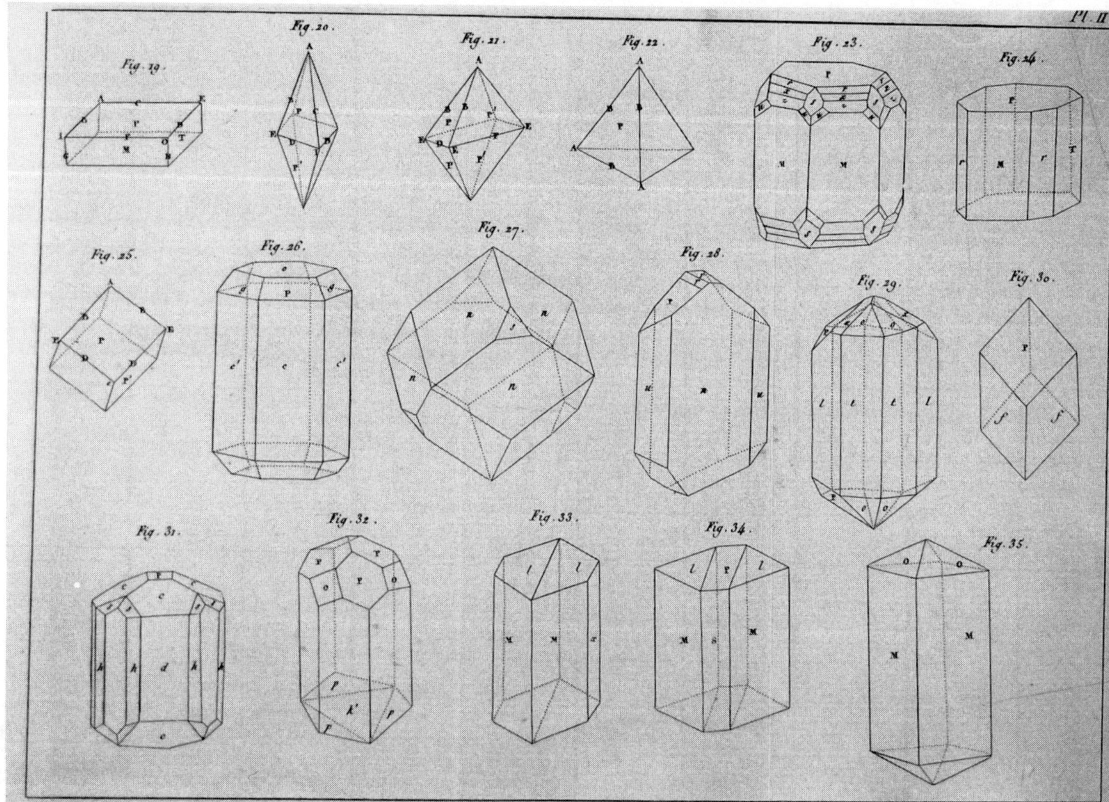

FIG. 3.3 René Just Haüy's geometrical forms of crystals, from his *Tableau comparatif des résultats de la cristallographie et de l'analyse chimique relativement à la classification des minéraux* (1809). See chapter 3, page 92.

GENRE VII. *Héméralopie*. État de la vue tel qu'on ne peut voir que pendant qu'on est exposé au grand jour.

ESPÈCE SIMPLE.

ESPÈCES COMPLIQUÉES.

GENRE VIII. *Nyctalopie*. État de la vue tel qu'on ne peut voir que dans l'obscurité.

ESPÈCE SIMPLE.

ESPÈCES COMPLIQUÉES.

GENRE IX. *Amaurose*. Privation complète de la faculté de voir.

ESPÈCE SIMPLE.

ESPÈCES COMPLIQUÉES.

ORDRE II. *Névroses des fonctions cérébrales.*

PREMIER SOUS-ORDRE. *Comata.*

GENRE X. *Apoplexie*. Suspension plus ou moins complète et subite de l'action des sens, de l'entendement, de la locomotion; respiration plus ou moins stertoreuse; continuation de l'action du cœur.

ESPÈCE SIMPLE. 1re variété. *Apoplexie forte.*
2e variété. *Apoplexie faible.*

ESPÈCES COMPLIQUÉES.

GENRE XI. *Catalepsie*. Suspension totale du sentiment et du mouvement, pouls et respiration à peine perceptibles, membres conservant leur position antérieure, ou celle qu'on leur donne.

ESPÈCE SIMPLE.

ESPÈCES COMPLIQUÉES.

GENRE XII. *Epilepsie*. Perte de connaissance avec mouvemens convulsifs et spasmodiques.

ESPÈCE SIMPLE.

Variétés. *Epilepsie idiopathique.*
sympathique.

ESPÈCES COMPLIQUÉES. Complication avec la manie, l'idiotisme, la démence.

DEUXIÈME SOUS-ORDRE. *Vésanies.*

GENRE XIII. *Hypochondrie*. Tension spasmodique dans diverses parties, flatuosités incommodes, maux imaginaires.

ESPÈCE SIMPLE.

FIG. 3.4 Classification of nervous disorders showing orders, genera, species, and varieties, from Philippe Pinel, *Nosographie philosohique,* 6th ed. (1818). See chapter 3, page 99.

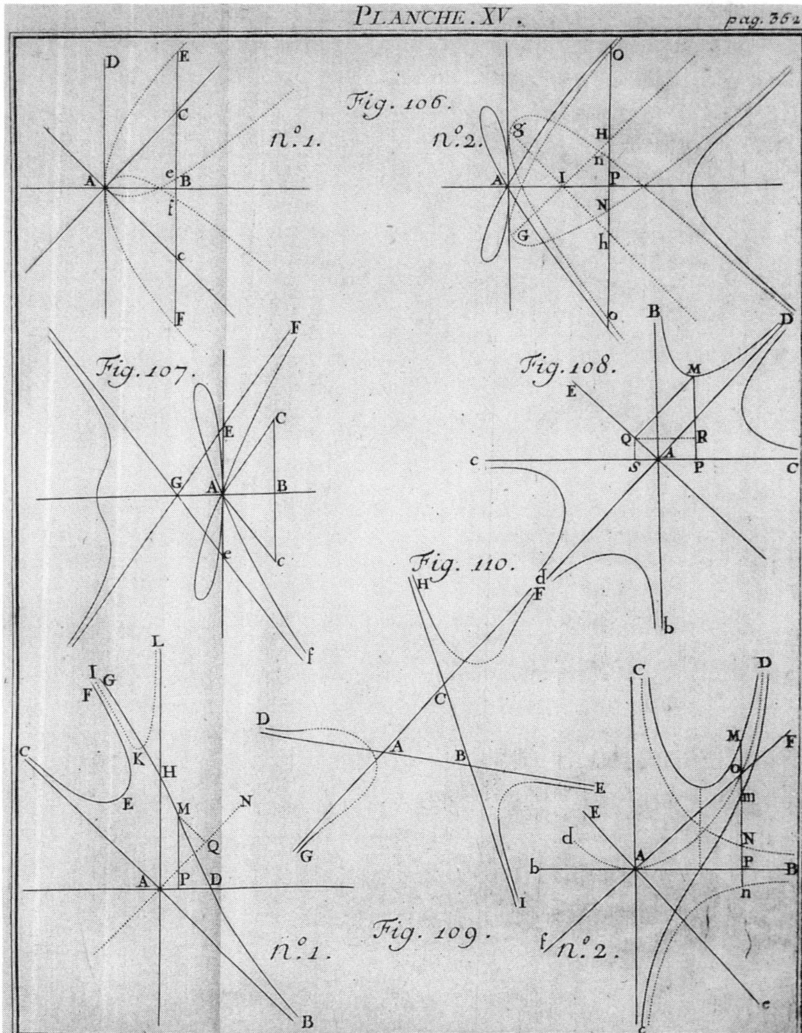

FIG. 3.5 Illustration of curves classified by Gabriel Cramer in his *Introduction à l'analyse des lignes courbes algébriques* (1750). See chapter 3, page 102.

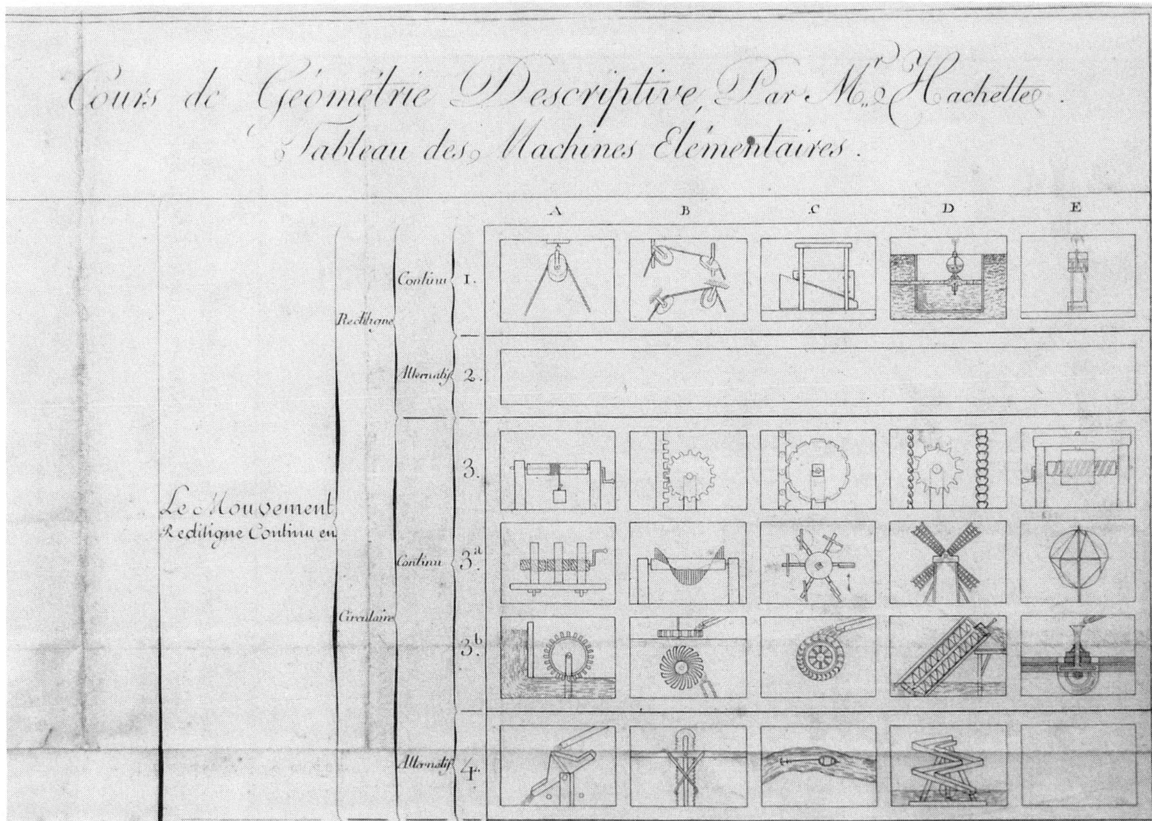

FIG. 3.6 Portion of Jean Nicolas Pierre Hachette's classification of simple machines, from his *Traité élémentaire des machines* (1811). See chapter 3, page 106.

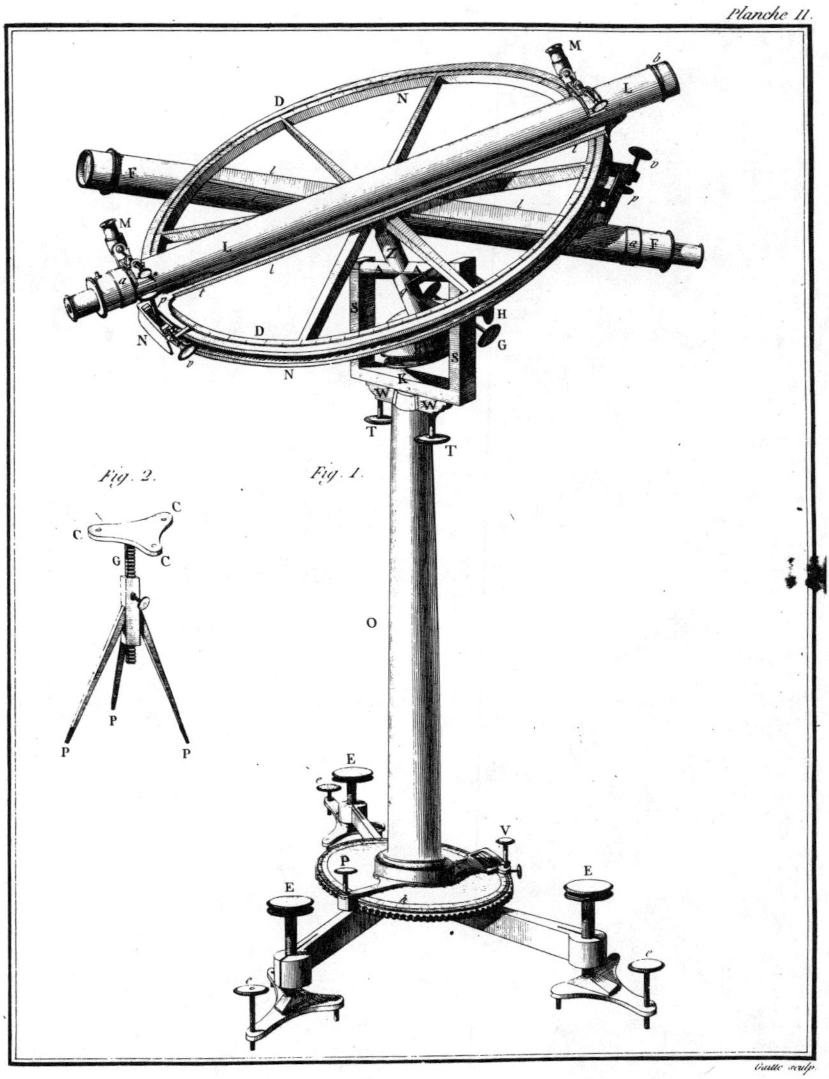

FIG. 6.1 The repeating circle, from Jacques Cassini, Pierre François Méchain, and Adrien-Marie Legendre, *Exposé des opérations faites en France* [1791]. By courtesy, National Land Survey (*Lantmäteriverket*), Gävle.

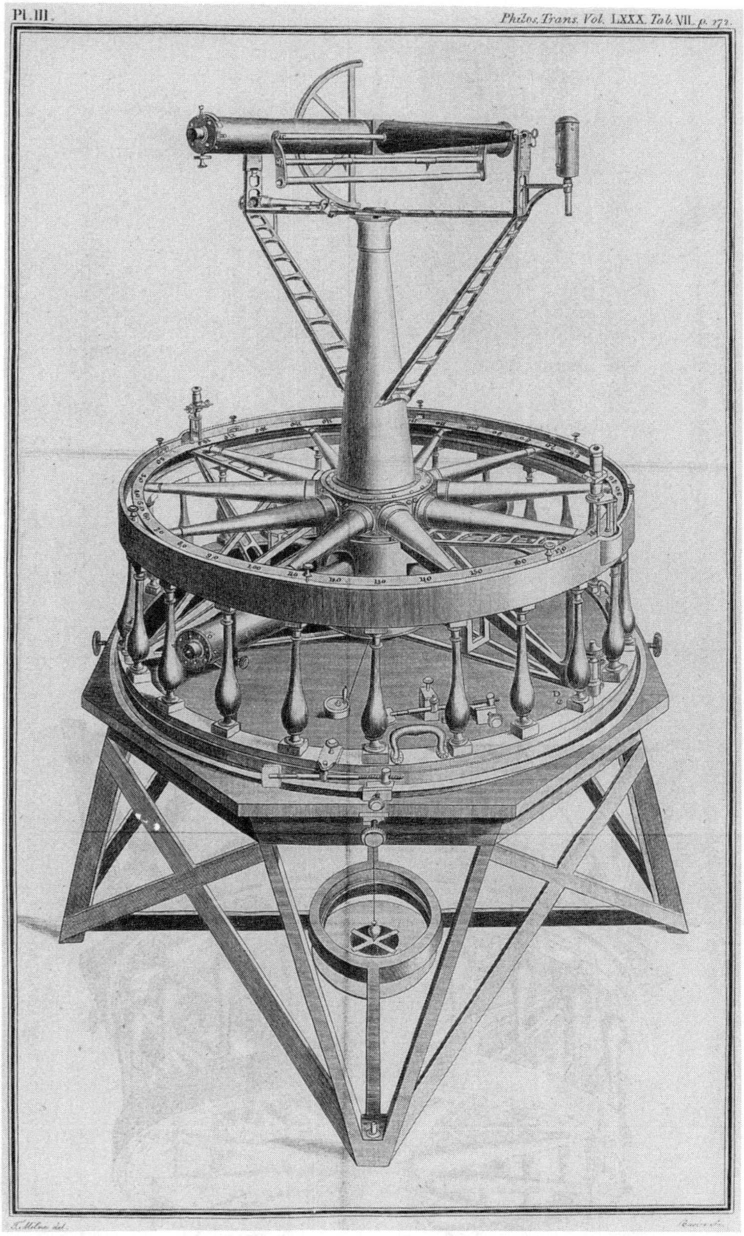

FIG. 6.2 According to William Roy, "It is a brass circle, three feet in diameter, and may be called a theodolet [theodolite], rendered exremely perfect." From Roy's "Account of the trigonometrical operation" in the *Philosophical transactions* for 1790.

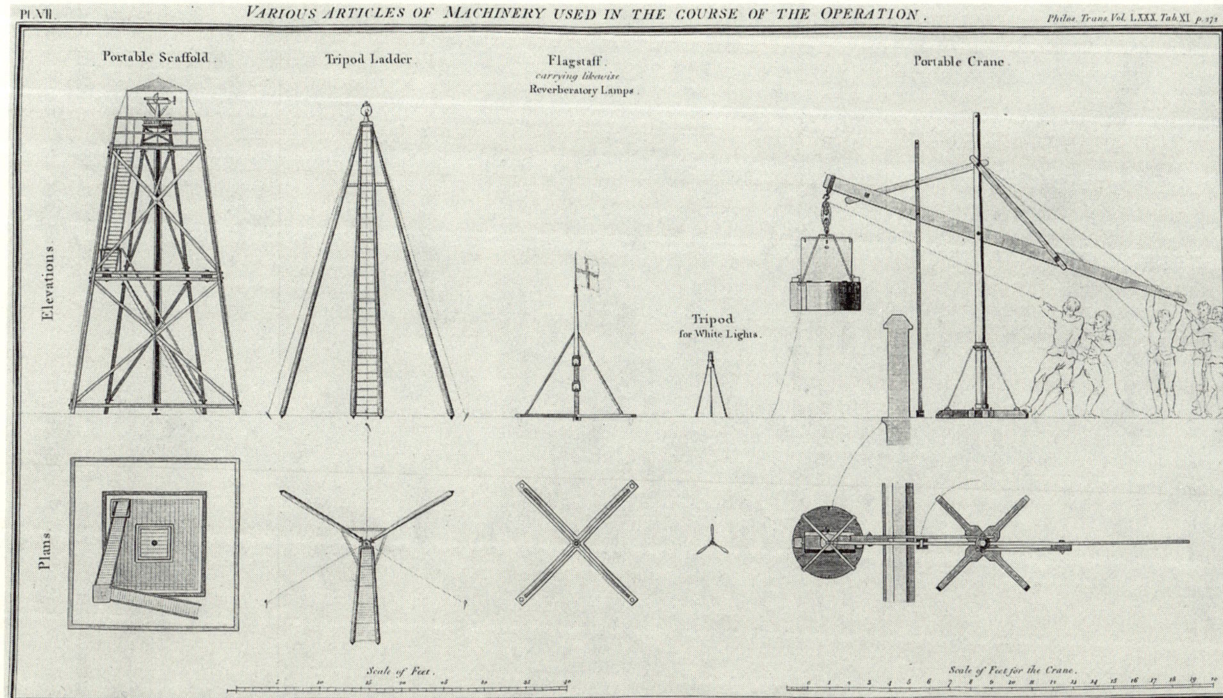

FIG. 6.3 The theodolite was used with a portable scaffold and crane in the high-tech military surveying operations of the late 18th century. From Roy's "Account of the trigonometrical operation" (1790).

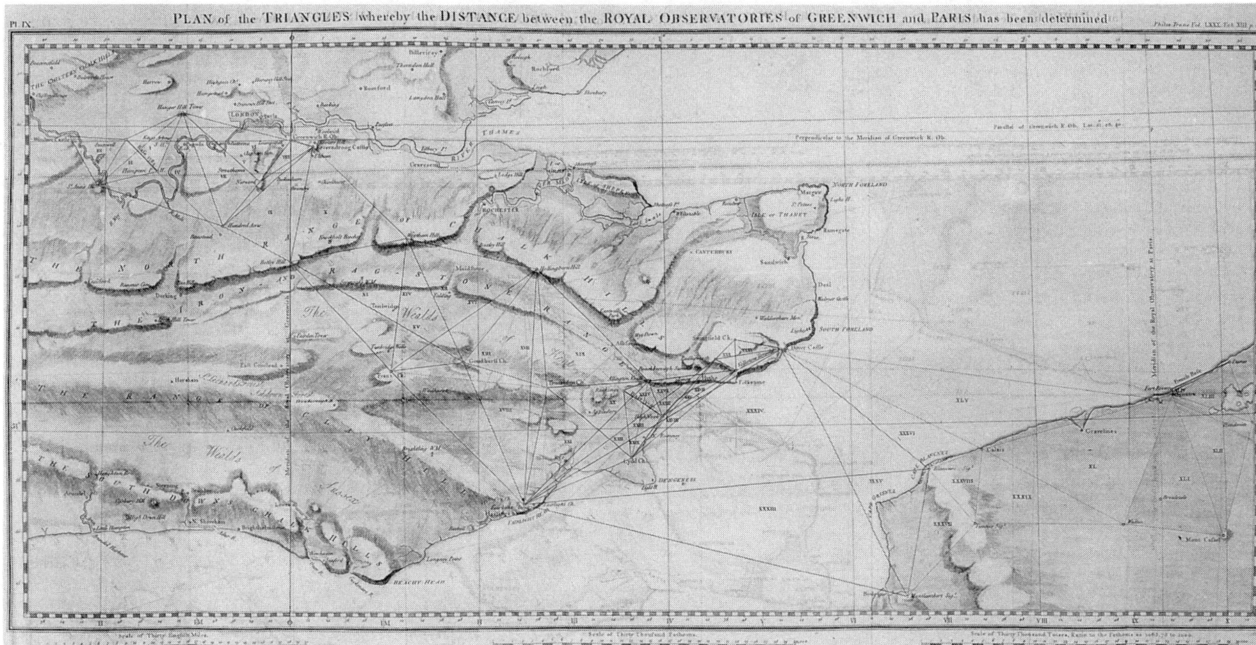

FIG. 6.4 The cross-channel connection, from Roy's "Account of the trigonometrical operation" (1790).

FIG. 8.1 The close connection between economics, technology, and chemistry in "die ganze Historie vom Kiess," from F.J. Henckel, *Pyrotologia* (1725). Instruments in the small buildings illustrate the role of chemistry in the production of sulfur, vitriol, and arsenic from pyrites.

FIG. 8.2 Ädelfors mine in Sweden as seen in 1767 by G. Jars. From his *Voyages métallurgiques, 2* (1780).

Tabell, som visar hvarje Milas förhållande af de vid Furudahls Bruk verkställde Prof-Kolningar.

Tab. 1.

		Kolningstid. Dygn.	Vedens vigt i Skepp:d B. V.	Milornes innehåll: efter Stock. mätning. Cub. alnar.	Milornes innehåll: efter yttre mätningen. Cub. alnar.	Milornes innehåll: Cubikfamnar Ved.	Milornes innehåll efter Kolning. Cub. aln.	Medel vigt af cub. aln Ved Lispund B. V.	Procent af verklig Ved.	Irandar: efter Stock. mätning. Cub. aln.	Irandar: efter yttre mätning. Cub. aln.	Kol: T:or à 36 k:pr.	Kol: Stigar à 12 T:or	Kolens vigt vid utrifningen. Skepp:d B. V.	Medel-vigt af en T:na Kol Lispd.	Kolens vigt emot veden. Proc.	Kolens rymd efter Stock. mätning. Proc.	Kolens rymd efter yttre mätning. Proc.	Cub. aln. Ved till en T:a Kol: efter Stock. mätning.	Cub. aln. Ved till en T:a Kol: efter yttre mätningen.
1811.																				
Resmilor	N:o 1	21	287,7	392,425	730,27	27,04	329,16	14,7	53,73	29,34	63	512	42,66	72,35	2,826	25,147	111,05	60,45	0,709	1,303
	2	30	374,05	513,688	1048	38,81	543,22	14,5	49	53,506	181,1	556	46,33	77,1	2,782	20,66	95,16	50,5	0,827	1,559
	3	34	343,26	474,745	968,28	35,86	553,77	14,6	49,105	25,817	58,27	658	54,83	90,6	2,755	26,39	115,42	56,93	0,682	1,383
	4	22	316,79	455,079	759,22	28,11	501,079	14,7	59,91	59,751	107,9	452	37,66	63,45	2,807	20,03	90,039	54,64	0,874	1,440
Liggmilor	1	22	256,55	348,682	571,394	21,163	356,97	14,2	61,023	12,49	20,475	467	38,91	72	3,085	28,06	109,4	66,754	0,719	1,179
	2	27	332,15	487,784	703,858	26,068	401,369	13,6	69,3	20,19	29,137	590	49,17	89,6	3,089	26,97	99,364	68,861	0,792	1,143
	3	24	304,35	422,297	639,067	23,67	423,079	14,4	66,08	53,079	80,325	504	42	74,10	2,976	24,64	107,5	71,035	0,732	1,108
1812.																				
Liggmilor	1	26	—	—	682,941	25,29	—	—	—	—	45,79	528	44	—	—	—	—	65,66	—	1,206
	2	24½	295,95	399,943	648,367	24,013	—	14,8	61,68	26,91	43,62	460	38,33	67,6	2,9[illegible]	24,48	97,10	59,90	0,810	1,311
	3	22	—	394,410	589,782	21,843	—	—	66,87	26,35	43,71	498	41,5	—	—	—	106,55	71,69	0,739	1,096
	4	25	—	379,820	556,642	20,616	—	—	68,24	26,72	43,31	484	40,33	—	—	—	107,92	74,98	0,727	1,060
Granved	5	20	262,35	381,385	519,90	19,255	—	13,[illegible]	73,35	25,66	41,59	560	46,66	70,075	2,[illegible]	28,63	124,2	92,20	0,635	0,894
	6	25	—	373,589	519,257	19,231	—	—	71,95	10,166	14,13	498	41,5	—	—	—	107,91	77,63	0,729	1,014
	7	21	—	457,997	649,175	24,043	—	—	70,55	13,8	19,56	593	49,4	—	—	—	105,13	74,21	0,749	1,059
	8	30	307,05	421,811	651,185	24,118	—	14,[illegible]	64,78	30,133	46,52	544	45,33	80,25	2,94	28,23	109,38	70,91	0,720	1,111
	9	32	—	—	641,355	23,753	—	—	—	—	22,8	552	46	—	—	—	—	70,26	—	1,120
	10	31	—	—	623,915	23,108	—	—	—	—	17,8	584	48,66	—	—	—	—	75,86	—	1,038
	11	31	—	499,868	673,411	24,941	—	—	[illegible]4,23	20,561	27,7	592	49,33	—	—	—	97,265	72,2	0,809	1,090
Resmilor	1	29	—	541,874	782,45	28,98	—	—	[illegible]9,25	69,105	99,8	592	49,33	—	—	—	94,38	68,293	0,798	1,153
	2	22	—	240,042	471,071	17,446	—	—	[illegible]3,079	24,831	46,78	336	28	—	—	—	117,49	62,363	0,670	1,262
	3	29	—	405,214	657,942	24,368	—	—	[illegible]1,588	76,369	124	400	33,33	—	—	—	95,789	58,995	0,822	1,334
Granved	4	27	189,1	271,746	465,65	17,244	—	13,9	[illegible]3,358	36,765	63	384	32	46,55	2,[illegible]	27,97	128,69	75,102	0,611	1,048
Liggmilor utan balkro	1	22	—	400,842	566,037	20,964	—	—	[illegible]0,82	59,775	84,41	480	30	—	—	—	110,82	78,504	0,710	1,003
	2	25	—	615,098	1114,251	41,26	—	—	5[illegible],20	81,978	148,5	720	60	—	—	—	106,35	58,7[illegible]	0,740	1,342
	3	30	—	355,297	544,551	20,168	—	—	6[illegible],39	83,384	127,8	384	32	—	—	—	111,21	72,56	0,708	1,085
Liggmila. Kolningen drifven efter längden af veden.		28	—	—	672,734	24,916	—	—	—	—	57,4	560	46,66	—	—	—	—	71,668	—	1,098
1813.																				
Liggmilor	1	29	—	—	579,75	21,47	—	—	—	—	5,79	504	42	—	—	—	—	69,15	—	1,138
	2	28	—	—	614,17	22,74	—	—	—	—	11,39	600	50,3	—	—	—	—	78,38	—	1,000
	3	26	—	397,76	600,79	22,28	282,05	—	66,20	6,75	9,19	544	45,33	—	—	—	109,56	72,41	0,731	1,087
	4	29	—	—	504,1	18,69	—	—	—	—	6,97	488	40,66	—	—	—	—	77,30	—	1,018
	5	31	—	—	560,19	20,74	—	—	—	—	11,72	480	40	—	—	—	—	68,91	—	1,142
	6	32	—	—	554,39	20,53	—	—	—	—	12,78	448	37,33	—	—	—	—	65,13	—	1,208
	7	34	—	440,431	661,22	24,48	309,825	—	66,61	12,29	20,79	568	47,33	—	—	—	107,39	69,84	0,753	1,127
	8	36	—	—	661,99	24,51	—	—	—	—	25,36	480	40	—	—	—	—	59,37	—	1,326
	9	34	—	—	601,71	22,28	—	—	—	—	19,75	448	37,33	—	—	—	—	60,62	—	1,299
Resmilor, Granved	1	40	—	—	686,46	25,42	—	—	—	—	51	560	46,66	—	—	—	—	69,40	—	1,134
	2	35	—	432,4	726,665	26,91	360,75	—	59,50	44,49	72	576	48	—	—	—	116,9	69,43	0,673	1,136
	3	45	—	504,984	807,02	29,89	416,957	—	62,57	49,44	80	600	50	—	—	—	103,7	64,99	0,759	1,211
Liggmilor utan balkro	1	37	—	370,417	559,411	20,71	269,45	—	66,21	2,75	4,45	552	46	—	—	—	118,2	78,33	0,671	1,005
	2	43	—	544,43	804,128	30,37	387,919	—	67,70	13,17	21,29	704	58,75	—	—	—	104,3	68,94	0.773	1,142

FIG. 10.1 Tabular display of experiments in charcoal-burning in Dalecarlia, from Carl David af Uhr, *Berättelse om kolnings-försök åren 1811, 1812 och 1813.*

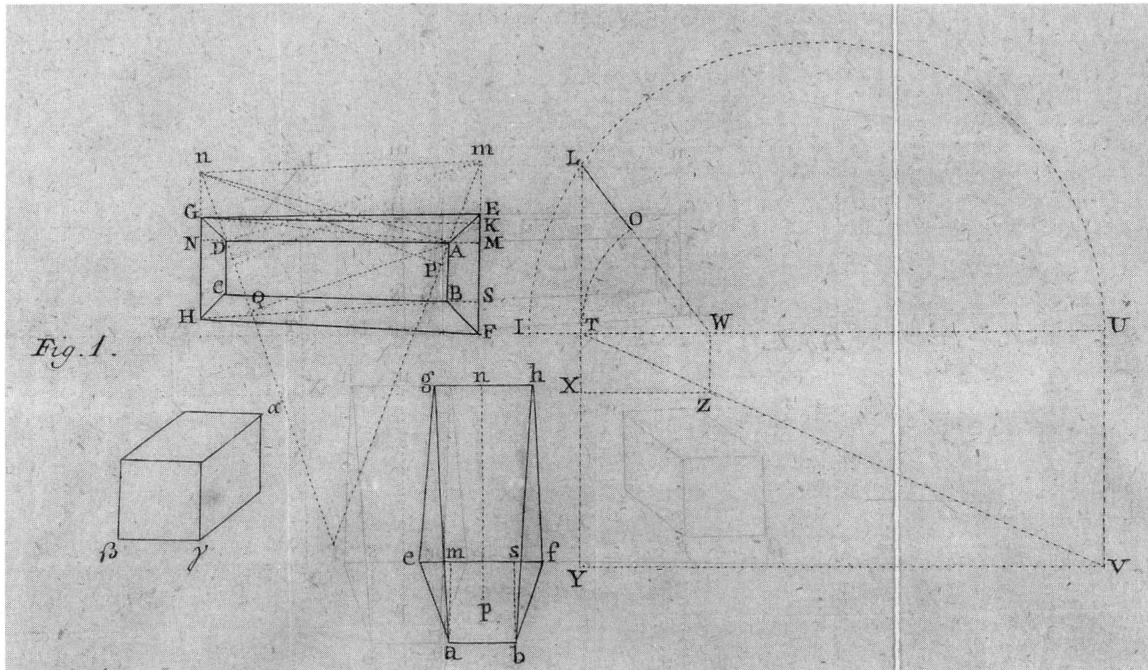

FIG. 10.2 Illustrations for Zacharias Plantin's studies of the volume of rhombic sledges of charcoal, from the *Kungl. Vetenskapsakademiens Handlingar* for 1778 and 1784.

Tab. N:o 1.

Journal, hållen vid Pumpningen för Fregatten *Hök*, den 5:te Augusti 1773. Djupgående för 7 Fot. Dito Achter 12 Fot 5 tum. Watnets högd ifrån bruns rännans botn 23 fot.

Gång-åſens Omlops antal.	Upfordrade Kiſtors antal, à 6 Cub. fot.	Den Upfordrade vattu Maſſans quantitet.	Länſat i Dockan.		Vattu Högden.	Manſkapets antal.	Tiden.	Kraftens väg.	Kraftens Haſt i Secunden.		Hvarje Karls använde ſtyrka.	Effecternes relativa förhållande.
		Cubic-fot	Fot	tum			Min.	Fot	Fot	tum	ſkålp.	
90	900	5400	—	6	96	V. 90	14	5706	6	9	10,8	8748
100	1000	6000	—	7	103	M. 90	18	6340	6	1	11,6	8468
100	1000	6000	—	7	110	B. 90	17	6340	6	3	12,4	9300
100	1000	6000	—	7	117	V. 90	18	6340	6	1	13,2	9636
100	1000	6000	—	7	124	M. 90	19	6340	5	6	14,0	9240
100	1000	6000	—	7	131	B. 90	17	6340	6	3	14,8	11100
100	1000	6000	—	7	138	V. 90	19	6340	5	6	15,6	10296
100	1000	6000	—	8	146	M. 90	20	6340	5	3	16,5	10395
100	1000	6000	—	7	153	B. 90	17	6340	6	3	17,3	12975
100	1000	6000	—	7	160	V. 90	21	6340	5	—	18,1	10860
100	1000	6000	—	8	168	M. 90	26	6340	4	1	18,5	9065
100	1000	6000	—	8	176	B. 90	19	6340	5	6	19,4	12804
100	1000	6000	—	9	185	V. 90	23	6340	4	7	20,9	11495
100	1000	6000	—	9	194	M. 90	29	6340	3	8	21,9	9636
100	1000	6000	—	9	203	B. 90	20	6340	5	3	22,9	14427
100	1000	6000	—	10	213	V. 90	25	6340	4	3	24,1	12291
100	1000	6000	1	—	225	M. 90	30	6340	3	6	25,5	10710
100	1000	6000	—	10	235	B. 90	25	6340	4	3	26,6	13566
100	1000	6000	—	10	245	V. 90	30	6340	3	6	27,7	11634
100	1000	6000	1	—	257	M. 90	43	6340	2	5	29,1	8439
100	1000	6000	—	10	267	B. 90	22	6340	4	8	30,2	16912
100	1000	6000	—	10	277	V. 98	32	6340	3	4	28,7	11480
100	1000	6000	1	—	289	M. 98	42	6340	2	6	30,0	9000
100	1000	6000	—	9	298	B. 90	21	6340	5	—	33,8	20280
50	500	3000	—	5	303	V. 98	20	3170	2	8	31,1	9952
50	500	3000	—	6	309	M. 98	19	3170	2	9	32,1	10593
50	500	3000	—	5	314	B. 90	14	3170	3	9	35,6	16020
50	500	3000	—	6	320	V. 106	20	3170	2	8	30,8	9856
50	500	3000	—	7	327	M. 106	27	3170	1	11	31,4	7222
50	500	3000	—	8	335	B. 98	15	3170	3	6	34,8	14616
50	500	3000	1	—	347	V. 106	17	3170	3	1	33,4	12358
50	500	3000	1	7	366	M. 106	16	3170	3	4	35,2	14080
Summa 2790	27900	167400						176886				

FIG. 10.3 Systematic display of data from Johann Eric Norberg's study (1772–3) of human muscle power, from the *Kungl. Vetenskapsakademiens Handlingar* for 1799.

H.

General-Tabelle

zur Uebersicht des periodisch-jährlichen Holzertrages im Forstreviere Jägerthal, unter der Forstinspektion Hubertusburg, Regierungsbezirks Friedrichstadt.

Jährlicher Holzertrag.

Perioden.	Block.		Eichenholz. Nutzholz.	Eichenholz. Kloben.	Eichenholz. Knüppel.	Eichenholz. Reiser.	Büchenholz. Nutzholz.	Büchenholz. Kloben.	Büchenholz. Knüppel.	Büchenholz. Reiser.	Birkenholz. Nutzholz.	Birkenholz. Kloben.	Birkenholz. Knüppel.	Birkenholz. Reise.
			Klftr.	Klftr.	Klftr.	Schck.	Klftr.	Klftr.	Klftr.	Schck.	Klftr.	Klftr.	Klftr.	Schck.
Iste Periode von 1820 bis 1839.	I.	Aus Hochwald haubar	76	75	31	—	—	—	—	—	—	—	—	—
	II.	desgl. Durchforstung	—	—	6	—	—	—	—	—	—	—	—	—
		Aus Niederwald	—	—	—	—	—	—	—	—	—	—	—	—
IIte Periode von 1840 bis 1859.	I.	Aus Hochwald haubar	50	44	19	—	—	—	—	—	—	—	—	—
	II.	desgl. Durchforstung	—	—	6	—	—	—	—	—	—	—	—	—
		Aus Niederwald	—	—	—	—	—	—	—	—	—	—	—	—
IIIte Periode von 1860 bis 1879.	I.	Aus Hochwald haubar	65	37	20	—	—	—	—	—	—	—	—	—
	II.	desgl. Durchforstung	—	—	23	—	—	—	—	—	—	—	—	—
		Aus Niederwald	—	—	—	—	—	—	—	—	—	—	—	—
IVte Periode von 1880 bis 1899.	I.	Aus Hochwald haubar	—	—	—	—	—	—	—	—	—	—	—	—
	II.	desgl. Durchforstung	—	4	46	—	—	—	—	—	—	—	—	—
		Aus Niederwald	—	—	—	—	—	—	—	—	—	—	—	—
Vte Periode von 1900 bis 1919.	I.	Aus Hochwald haubar	—	—	—	—	—	—	—	—	—	—	—	—
	II.	desgl. Durchforstung	—	15	62	—	—	—	—	—	—	—	—	—
		Aus Niederwald	—	—	—	—	—	—	—	—	—	—	—	—
VIte Periode von 1920 bis 1939.	I.	Aus Hochwald haubar	—	—	—	—	—	—	—	—	—	—	—	—
	II.	desgl. Durchforstung	—	24	103	—	—	—	—	—	—	—	—	—
		Aus Niederwald	—	—	—	—	—	—	—	—	—	—	—	—
VIIte Periode von 1940 bis 1959.	I.	Aus Hochwald haubar	76	25	20	—								
	II.	desgl. Durchforstung	—	59	109	—								
		Aus Niederwald	—	—	—	—								
VIIIte Periode von 1960 bis 1979.	I.	Aus Hochwald haubar	—	—	—	—								
	II.	desgl. Durchforstung	—	94	84	—								
		Aus Niederwald	—	—	—	—								
IXte Periode von 1980 bis 1999.	I.	Aus Hochwald haubar	—	—	—	—								
	II.	desgl. Durchforstung	—	135	62	—								
		Aus Niederwald	—	—	—	—								
Xte Periode von 2000 bis 2019.	I.	Aus Hochwald haubar	118	39	32	—								
	II.	desgl. Durchforstung	—	136	53	—								
		Aus Niederwald	—	—	—	—								

Perioden.	Block.		Erlenholz. Nutzholz.	Erlenholz. Kloben.	Erlenholz. Knüppel.	Erlenholz. Reiser.	Erlen- und Birkenholz vermischt. Nutzholz.	Kloben.	Knüppel.	Reiser.	Kiefernholz. Nutzholz.	Kiefernholz. Kloben.	Kiefernholz. Knüppel.	Kiefernholz. Reiser.	Fichtenholz. Nutzholz.	Fichtenholz. Kloben.	Fichtenholz. Knüppel.	Fichtenholz. Reiser.
			Klftr.	Klftr.	Klftr.	Schck.	Klftr.	Klftr.	Klftr.	Schck.	Klftr.	Klftr.	Klftr.	Schck.	Klftr.	Klftr.	Klftr.	Schck.
Iste Periode von 1820 bis 1839.	I.	Aus Hochwald haubar	—	—	—	—	—	—	—	—	1209	821	268	—	—	—	—	—
	II.	desgl. Durchforstung	—	—	—	—	—	—	—	—	—	—	81	—	—	—	—	—
		Aus Niederwald	—	250	255	—	—	—	—	—	—	—	—	—	—	—	—	—
IIte Periode von 1840 bis 1859.	I.	Aus Hochwald haubar	—	—	—	—	—	—	—	—	1150	833	188	—	—	—	—	—
	II.	desgl. Durchforstung	—	—	—	—	—	—	—	—	—	—	281	—	—	—	—	—
		Aus Niederwald	—	241	241	—	—	—	—	—	—	—	—	—	—	—	—	—
IIIte Periode von 1860 bis 1879.	I.	Aus Hochwald haubar	—	—	—	—	—	—	—	—	1227	672	204	—	—	—	—	—
	II.	desgl. Durchforstung	—	—	—	—	—	—	—	—	—	21	377	—	—	—	—	—
		Aus Niederwald	—	238	238	—	—	—	—	—	—	—	—	—	—	—	—	—
IVte Periode von 1880 bis 1899.	I.	Aus Hochwald haubar	—	—	—	—	—	—	—	—	913	933	199	—	—	—	—	—
	II.	desgl. Durchforstung	—	—	—	—	—	—	—	—	—	19	564	—	—	—	—	—
		Aus Niederwald	—	238	238	—	—	—	—	—	—	—	—	—	—	—	—	—
Vte Periode von 1900 bis 1919.	I.	Aus Hochwald haubar	—	—	—	—	—	—	—	—	868	868	179	—	—	—	—	—
	II.	desgl. Durchforstung	—	—	—	—	—	—	—	—	—	43	716	—	—	—	—	—
		Aus Niederwald	—	238	238	—	—	—	—	—	—	—	—	—	—	—	—	—
VIte Periode von 1920 bis 1939.	I.	Aus Hochwald haubar	—	—	—	—	—	—	—	—	871	872	158	—	—	—	—	—
	II.	desgl. Durchforstung	—	—	—	—	—	—	—	—	—	61	673	—	—	—	—	—
		Aus Niederwald	—	238	238	—	—	—	—	—	—	—	—	—	—	—	—	—

Perioden.	Block.		Summa des haubaren Holzes. Nutzholz.	Kloben.	Knüppel.	Reiser.	Summarum des haubaren Holzes. Klftr.	Schck.	Summa des Durchforstungsholzes. Nutzholz.	Kloben.	Knüppel.	Reiser.	Summarum des Durchforstungsholzes. Klftr.	Schck.	Summarum des haubaren und Durchforstungsholzes. Klftr	Schck.
			Klftr.	Klftr.	Klftr.	Schck.	Klftr.	Schck.	Klftr.	Klftr.	Klftr.	Schck.	Klftr.	Schck.	Klftr	Schck.
Iste Periode von 1820 bis 1839.	I.	Aus Hochwald haubar	1285	896	299	207										
	II.	desgl. Durchforstung	—	—	—	—	2985	333	—	—	87	7	87	7	3072	340
		Aus Niederwald	—	250	255	126										
IIte Periode von 1840 bis 1859.	I.	Aus Hochwald haubar	1200	877	207	190										
	II.	desgl. Durchforstung	—	—	—	—	2766	310	—	—	287	24	287	24	3053	334
		Aus Niederwald	—	241	241	120										
IIIte Periode von 1860 bis 1879.	I.	Aus Hochwald haubar	1292	709	224	185										
	II.	desgl. Durchforstung	—	—	—	—	2701	304	—	21	400	35	421	35	3122	339
		Aus Niederwald	—	238	238	119										
IVte Periode von 1880 bis 1899.	I.	Aus Hochwald haubar	913	933	199	170										
	II.	desgl. Durchforstung	—	—	—	—	2521	289	—	23	610	53	633	53	3154	342
		Aus Niederwald	—	238	238	119										
Vte Periode von 1900 bis 1919.	I.	Aus Hochwald haubar	868	868	179	160										
	II.	desgl. Durchforstung	—	—	—	—	2391	279	—	58	778	70	836	70	3227	349
		Aus Niederwald	—	238	238	119										
VIte Periode von 1920 bis 1939.	I.	Aus Hochwald haubar	871	872	158	158										
	II.	desgl. Durchforstung	—	—	—	—	2377	277	—	85	776	72	861	72	3238	349
		Aus Niederwald	—	238	238	119										

Diese Berechnung dient bloß zur Uebersicht des künftig wahrscheinlichen Holzertrages der Eichenbestände.

Nota. Für die Klafter Eichen und Nadelholz sind 5 Büschel Reiser, und für die Klafter Erlenholz „ 15 „ „ gerechnet.

FIG. 11.1 Georg Hartig's 200-year plan for the Jägerthal forest district, from his *Neue Instructionen* (1819). By courtesy, Department of Special Collections, Yale University.

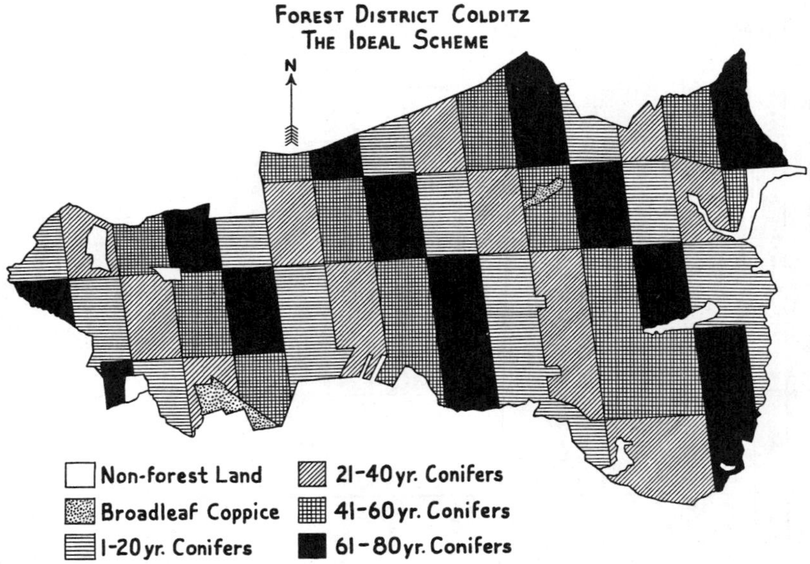

FIG. 11.2 Transformation of the Colditz forest district after 100 years of managements according to Cotta's ideal scheme. From Heske, *German forestry*. By courtesy, Stanford University Libraries.

Forest District Colditz (Saxony)
Situation in 1821

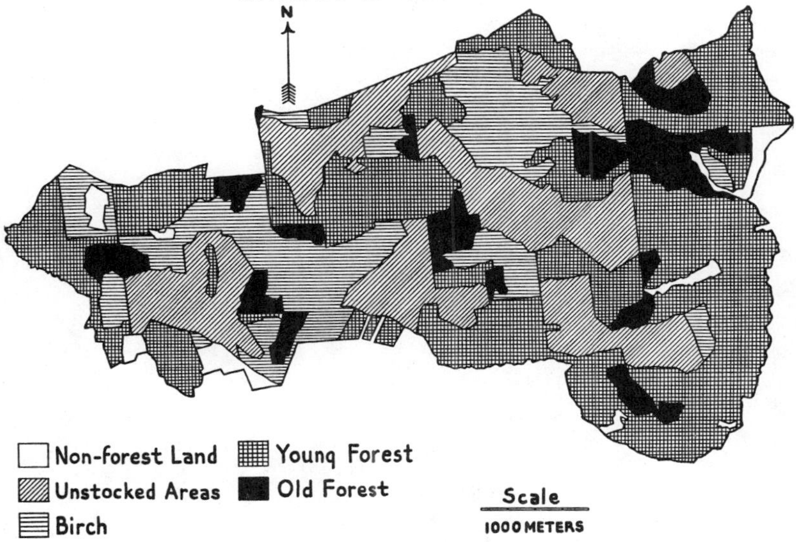

Forest District Colditz
Situation in 1921

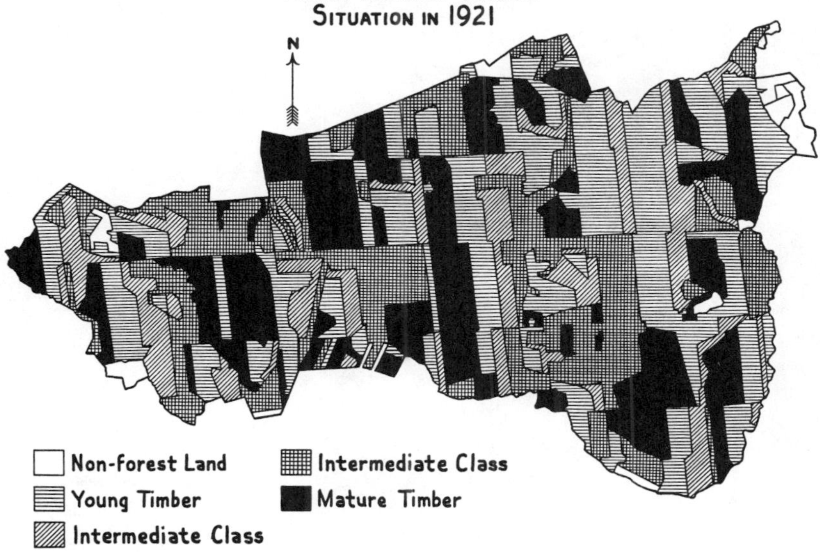

PART 3

WIDER APPLICATIONS

8 The Changing Role of Numbers in 18th-Century Chemistry

By Anders Lundgren

According to a well-established tradition in the history of chemistry, the chemical revolution at the end of the 18th century was a product of the increased use of quantitative methods and of the balance, leavened by the law of the conservation of matter. Despite this emphasis on quantification in accounts of the transformation of 18th-century chemistry, most historians of chemistry have dismissed the topics of quantification, mathematics, and measurements in a few words.[1] To redress the balance, I propose two approaches for studying the changing use of numbers in 18th-century chemistry.

The first approach stresses the influence of economic interests. During the 18th century, chemistry was often classified as an applied science. Mineralogy, metallurgy, and assaying, as aspects of applied chemistry, gave rise to an increased use of the balance and thereby a new role for numbers in chemistry. The second approach assays the influence of experimental physics. Before the beginning of the chemical revolution, important parts of experimental physics had been subjected to quantification.[2] Many chemists during the 18th century

I should like to thank Marco Beretta, Christoph Meinel, and Evan Melhado, as well as the other authors of this volume, for comments and criticism of earlier versions.

1. See Henry Guerlac, "Lavoisier and his biographers," *Isis, 45* (1954), 51–62, and *Essays and papers in the history of modern science* (Baltimore: Johns Hopkins University Press, 1977), 314–26; Bernadette Bensaude-Vincent, "A founder myth in the history of modern sciences? The Lavoisier case," in L. Graham et al., eds., *Functions and uses of disciplinary histories* (Dordrecht: D. Reidel, 1983), 53–78; J.R. Partington, *A history of chemistry*, 4 vols. (London: Macmillan, 1962), *3*, 674–88; Henry Leicester, *The historical background to chemistry* (New York: Dover Publications, 1971; reprint of 1956), 151.

2. J.L. Heilbron, *Elements of early modern physics* (Berkeley: University of California Press, 1982). The influence of the mechanical world view, which led to a shift from organic to mechanical analogies in mineralogy, invoked no new use of numbers. Cf. D. Oldroyd, "Mechanical mineralogy," *Ambix, 21* (1974), 176–8.

invoked physical concepts in explaining their subject matter: witness the assimilation of Newtonian forces and chemical affinities. Lavoisier was only one of a company of chemists who saw themselves as physicists.[3] Consideration of the influence of experimental physics on laboratory instruments and chemical theories will also speak to the use of numbers in chemistry.

THE BALANCE

The close association of chemistry with practical arts has figured prominently in the history of technology. In the index to the eight-volume *History of technology* edited by Charles Singer et al., for example, references to chemistry fill almost one page, whereas there is no reference at all to physics.[4] Historians of the chemical revolution have not dwelt on this association with technology, presumably because they have regarded the revolution as a transformation in theory.[5]

In the beginning of the 18th century, traditional chemical theory had little to do with the daily practice of chemistry. Existing theory was antiquated, almost entirely qualitative, and infused with compounds of Aristotelian elements and Paracelsian principles. As chemical descriptions of processes and substances were refined, old theories

3. Henry Guerlac, "Chemistry as a branch of physics: Laplace's collaboration with Lavoisier," *Historical studies in the physical sciences, 7* (1976), 193–276, and *Antoine-Laurent Lavoisier: Chemist and revolutionary* (New York: Scribners, 1975), esp. 74; Evan Melhado, "Chemistry, physics, and the Chemical Revolution," *Isis, 76* (1985), 195–211; Carleton E. Perrin, "The triumph of the antiphlogistians," in H. Wolff, ed., *The analytic spirit: Essays in the history of science in honor of Henry Guerlac* (Ithaca: Cornell University Press, 1981), 40–63; A.L. Donovan, "Lavoisier and the origins of modern chemistry," *Osiris, 4* (1988), 214–31.

4. Charles Singer et al., eds., *A history of technology*, 8 vols. (Oxford: Clarendon Press, 1955–84), *8, Consolidated index*, comp. Richard Raper, 150. Misunderstanding of the role of physics in the development of technology may also play a part in this imbalance.

5. See, however, David Oldroyd, "Some phlogistic mineral schemes, illustrative of the evolution of the concept 'earth' in the 17th and 18th centuries," *Annals of science, 31* (1974), 269–305; Theodore M. Porter, "The promotion of mining and the advancement of science: The chemical revolution in mineralogy," ibid., *38* (1981), 543–70.

lost their empirical foundation. Theories capable of organizing the growing body of empirical observations were in order. Mining practice, one of the most important practical fields and one that contributed to a changing use of numbers in chemistry, will be a focus here. The shape of the resulting systems bore the imprint of mining practice.

By 1700, the balance had long been in use in both metallurgy and assaying.[6] The hydrostatic balance determined density and controlled purity of different substances, especially the noble ones, and assisted in the control of less than noble practices. After introducing the balance in testing gold sand imported from Guinea, the British noticed a marked decrease in "the swindling the natives practiced."[7] But the method could be used only to determine mixtures of metals, never to decide chemical composition. In fact, the hydrostatic balance was afflicted by so many sources of error that it gave results scarcely better than assayer's needles and touchstones. (The needles were gold-silver mixtures of known composition; by matching the color of an unknown sample to that of a needle, the assayer could quickly estimate its makeup.)[8] Despite the fact that the balance played a modest role in practical metallurgy, knowledge of density did not bring the chemist to a better understanding of chemical processes or of the chemical characteristics of a given substance.

The chemical (as opposed to the hydrostatic) balance does not appear in illustrations of laboratories of the 17th and early 18th centuries.[9] Chemists did not use it in their daily work. Only in

6. Not much has been written on the balance in chemistry; see the recent book edited by John T. Stock and Mary Virginia Orna, *The history and preservation of chemical instrumentation* (Dordrecht: Reidel, 1986). See, however, R. Multhauf, "On the use of the balance in chemistry," American Philosophical Society, *Proceedings, 106* (1962), 210–8; John T. Stock, *Development of the chemical balance* (London: Science Museum, 1969); Peta D. Buchanan, *Quantitative measurement and the design of the chemical balance 1750–ca. 1900* (Ph.D. dissertation, University of London, 1982).

7. Daniel Menlös, *Kort beskrifning af den hydrostatiske wåg-balcken* (Stockholm: Andr. Biörkman, 1728), "Företal."

8. Agricola, *De re metallica*, transl. Herbert Clark Hoover and Lou Henry Hoover (London: The Mining Magazine, 1912), 252–60.

9. John Read, *The alchemist in life, literature and art* (London: Nelson, 1947), and F. Ferchtl and A. Süssenguth, *Kurzgeschichte der Chemie mit 200 Abbildungen* (Mittenwalde: Arthur Neumayer, 1936), provide many examples.

commercial mining, which typically involved amounts of material far larger than anything of interest to the chemist or the assayer, was the balance at home. The most sensitive chemical balances were used exclusively for the weighing of noble metals. In *De re metallica* (1556), Georgius Agricola treats the balance in a section on the assayer's work and the purity of metals, and emphasizes that the most sensitive must be confined to weighing "the bead of gold and silver," since ores and other large weights would injure it.[10] The balance played no part in the production of gold, but only in the measurement of the final, refined result.

Echoes of Agricola's attitude toward sensitive balances can be heard to the end of the 18th century. In 1689 J.J. Becher distinguished three types of balances with respect to sensitivity; his categories recurred in the writings of Johann Cramer in the mid, and of Sven Rinman in the late, 18th century. According to Rinman, the most sensitive balances, which could register changes as small as 1/128 *ass* (about 0.4 milligram), should be used only to weigh "the smallest bead...of the noble metals."[11] Agricola, Becher, Cramer, and Rinman all assigned the same tasks to the balance. They did so independently of any theoretical commitments. Theories in chemistry retained their qualitative character until the end of the 18th century. However, from about 1750 the balance began to take on importance in the shaping of chemical theory.

10. Agricola, *De re metallica*, 264ff.

11. Becher's three types were the *Statera docimastica*, *Bilanx pro ponderando auro*, and *Bilanx civilis*, in Becher, *Tripus hermetica fatidicus* (Frankfurt an Moen, 1689), 30. Johann Cramer, *Anfangsgründe der Probierkunst* (Stockholm, 1746), 193, distinguishes between the most sensitive balance, "Probier- oder Korn-waage," the "Erz-waage," and the "Bleywaage, worauf man, ohne ihr einen Schaden zuzufügen, dreyssig bis vierzig Probiercentner abwiegen kann." Sven Rinman makes a similar distinction in his *Bergwerks-Lexicon*, 2 vols. (Stockholm, 1789), *2:1*, 330–1. For the metric equivalent of the *ass*, see S.O. Jansson, *Måttordbok* (Stockholm: Nordiska Muséet, 1950).

SYNTHESIS AND ANALYSIS QUANTIFIED

By synthetic quantification, I mean a recipe that states the amount of each ingredient in numbers and the relative proportion needed to produce a substance with specified properties. In analytical quantification, the chemist determines the proportions among the constituent parts of a given compound substance. Synthetic quantification had a long tradition in pharmaceuticals, mineralogy, and metallurgy, as in the production of different kinds of brass.[12] Analytic quantification scarcely existed before the middle of the 18th century.

Synthetic quantification did not demand exact balances. The figures given in recipes were only approximate: ingredients did not come pure and the recipes were the result of trial and error. A chemist tried one part of A and one of B, then two of A and one of B, and so forth, and then, by comparing the different products, chose the better recipe. A case in point is Johann Heinrich Pott, whose *Chymische Untersuchungen* of 1754 included extensive "Tabellen von denen Würckungen der verschiedenen Mischungen derer Erden." The tables recorded Pott's attempts to obtain a certain iron product able to withstand high heat. He mixed ingredients in various proportions,[13] but the abundance of expressions like "about 4 parts to 1," "a little more was added," "a greater part," "most of it," in his tables suggest that we should not ascribe much weight to the numbers offered. Pott and his contemporaries resorted to these imprecisions because they lacked standards for measurements and uniform scales and weights.[14] Compounding this lack was a plethora of impurities, which could cause successive weighings to give very different results even if carried out in exactly the same way. This last problem particularly afflicted the medical branch of applied chemistry, pharmacy.[15]

12. For the example of Lazarus Ercker, see Partington, *A history of chemistry, 2*, 104–7.

13. Johann Heinrich Pott, *Chymische Untersuchungen*, 3 vols. (Berlin: Christian Friedrich Voss, 1746–54), *3*, 15, 35–51.

14. Cramer, *Anfangsgründe der Probierkunst*, 193.

15. J.K. Crellin and S.R. Scott, "Pharmaceutical history and its sources in the Wellcome Institute, II. Drug weighing in Britain c. 1700–1900," *Medical history, 13* (1969), 51–67, esp. 56.

In pharmacies balances played much the same roles as in metallurgy and assaying.[16] A pharmacist sold substances by weight, as indicated by the prominent position of the balance on the shop counter, and compounded drugs by weight according to the recipes in the pharmacopoeias. As in the case of the assayer, however, the pharmacist's method of measuring was independent of theory. The role of chemistry in the business was confined to qualitative aspects and in practice influenced the making of very few pharmaceuticals.

As practitioners of an art, apothecaries were trained by apprenticeship.[17] The theory of their business was reserved for philosophical chemists who hobnobbed with savants. Carl Wilhelm Scheele confined his chemical work to his spare time, and Torbern Bergman explicitly reserved the theoretical part of pharmacy for the chemist.[18] In all this pharmacy resembled metallurgy. Many metallurgists in Germany had either practiced as or trained under an apothecary.[19] However, commercial demands on pharmacy did not lead, as in mining, to an increased use of analytical quantification. Synthesis, not analysis, was important for the apothecary; whose financial well-being did not depend on successful analysis. Nor was quantitative analysis important in medicine: with the techniques then available, it would have been impossible to subject the concerns of physicians to quantitative analysis.[20]

The increasing use of analytical quantification in assaying during the 18th century was a consequence of the expanding number of known ores and minerals. In Sweden, for example, where mining

16. Ferchtl and Süssenguth, *Kurzgeschichte*, 38, 41, 73, 104; *Kermer and Urdang's history of pharmacy*, 4th ed., revised by G. Sonnendecker (Philadelphia: Lippincott, 1976), 82, 90.

17. Dieter Pohl, *Zur Geschichte der pharmazeutischen Privatinstitute in Deutschland von 1779 bis 1873* (Ph.D. dissertation, Marburg on Lahn, 1972), 14f.

18. Wolfgang Schneider, *Geschichte der pharmazeutischen Chemie* (Weinheim: Verlag Chemie, 1972), 133–4. For Scheele, see Georg Urdang, *Pictorial life history of the apothecary chemist Carl Wilhelm Scheele* (Madison, Wisconsin: American Institute of the History of Pharmacy, 1943); for Bergman, see University Library, Uppsala, ms D 1459f.

19. Karl Hufbauer, *The formation of the German chemical community (1720–1795)* (Berkeley: University of California Press, 1982), 53–7.

20. The place of practical medicine and practical pharmacy in the development of chemistry requires further investigation.

products were the backbone of the economy, discovering new sources of ores took high priority, especially after the failure of the Falun Copper Mine at the beginning of the century. This failure prompted search for substitutes that might have the same economical importance as copper.[21]

In this process chemistry played a central part. Laboratories were built close to the different mines, in places like Ädelfors, Falun, and the little-known Skisshyttan, where Axel Fredrik Cronstedt organized a small chemical research center. The nature and activities of these relatively unknown laboratories need to be examined. They proved equal to their task; they helped solve the problems raised by mining and the iron industry—for example, how carbon influenced the quality of steel.[22]

The significance of chemistry for metallurgy was further emphasized by the fact that the government Board of Mines employed a laboratory worker from the late 1630s, and regularly operated a *Laboratorium chymicum* from the 1680s.[23] The main task of the laboratory was to prepare pharmaceuticals for the mining industry. From the beginning of the 18th century, it concentrated increasingly on mineral analysis, assaying, and mineralogy. In keeping with its original assignment, the first directors of the laboratory were physicians; but in 1718, when Georg Brandt became its director, the responsibility of running it was transferred to a mineralogist.

In German-speaking central Europe, interest in chemistry likewise grew with the economic importance of mining. Many German chemists, like Becher and Johann Heinrich Gottlob von Justi, also occupy distinguished places in the history of German economic thought.[24]

21. For the Falun copper mine, see E.H. Hekscher, *Svenskt arbete och liv* (Stockholm: Bonniers, 1980; 1st ed., 1941), 101–4, 195–7; and Sten Lindroth, *Gruvbrytning och kopparhantering vid Stora Kopparberget intill 1800-talets början*, 2 vols. (Uppsala: Almkvist and Wiksell, 1955), *1*, 410–1. For the Ädelfors mine, see Heckscher, 195.

22. Anders Lundgren, "Bergshantering och kemi i Sverige under 1700–talet," *Med hammare och fackla, 24* (1985), 89–124.

23. For the early history of the laboratory, see Sten Lindroth, "Urban Hiärne and the *Laboratorium Chymicum*," *Lychnos*, 1946–7, 51–112; also J.A. Almquist, *Bergskollegium och bergslagsstaterna 1637–1857* (Stockholm: Riksarkivet, 1909).

24. H. Hassinger, *Johann Joachim Becher 1635–1692: Ein Beitrag zur Geschichte des Merkantilismus*, Veröffentlichungen der Kommission für neuere Geschichte

When the French government wanted information on mining from the rest of Europe, they sent out a mining engineer with a thorough training in chemistry, Gabriel Jars.[25]

Jars' appointment symbolized an important development in our history. Around 1700, empirical knowledge was sufficient to permit miners to decide the value of well-known minerals: a trained eye could sort samples by sight. But traditional knowledge was inadequate to manage the many new minerals discovered during the 18th century. Assayers found themselves in need of chemical knowledge. Only chemists armed with the methods of analytic quantification could answer the question, "How much iron does this ore contain, and how much of it can be made available?" In his *Anfangsgründe der Probierkunst* (1746), Johann Cramer declared that the assayer must be able to decide the composition of different substances "[i]n order to know what and how much of a [constituent] might be found in the substance under study, or could profitably be obtained from it."[26]

The growing need for analytical quantification raised the balance to a more prominent position in the assayer's workplace, the forerunner of the chemical laboratory. When Cramer described his laboratory, or "Arbeitsstätte," in 1746, ovens and cupels were still the most important instruments, but balances also received attention; Cramer claimed that his newest balance could give "the weight of the smallest body exactly." Still, weight is not everything, and Cramer acknowledged that even the most sensitive balance had to be supplemented by needles and touchstones in order to determine the composition of important substances.[27]

Österreichs, 38 (Vienna, 1951); Johann Heinrich Gottlob von Justi, *Gesammelte chymische Schriften, 1* (Berlin: Verlag des Buchladens der Real-Schule, 1760), 67; Christoph Meinel, "Zur Socialgeschichte des chemischen Hochschulfaches im 18. Jahrhundert," *Berichte zur Wissenschaftsgeschichte, 10* (1987), 147–68.

25. Gabriel Jars, *Voyages métallurgiques...en Allemagne, Suède, Norwège, Angleterre et Escosse*, 3 vols. (Lyon: Gabriel Regnault, 1774–81). Cf. *Dictionary of scientific biography* (*DSB*), s.v. "Jars, Antoine Gabriel."

26. Cramer, *Anfangsgründe der Probierkunst*, 3–4.

27. Ibid., 182, 183. The expression "kleinsten Körper" has no theoretical meaning, but is used to suggest the smallest quantity that can be weighed.

From mineralogy, analytical quantification spread into other fields of chemistry, including wet analysis. Mineral water was a favorite subject of study, even though its constituents were not intended for individual sale. Analysis facilitated subsequent synthetic quantification of naturally occurring mineral waters deemed to be especially valuable.[28] Another sort of wet analysis, titrimetry, also became more important during the second half of the 18th century. Here, too, commercial concerns were at work, since analysis helped in determining the purity of sulfuric acid, among other substances.[29]

Toward the end of the century analytical quantification seeped into ordinary textbooks of chemistry. Bergman's edition of H.T. Scheffer's lectures, *Chemiske föreläsningar* (Uppsala, 1775), is an instructive example. The material Bergman added almost doubled the size of the original, published thirty years before. In Bergman's additions, the composition of chemical substances is given in the form "100 parts of A contains *x* parts of B, *y* parts of C," and so on;[30] no such formulas appeared in Scheffer's original text.

By Bergman's time it was common for chemists to describe composition quantitatively.[31] Formulas did not necessarily afford great precision, however. Analytical quantification was a more difficult assignment than synthetic quantification; the same analytical procedure could yield very different results even when applied twice by the same chemist to the same substance. Insensitivity of the balances and, more importantly, difficulty in procuring and identifying pure substances, contributed to the variations. Early analytical quantifica-

28. Gmelin, *Geschichte der Chemie*, 3 vols. (Göttingen: Johann Georg Rosbuch, 1797–99), *2*, 740–90, and *3*, 718–52, contains a rich catalogue.

29. W.A. Campbell, "Analytical chemistry," in C.A. Russell, ed., *Recent developments in the history of chemistry* (London: The Royal Society of Chemistry, 1985), 180–1; E. Rancke Madsen, *The development of titrimetric analysis till 1806* (Copenhagen: G.E.C. Gad Publishers, 1958). The developments leading to the attempts by Wenzel, Kirwan, and later Ritter to measure attractive forces by the study of properties of solutions indicates the importance of wet analysis for the changing use of numbers in chemistry. It is unlikely, however, that investigation of these developments would contradict the importance of economic factors in the rise of analytical quantification as discussed in the text for the case of mineralogy.

30. Parts are usually given without decimal places.

31. Guerlac, "Quantification in chemistry," *Isis, 52* (1961), 201–2.

tion thus represented a new use of numbers—whatever their exactness—but did not at first assist in the development of chemical theory.

Constant interaction between chemistry and mineralogy would change the situation. Analytical quantification led to important changes in mineralogical classification systems. Up to the middle of the century, these classifications remained a species of natural history—relatively simple and qualitative, based on the external characteristics of minerals. The more substances mineralogists needed to classify, however, the greater the difficulties they experienced in relying only on external characteristics.[32] Classification by internal characteristics—that is, by chemical composition—offered an attractive alternative or supplement. The shift from classification on the basis of external physical properties to that based on internal, chemical composition began around 1750. Cronstedt's system of 1758 is an early example.[33] He denied the value of description from external factors and plumped for one based on chemical composition. His program gave chemistry a new role as a describer of mineralogical species. Although Cronstedt did not use numbers to express chemical composition in numbers, the growing importance of analytical quantification made it just a matter of time before others did so as a natural part of chemical description.

The determination of composition by weight consequently was not inspired primarily by attempts to formulate a classificatory system, but by the need to describe individual species more accurately. The balance and analytical quantification contributed to improved description, but only within a basically qualitative scheme of classification.

32. Crystallography can be viewed as quantitative description of external properties. However, no methods correlated crystal form with chemical composition, and crytallography did not assume significance for chemistry until the early 19th century. See John G. Burke, *Origins of the science of crystals* (Berkeley and Los Angeles: University of California Press, 1966). Hans-Werner Schütt argues (in *Die Entdeckung des Isomorphismus* [Hildesheim: Gerstenberg Verlag, 1984], 51) that Haüy's development of crystallography was dependent on the quantitative chemistry of Lavoisier.

33. Axel Fredrik Cronstedt, *Försök til Mineralogie, eller Mineral-Rikets Upställning* (Stockholm: Wildisksa tryckeriet, 1758), translated into German (1760), English (1770), and French (1771). Cf. David Oldroyd, "A note on the status of A.F. Cronstedt's simple earths and his analytical methods," *Isis, 65* (1974), 506–12.

(The order in a system of mineralogical classification, for example, continued to depend on the qualitative properties of the substances to be classified.) It was only with Joseph Louis Proust's system of definite proportions in the 1790s that improved descriptions were incorporated into a quantitative classificatory scheme. Proust arrived at his basic concepts while working as a chemist and mineralogist for the Spanish Government.[34] The research that inspired the theory of definite proportions, which immediately found its way into the mineralogical classification systems and gave a forceful impetus to the change from qualitative to quantitative description, concerned economically important oxides of iron.

Mineralogy thus began to transform itself from a technical art into a science. In it chemistry played important roles, providing both the overall qualitative pattern for classification and exact, quantitative descriptions of mineral species. This dual role finds a parallel in the tension between practice (economic interest) and theory (scientific interest) in the development of mineralogical classifications. Economic interest argued for a system based on monetary value; chemists preferred a system based on composition. Bergman even tried to amalgamate the two systems into one.[35]

Bergman envisioned a scientific mineralogical system whose classes were determined by the most dominant ingredient of the mineral where dominance was defined quantitatively. Value can also be quantified, however, and it was therefore both possible and advantageous in practice to include in one class all minerals containing gold and silver regardless of their content of noble metal.[36] Similarly, any mineral that derived its economic or metallurgical properties from one particular component might be placed in the class of that component, not in the class of the quantitatively dominant component.

Every systematizer used both qualitative and quantitative methods in the laboratory. The blowpipe, for example, never ceased to be an important tool for the mineralogist. According to Berzelius, miner-

34. For Proust, see *DSB*, s.v. "Proust, Joseph Louis."

35. Torbern Bergman, *Outlines of mineralogy*, transl. William Wihtering (Birmingham: Piercy and Jones, 1783), 10–1.

36. Ibid., 11.

alogical classification approached "mathematical certainty" thanks to the theory of chemical proportions.[37] But he also insisted that the constituents of a compound must be found "in their nature as well as in their quantity." He combined the results of his undisputed mastery with the blowpipe with a consistent application of numbers. The game can also be played on pharmaceuticals. Berzelius was the first to do so systematically.[38]

EXPERIMENTAL PHYSICS AND CHEMISTRY

During the 18th century, the word "physics" meant both general knowledge of nature and "experimental philosophy." In the second, newer sense, physics was quantified or quantifiable—it attempted to formulate mathematical laws from experimental results obtained by means of specially constructed instruments.[39] It is the influence on chemistry of experimental physics that concerns us here.

Around 1700, there were few connections between physics and chemistry. John Keill's and John Freind's efforts to use Newtonian concepts in calculating chemical affinity did not attract many followers.[40] The influential 31st Query to Newton's *Optics* referred affinity to atoms interacting by Newtonian forces. No one succeeded in handling the subject mathematically, however, and at the end of the century some chemists working in the Newtonian tradition, like Bergman, declared it to be impossible. Newtonian views of atomism and affinity had little or no effect on concrete chemical work. A common approach to the relations between matter theory and chemical thinking may be seen in the work of Ernst Stahl, who accepted a corpuscular philosophy but based his chemical understanding of

37. Berzelius, *Försök att genom användandet af den electrokemiska theorien och de kemiska proportionerna grundlägga ett rent vettenskapligt system för mineralogin* (Stockholm: A. Gadelius, 1814), 7–8.

38. Maurice Crosland, *Historical studies in the language of chemistry* (London: Heinemann, 1962), 216.

39. See Heilbron, *Elements of early modern physics*, 1–5.

40. See Arnold Thackray, *Atoms and powers. An essay on Newtonian matter-theory and the development of chemistry* (Cambridge, Mass.: Harvard University Press, 1970), 57–67.

matter on laboratory work. He made a clear distinction between chemistry and physics, saw chemical elements as compounds, and did not consider it the task of the chemist to penetrate deeper.[41] Throughout the 18th century, the atomic theory hovered in the philosophical background against which chemists carried out their work.

The difficulty of applying physical theory to chemical phenomena derived from the fundamental difference between the two disciplines: physics dealt with properties common to all substances, whereas chemistry dealt with their special characteristics. The chemist described unique substances and did not aim at discovering general laws. Even Proust's theory of definite proportions did not explain why these proportions existed. Proust is said to have had little interest in any sort of theory.[42] As late as 1810, when Berzelius considered definite chemical proportions, he could not decide whether the proportions "obey laws, common for all substances, or depend on circumstances unique for each substance."[43]

Working chemists saw physics as very different from chemistry.[44] Johann Friedrich Henckel, in his influential *Pyrotologia oder Kiess-Historie* (1725), mentioned "gravitas specifica" as a property to be studied with the hydrostatic balance, but the insights thereby gained would be physical rather than chemical. Even though determination of specific weights was still part of metallurgy, Henckel did not own a hydrostatic balance; the remarks about specific weight in his book were instead contributed by his less well-known colleague, Dr. Meuder. To learn chemical properties, Henckel argued, there was no way to avoid tedious laboratory work, despite the opinions of philosophers "who did not like to dirty their hands with coal."[45] Pott liked

41. Ibid., 171–6; Martin Carrier, "Zum korpuskularen Aufbau der Materie bei Stahl und Newton," *Sudhoff's Archiv, 70* (1986), 1–17; J. Gough, "Lavoisier and the fulfillment of the Stahlian revolution," *Osiris, 4* (1988), 23.

42. See *DSB*, s.v. "Proust," 168.

43. Berzelius, "Försök rörande de bestämda proportioner, hvari den oorganiska naturens beståndsdelar finnas förenade," *Afhandlingar i fysik, kemi och mineralogi, 3* (1810), 162.

44. For the example of Gabriel François Venel, see Evan Melhado, "Chemistry, physics, and the Chemical Revolution," *Isis, 76* (1985), 196–9.

45. Johann Friedrich Henckel, *Pyrotologia oder Kiess-Historie* (Leipzig: Johann Christian Martini, 1725), 878, 999, 1001. Cf. Partington, *A history of chemistry, 2*, 709.

to say that physics was the study of the superficial, far from the reality examined in chemistry: "it is an important difference, that superficial physics only describes the external and largely changeable shape of objects, while a reasonable chemistry can discover and bring to light through its experiments the inner forces and characteristics, fundamental composition, and *partes constituantes* of objects."[46]

In spite of the disparaging rhetoric, ideas borrowed from physics took hold in chemistry, particularly in the English development of pneumatic chemistry with its emphasis on empiricism, instrumentalism, and, ultimately, measurement.[47] Here the influence of Stephen Hales and Joseph Priestley on continental chemistry, and especially on Lavoisier, deserves particular emphasis.[48] Experimental physics made crucial contributions to chemistry—among them, a new attitude toward instruments and the quantitative facts they yielded, and a new methodology, which included an instrumentalist interpretation of theories.

Instruments and Facts

As the use of instruments characteristic of experimental physics spread to chemistry, new sorts of facts seized the chemist's attention. In 1750, Pott could say of fire as a chemical substance: "Although in its subtlety it cannot be investigated by number, measure, or weight, yet chemistry discovers a goodly number of its attributes."[49] The impact of experimental physics changed matters. The means of production of chemical facts in themselves remained much the same—distillation, vaporization, and precipitation, according to the tradi-

46. Pott, *Chymische Untersuchungen, 1*, 60.

47. See John Robison, "Preface," in J. Black, *Lectures on the elements of chemistry*, 2 vols. (Edinburgh: Mundell and Son, 1803), *1*, lxiv, which uses the term "pneumatic science."

48. Henry Guerlac, "The continental influence of Stephen Hales," *Archives internationales d'histoire des sciences, 15* (1951), 393–404, also in *Essays*, 275–84; and *Lavoisier—the crucial years* (Ithaca: Cornell University Press, 1961), 25–35. The methodological parallels between English pneumatics and Lavoisier have been stressed by J.R.R. Christie, "Ether and the science of chemistry," in G.N. Cantor and M.J.S. Hodge, eds., *Conceptions of ethers. Studies in the history of ether theories, 1740–1900* (Cambridge: Cambridge University Press, 1981), 104.

49. Pott, *Chymische Untersuchungen, 1*, 61.

tional practices of assayers, pharmacists, and others in chemical trades—but, thanks to the influence of physics, facts yielded up by the balance assumed greater importance.

The major innovation at midcentury was not high accuracy in measurements, but rather numerical measurement per se. Exactness was not essential to the formulation of the theory of definite proportions. Proust derived his ideas about the chemical significance of proportions from his work in ordinary practical metallurgy, the inaccuracy of which left plenty of room for the debates between himself and Berthollet over the nature of chemical combination.[50] The arguments central to Lavoisier's classical investigations on the supposed conversion of water to earth did not depend on great accuracy;[51] they did, however, rest on a numerical base. Nor did his studies of fermentation indicate the importance of exact measurement in the concrete study of chemical processes. To be sure, Lavoisier gave the law of the conservation of mass in mathematical form in order to demonstrate its exactness, but he never came close to exactness in actual experiments.[52] None was needed. The balance merely gave a gravimetric criterion for identifying and describing a unique chemical substance.[53]

Lavoisier thus relied on a rhetoric of numbers. The complication of chemical reality, which could not be idealized, might have compromised the rhetoric. But Lavoisier and others explained away

50. Cf. Seymour H. Mauskopf, "The Anales del Real Laboratorio de Quimica de Segovia of J.L. Proust," paper read at the XVIIth International Congress of History of Science (Berkeley, California, 1985); Satish C. Kapoor, "Berthollet, Proust, and proportions," *Chymia, 10* (1965), 84–5.

51. The balance Lavoisier used was more exact than necessary: its precision was one-tenth of a grain (0.06 gram), but the "earth" in the pelican weighed 3 grains (0.15 gram) more than the loss of the pelican. See Partington, *A history of chemistry, 3*, 377–80; Maurice Daumas, *Scientific instruments of the seventeenth and eighteenth centuries and their makers* (London: Batsford, 1972), 221–7.

52. F.L. Holmes, *Lavoisier and the chemistry of life* (Madison: University of Wisconsin Press), 269, 281–3, 394–7, emphasizes that Lavoisier never tried to verify the law of conservation of mass, but that he assumed it. The significance of that law for a changing use of numbers in chemistry is therefore minimal.

53. J.B. Gough, "Lavoisier and the fulfillment of the Stahlian revolution," *Osiris, 4* (1988), 15–32; John G. McEvoy, "Continuity and discontinuity in the chemical revolution," ibid., 195–213.

large numerical discrepancies by invoking unknown chemical processes. In the water conversion experiment, for example, the numerical shortfall was blamed on a chemical reaction between the glass and the water.[54] Still, the precision balance and the law of the conservation of matter conferred upon numbers a rhetorical value similar to what they enjoyed in physics and other fields during the late 18th century. Lavoisier made good use of the eloquence of the balance when arguing for the new chemistry.

Imponderables posed special technical problems. Lavoisier and Joseph Black wished to subject imponderables to quantitative study, but the usual array of chemical instruments offered no help. Other experimental devices were called for, such as the thermometer, which had not been an instrument for the chemist, and above all the calorimeter, recently constructed. These new instruments, introduced into chemistry from physics, became central to the study of chemically important substances. Deliberately constructed to yield quantitative results, they contributed to the introduction of numbers into chemistry. The study of heat stood at the intersection between physics and chemistry. Lavoisier and his physicist colleague Laplace met the problem of measuring the amount of heat participating in a chemical reaction by inventing the ice calorimeter.[55] Bergman, both physicist and chemist, found a way to measure the relative phlogiston content of two metals. He knew that a metal lost its phlogiston in acid solution but could regain it when another metal was added to the solution. He therefore dissolved a certain weight of one metal in acid and then weighed the amount of a second metal necessary to precipitate entirely the first from solution. In his chemistry, the amounts of phlogiston in the two metals were proportional to the weights so determined.[56]

54. Stock, *Development of the chemical balance*, 2. Scheele's qualitative proof of the impossibility of transmutation was considered as valid as Lavoisier's quantitative demonstration. Cf. Partington, *A history of chemistry, 3*, 380.

55. Guerlac, "Chemistry as a branch of physics."

56. Hugo Olsson, *Kemiens historia i Sverige intill 1800* (Uppsala: Almqvist and Wiksell, 1971), 227f.

Instruments and Theory

The influence of experimental physics on chemical theory was still negligible in the middle of the century. Although interest in Newtonian ideas about affinity then began to increase, and although affinity was supposed to be a distance force, the tables remained descriptions of empirical facts, in practice irrelevant to any theory of affinity. This generalization holds for the work of the thoroughgoing Newtonian Etienne François Geoffroy and also for Bergman, who brought the affinity tables to their fullest form.[57] Also, the Newtonian concept of the ether did not attract attention during the heyday of phlogiston and affinity studies. It later influenced Lavoisier.[58]

The influence of physics on chemistry was most evident on the continent in France, which, perhaps not coincidentally, had a relatively weak tradition in mining.[59] In Germany and in Sweden, chemists took less interest in physics because of the difficulties of applying it to mineralogy and because of the tendency of Stahl and his followers to keep Newtonian mechanics away from practical chemical work. Chemists who did show interest in physics in the last decades of the 18th century typically had some attachment to the universities. That in any case was true of Sweden.[60]

Relations between mathematics and chemistry were strained by the inability of the one to calculate anything of interest for the other. This was especially true in atomic theory. As Joseph Black put it, the assumption that a certain attractive force existed between certain atoms was void, since "all the mathematicians of Europe are not

57. For Geoffroy, see I.B. Cohen in *Melanges Alexandre Koyré, 1*, 102; for Bergman, A. Duncan, "Introduction," in Torbern Bergman, *A dissertation on elective attractions*, 2d ed. (London: Cass, 1970).

58. Evan M. Melhado, "Oxygen, phlogiston, and caloric: The case of Guyton," *Historical studies in the physical sciences, 13* (1983), 311–34, esp. 311–20.

59. Henry Guerlac, "Some antecedents of the Chemical Revolution," *Chymia, 5* (1959), 73–112; also in *Essays*, 340–74.

60. Karl Hufbauer, *The formation of the German chemical community*, 97; A. Lundgren, "The new chemistry in Sweden: The debate that wasn't," *Osiris, 4* (1988), 146–68.

qualified to explain a single combination by these means."[61] Macquer, though appreciative of Newtonian methods, thought that some mathematics was needed to formulate a general theory of chemistry; "but that [he said] does not fall into our line of work." As a pharmacist he recognized the complicated reality of the chemist: "Perhaps chemistry is not yet sufficiently advanced to be made the subject of calculation, perhaps it will never be [since] the problems that it will present mathematicians might be so complicated that they would be beyond all human effort."[62] Bergman, who was close to Pierre-Joseph Macquer and Guyton de Morveau, had a thorough knowledge of physics, admired Newtonian methods, and was capable in mathematics. None of these tools seemed useful for the study of the atom. For Bergman, empirical knowledge of atoms was itself impossible; certainly they could not be studied quantitatively.[63]

What chemists did take from experimental physics was an instrumentalist attitude toward theories. William Cullen and Joseph Black, following one methodological approach inherited from Newton, insisted that empirical knowledge and theoretical explanations should be kept separate.[64] The first part of Black's classical treatise, ***Experiments on magnesia alba*** (1750), is given over to experiments; the second, to their interpretation and theoretical explanation.[65] The instrumentalist approach fit well with a new view of theories during the late 18th century. English chemists wrote about the caloric

61. Black, *Chemical lectures, 1*, 283. Although it is difficult to separate Robison's views from Black's, at issue here is the attitude itself.

62. Melhado, "Oxygen, phlogiston, and caloric," 320; J.P. Macquer, *Dictionaire de chymie*, 4 vols., 2d ed. (Paris: P. Fr. Didot, 1778), *3*, 88, s.v. "pesanteur." Cf. Thackray, *Atoms and powers*, 208f.

63. For Bergman's Newtonianism, see "Introduction," in his *A dissertation on elective attractions*, translated from Latin (London: J. Murray and Charles Elliot, 1785); for the impossibility to measure atoms, see his "Företal," in H.T. Scheffer, *Chemiske föreläsningar* (Uppsala: M. Swederi, 1775), 6.

64. Stephen Hales, *Statical essays: Containing vegetable statics*, 3d ed. (London: Innys and Manby, 1738), 170–2; A.L. Donovan, "Pneumatic chemistry and Newtonian natural philosophy in the eighteenth century. William Cullen and Joseph Black," *Isis, 67* (1976), 217–28, and *Philosophical chemistry in the Scottish Enlightenment* (Edinburgh: Edinburgh University Press, 1975), 24–5.

65. See Donovan, *Philosophical chemistry*, for a fuller analysis. Priestley does not break the general scheme: cf. S. Schaffer, "Priestley's questions: An historiographic study," *History of science, 12* (1984), 151–83.

theory of heat in an instrumentalist way and did not commit themselves about its absolute truth.[66]

Instrumentalism triumphed in chemistry with Lavoisier's definition of an element.[67] It was perhaps his most important contribution to a new theoretical role for numbers in chemistry. The definition of a chemical element as the simplest substance available in the laboratory ignored philosophical questions concerning the structure of matter and denied elementary status to the old elements and principles. Lavoisier's definition turned the concept "element" into empirical operations independent of any hypothesis about the structure of matter; at the same time he made the definition the starting point for the construction of theories.

Lavoisier's view of chemical elements represented a break from the earlier concept of chemical facts as so many benchmarks in the search for absolute truth.[68] Lavoisier's definition augmented by the atomic theory, gave numbers a new function in chemistry. This theory can be regarded as a combination of definite chemical proportions, derived from practical chemistry, with the instrumentalist notion of a simple body, derived from experimental physics. Dalton's atomic theory used numbers to express its central concept—atomic weight. With the help of Dalton's rules of simplest combination and the assumption of definite proportions, the different weights could be interrelated. The balance thereby acquired a definitive role in the construction of chemical theory.

Although the atomic theory gave to chemistry numbers invested with theoretical significance, it did not provide generalized laws in mathematical form. The study of discrete atomic weights thus made it possible to combine theory and practice in a quantitative way without commitment to the existence of atoms. The atomic theory

66. Robert Fox, "Dalton's caloric theory," in D.S.L. Cardwell, ed., *John Dalton and the progress of science* (Manchester: Manchester University Press, 1968), on 92; Russell McCormmach, "Henry Cavendish on the theory of heat," *Isis, 79* (1988), 37–67; Hankins, 107; Christie, "Ether," 101–3.

67. Guerlac, in "Quantification in chemistry," considers quantitative descriptions of individual facts an essential part of quantification in chemistry, but does not discuss its significance for chemical theories.

68. See Hufbauer, *The formation of the German chemical community*, 8–10.

eventually overcame philosophical opposition by its success in explaining experimental facts and by distancing itself from physical atomic theory.[69]

Phlogiston Modified

It might seem inappropriate to treat the phlogiston theory here, since it is rightly considered a qualitative system. But as the dominant theory of 18th-century chemistry, and as a theory undermined by Lavoisier's brand of "quantification," the modifications of phlogiston theory in the face of quantitative facts have a claim on our attention.

Phlogiston functioned as a qualitative classifier and offered a rational explanation of the behavior of combustible substances. Like Aristotelian elements and Paracelsian principles, phlogiston could not be isolated in pure form, and it could not easily explain details of chemical processes. Yet there was one important difference: the existence and properties of phlogiston had been inferred from many and repeated empirical observations. It was easy to demonstrate that smoke, heat, fire, and perhaps also matter escaped from burning bodies. Moreover, phlogiston theory was developed by chemists who considered chemistry a practical science and who had a deep knowledge of metallurgical and mineralogical practice.

Swedish and German chemists close to mineralogy generally adhered to the theory and pushed it to its limits. Bergman's attempts to calculate the phlogiston content of different metals is especially interesting in showing how a chemist could combine Newton and Stahl.[70] Adherence to Newtonian method might thus go hand in hand with acceptance of the phlogiston theory. Bergman insisted on strong empirical foundations and on an instrumentalist interpretation

69. See Alan Rocke, *Chemical atomism in the nineteenth century. From Dalton to Cannizzaro* (Columbus: Ohio State University Press, 1984). Dalton's full place in the chemical revolution has not been sufficiently studied. Cf. Robert Siegfried, "The Chemical Revolution in the history of chemistry," *Osiris, 4* (1988), 34–50; J.B. Gough, "Lavoisier and the fulfillment," ibid., 33.

70. Bergman has been called a "Newtonian Stahlian." See Evan Melhado, *Jacob Berzelius. The emergence of his chemical system* (Stockholm: Almqvist and Wiksell, 1981), 63. There is no reason to assume that Bergman was unique in this respect.

of theories. He distinguished between *chemia vulgaris*, or traditional descriptive chemistry, and *chemia sublimior*, or transcendental chemistry; he compared the objects of the latter to the "fluxions and infinitesimals of the more sublime or transcendental geometry."[71] An instrumentalist interpretation of phlogiston theory helped its adherents to make the switch to the oxygen theory. Quantitative facts alone did not kill the phlogiston theory. Increased accuracy of measurements certainly played no important part in its demise. A very crude balance was sufficient to show that weight increased during combustion. The details of the shift deserve careful investigation, and the influence from experimental physics should be taken as a starting point for reinterpretation of the role of phlogiston in the history of chemistry.

CONCLUSIONS

During the 18th century an increasing body of chemical facts was expressed in quantitative form. Economic pressures in mineralogy and the general influence of experimental physics inspired the development of new mineralogical systems, in which quantitative descriptions of the units to be classified were an essential feature, and forced a change in the use of the balance for analytic as well as synthetic quantification, which eventuated in the theory of definite proportions.

Experimental physics did not influence chemistry by the direct application of basic physical concepts of affinity or the atom or by the direct use of mathematics. Rather, it encouraged the use of instruments yielding quantitative data: balance, thermometer, and calorimeter. On the epistemological level, it brought an instrumentalist view of theories and so contributed fundamentally to the reinterpretation of known facts that lay at the heart of the chemical

71. Bergman, *An essay on the usefulness of chemistry*, transl. F.X. Schwediauer (London: J. Murray, 1783), 16–7; cf. Thackray, *Atoms and powers*, 216. A well-known argument similar to Bergman's is that of Lavoisier, in *Traité élémentaire de chimie*, 3 vols. (Paris: Chez Cuchet, 1789), *1*, xxi–xxii. Cf. Lundgren, "The debate that wasn't."

revolution. Instrumentalism was at work, for example, in the phlogistic debates and in the attempts to define an element. Quantification in chemistry did not result in generalized laws until it was augmented by the atomic theory; even then the theory bore the mark of chemistry as a whole—the task of describing the unique. Finally the rhetoric of number played an important role in propagating Lavoisian chemistry.

The change can be summarized in the sorts of questions that were put to the balance. In the beginning of the 18th century, the main question was "How much is needed to produce this substance?"—a question of synthetic quantification. Later in the century, the question shifted to one of analytic quantification: "How many parts of different constituents make up this substance?" With the atomic theory, the question became "What is the atomic weight of this element?" The evolution of questions reflected the growing theoretical importance of the balance. No great increase in the accuracy of the balance prompted or accompanied this change. The exact measurements necessary for a quantitative physics were not necessary for a chemical revolution; what was required was increased awareness within chemistry of the significance of measurement.

9 The Most Confused Knot in the Doctrine of Reproduction

By James Larson

The systems and methods discussed by Gunnar Broberg and John Lesch in chapters 2 and 3 are among the most characteristic and enduring achievements of the Enlightenment. All of these compendia, from the most circumspect regional flora to the most ambitious inventory of the terraqueous globe, were inspired by the belief that nature's own plan is not only fixed, but intelligible, and known in outline. The physical splendor of these quarto and folio volumes proclaims the scientific and social importance of that belief.

This enormous production, focused entirely upon the stable, observable aspects of natural diversity, has always overshadowed another, smaller but equally important literature written by and for specialists. The transactions and memoirs of academies and universities of the second half of the 18th century are full of studies of degeneration, race, hybridization, metamorphosis, and monstrosity—the irrational back side of the picture devoted to stability and order. Naturalists recognized the existence of these other kinds of diversity with reluctance, and tried to limit the effect of their recognition upon the systems of fixed, natural forms. As knowledge about any one kind of divergence from established order accumulated, specialists segregated the subject and treated it as a separate set of problems. These separate studies did not constitute an integrated body of knowledge, not even at the end of the century, but they did move research in a new direction and they were inspired by a common belief. Naturalists were forced to confront the processes constitutive of living forms and to invent new methods for the study of these processes. If some kinds of diversity seemed to threaten the basic units of natural order with dissolution, it was still possible to believe that they could be made intelligible, and might even throw light upon the world of stable, established order.

This chapter analyzes one of these specialized studies, the experimental series undertaken by Joseph Gottlieb Koelreuter to discover the reproductive limits of plant species. Koelreuter cannot be seen as a typical or representative figure. The man was unknown outside a small circle of savants in Germany and St. Petersburg. The number of persons who understood his work could be counted on the fingers of one hand. His influence on contemporaries was negligible. In both his professional life and his research, Joseph Gottlieb Koelreuter was, in the exact meaning of the word, unique.

No other experimental sequence in natural history of the late 18th century equaled Koelreuter's in extent. According to Robert Olby, he carried out "more than 500 different hybridizations involving 138 species, and examined the shape, colour, and size of pollen grains from over 1,000 different plant species."[1] It is not just the number of experiments that is unusual: Koelreuter's use of counting and measurement in the analysis of data is unusual. His contemporaries were masters of impersonal, objective observation, and it would be possible to put together an anthology of their work on generation equal in precision to anything ever written. But throughout the 18th century naturalists continued to ignore or resist the use of mathematical methods and symbols in their studies of morphogenesis. The theory of generation, as they understood it, was observational, qualitative, and finalist. The central conceptual tradition was still Aristotelian, and Aristotle's depreciation of mathematics as irreconcilable with the study of final causes was so thoroughly a part of the science that naturalists did not need to justify their resistance to measuring and counting. Almost everyone considered the constitution of living forms so complicated and subject to so many conditions that their complexity was beyond the reach of mathematical analysis.

Koelreuter's concerns were limited, however: the boundaries imposed by nature on crosses between two plant species—"the most confused knot in the entire doctrine of reproduction." He did not

1. Robert C. Olby, *The origins of Mendelism* (London: Constable, 1966), 21.

believe he could untie the knot, something for which human wisdom would perhaps prove too weak, but he did intend to set forth the obvious features that lay at its basis. Since in any one cross only two plant species were involved, Koelreuter conceived his problem as analogous to a combinatory. Specific differentia were susceptible of measure and enumeration; combinations would result in the addition of a sum or the subtraction of a difference. Koelreuter's tools, measurement and counting, were the same as those of contemporary systematists. The difference lay in the consistency with which Koelreuter applied these simple tools. His experiments with plant bastards were set up on such a scale and carried out in such a way that he could determine the different forms and sizes under which hybrid offspring appear, arrange these according to their different generations, and ascertain roughly some of their numerical relations. Koelreuter's results were products of the stubborn persistence with which he aimed at the strict numerical determination of carefully limited aspects of plant morphology.

I have mentioned results. Let me admit that I could no more have understood these results than did Koelreuter's contemporaries had it not been for the work of Robert Olby. In two short papers, Olby has made Koelreuter's work accessible as no previous historian has managed to do.[2] However, Olby's insistence upon direct lines between Koelreuter and modern genetics is questionable and different from the approach taken here.

THE PROBLEM OF PLANT SEXUALITY

The competition on the question of plant sexuality proposed by the St. Petersburg Academy in 1759 provides an obvious point of entry into the study of plant hybrids and hybridization in the late 18th century. By the mid-18th century the question whether plants have sexual organs and reproduce sexually would seem to have been

2. See ibid., 17–36; Olby, "Joseph Koelreuter, 1733–1806," in Robert C. Olby, ed., *Late 18th-century European scientists* (Oxford: Pergamon Press, 1966), 33–65; and his "Joseph Gottlieb Koelreuter," *Dictionary of scientific biography* (*DSB*), *8*, 440–2.

decided by the many proofs offered by European naturalists. This was by no means the case, and many botanists continued to contest the idea.[3] To end the conflict the Academy of Sciences in St. Petersburg proposed a prize question: "Either to confirm or to deny, by means of new proofs and experiments, as well as those already known, the doctrine of the sex of plants, preceded by a history and an account of all the parts of the plant that play some role in fructification and the formation of seed and fruit."[4] The competition summed up received opinion on the subject of hybridization and initiated new research that marks the beginning of serious investigation of hereditary characters in plants. The Academy received three entries. Two were considered unsatisfactory; the third, however, was judged "praemio omnino dignus"; the prize was awarded to Carl von Linnæus, at the session of September 6, 1760.[5]

Linnæus' paper offered nothing not already familiar to his readers; he simply restated positions on the subject of plant sexuality that he had defended throughout his career. In plants there is no fruit without flower; since the flower is a necessary antecedent to fruit, it follows that sexual organs must exist in the flower. In fact, Linnæus wrote, "the flower consists of nothing but sexual organs." Those parts with the rudiments of fruit, by analogy with the animal kingdom, must be female.[6] Linnæus separated and combined male and female floral parts to show that seed developed only when pollen contacted the stigma, and argued that the possibility of fertilization of a female plant by a male of another species exhibited clearly the sexual duality of plants.[7]

Linnæus considered the production of hybrids the decisive proof of the sexuality of plants. He cited four hybrids, but offered little by way of proof that they were the offspring of two separate species. One (*Tragopogon hybridum*) had been hand-pollinated and marked

3. François Delaporte, *Nature's second kingdom*, transl. Arthur Goldhammer (Cambridge, Mass.: MIT Press, 1982), 91–148.

4. Carl Linnæus, "Nya bevis för sexualitet hos växterna," *Skrifter, 4* (Uppsala: Almqvist and Wiksell, 1908), 129.

5. Ibid., 128.

6. Ibid., 113–8.

7. Ibid., 119–26.

with a thread. (Koelreuter did not consider this cross above suspicion.) The other three plants (*Veronica spuria, Delphinium hybridum,* and *Hieracium hybridum*), which had been found in the wild or in gardens, exhibited characters intermediate between two known species. Linnæus assumed without demonstration that these plants would reproduce through seed to form "constant varieties." This would constitute a new kind of metamorphosis in plants. "It cannot be doubted that we have here new species brought forth through hybridization."[8] The bastard plant, although resembling the father outwardly, is the image of the mother with respect to the inner medullar substance and fructification. The numerous African *Gerania*, for example, led Linnæus to conclude that there are as many forms in one genus in the plant kingdom as have emerged from one species through the crossing of flowers. A genus is thus only the epitome of those plant forms that stem from a single mother and various fathers.[9]

These speculations contradicted the few established facts concerning plant crosses. However plausible Linnæus' hypothesis, especially for genera in parts of the world where nature produces a copious variety of species, crosses between species are difficult to achieve. They are, moreover, impossible to perpetuate: because of their absolute or relative infertility, or because of the regressive degeneration to which their issue was subject, the aid of one of the parental stocks is required for fertilization. Insurmountable barriers of sterility, degeneration, and distribution limit the effects of crossing far more stringently than Linnæus imagined. Thus, for each species known to be distinct and constant, most naturalists assumed a common origin and epoch.

The key word is, of course, "assumed." The constancy of species characters and the absolute or relative sterility of crosses between species were subjects about which most naturalists of the 18th century found themselves obliged to make assumptions, just as they assumed that the development of a new individual was simply the

8. Ibid., 127.
9. Ibid., 127–8.

gradual distension of a preformed being. Trained in the identification and analysis of external conformation, they were at a loss to appreciate properties and novelties that breeding experiments alone might reveal. They classified so-called bastards as a special kind of anomaly, outside the rules ordinarily followed in reproduction.[10] Each separate species is designed to function independently, yet to contribute "all those perfections towards the ends for which it had been determined."[11] Intermediate forms lack this teleological justification, and naturalists reckoned among the wiser provisions of nature both the rarity of crosses between species in their free natural state and the infertility of bastards. Crosses between parents with very different organizations were seen as flatly impossible, and conjecture about crosses between genera and orders was idle speculation.[12]

An adjunct of the St. Petersburg Academy, Joseph Gottlieb Koelreuter (1735–1806) had been a proponent of the sexual theory of plant reproduction since his early years as a student at Tübingen. One of his professors, J.G. Gmelin, had been among the first to recognize the significance of Linnæus' work on hybrids, and in 1749 had adopted the subject for his own inaugural lecture. In it, Gmelin called for experiments in hybridization.[13] Koelreuter completed his

10. Johann Friedrich Blumenbach, *Handbuch der Naturgeschichte*, 8th ed. (Göttingen: Dieterich, 1807), 21.

11. Joseph Gottlieb Koelreuter, *Vorläufige Nachricht von einigen das Geschlecht der Pflanzen betreffenden Versuchen und Beobachtungen, nebst Fortsetzungen* ([1761–66] Leipzig: Wilhelm Engelmann, 1893), 42–3. Hereafter *Vorläufige Nachricht* (1761) is cited as *Vorl. Nach.*; and the *Fortsetzungen* of 1763, 1764, and 1766, as *I. Forts.*, *II. Forts.*, and *III. Forts.*, resp.

12. Obviously these limitations did not apply to crosses between varieties of a single species, since it was well known that these often produced new races, even in a state of nature. Again assuming the fixity of species, botanists concerned themselves with establishing whether specimens were crosses between species or varieties. Observations and experiments were read over against the principle of fixity in order to arrive at the foreseen result—that species hybrids were absolutely or relatively sterile.

13. Carl-Otto von Sydow, "Linnæus and Gmelin," *Svenska Linnésällskapets Årsskrift*, 1978, 212–22. Cf. Olby, *The origins of Mendelism*, 17–36; see also his "Joseph Koelreuter, 1733–1806," and his *DSB* article on Koelreuter; Bentley Glass, "Heredity and variation in the 18th-century concept of species," in Bentley Glass et al., eds., *Forerunners of Darwin: 1745–1859* (Baltimore: Johns Hopkins Press, 1959), 158–63; Ernst Mayr, "Joseph Gottlieb Koelreuter's contributions to biology," *Osiris, 2* (1986), 135–76; H.F. Roberts, *Plant hybridization before Mendel* (Princeton: Princeton University Press, 1929), 34–61; Conway Zirkle, *The beginnings of plant hybridization* (Philadelphia: University of Pennsylvania Press, 1935), 197–200.

degree at Tübingen in 1755 and became an adjunct in natural history at St. Petersburg. When the Academy announced its competition on plant sexuality he recalled Gmelin's recommendation and set to work to produce plant hybrids.

Like Linnæus, Koelreuter considered the production of hybrids a decisive proof of the sexual duality of plants; unlike Linnæus, Koelreuter believed that nature limits this kind of anomaly, thereby preserving the order and harmony that had reigned in Eden. Two different species of animals living in a state of nature do not produce bastards; nature avoids disorder by means of natural instinct. She has equally certain methods for avoiding comparable disorder in plants.[14]

> Perhaps it has also been one of her intentions, in order to avoid just such a disturbing disorder, that she disposes one plant in Africa, and gives another its place in America. Perhaps it is partly on this account that she has confined within the limits of a certain region only those plants which in respect to structure have the least likeness with one another, and are consequently least likely to bring about disorder among themselves.

Bastards are products of the artifice of man, as exercised in botanical and zoological gardens.

> Here at any rate man gives plants of a certain kind the opportunity that he gives his animals, often assembled from widely separated parts of the world, which he keeps penned in a zoological garden, or in an even narrower space.[15]

Koelreuter was convinced that nature limits the potential for disorder, even under such unnatural conditions, and he set out to discover those limits.

Koelreuter produced his first plant bastards during the fall of 1760, after Linnæus won the Academy competition. The offspring of a cross between two tobacco species, *Nicotiana paniculata* and *N. rustica*, flowered the following March, and in the fall of 1761 Koelreuter published a brief account entitled *Vorläufige Nachricht von einigen das Geschlecht der Pflanzen betreffenden Versuchen und Beobachtungen*. He reported the results of continued experiments in

14. *Vorl. Nach.*, 29.
15. Ibid.

three *Fortsetzungen* (1763, 1764, and 1766). The *Vorläufige Nachricht* and the three *Fortsetzungen* offer a coherent account, not only of experiments in hybridization, but of the processes of pollination and fertilization. Koelreuter also published a number of individual papers on these subjects in the *Commentarii* of the St. Petersburg Academy.

The *Vorläufige Nachricht* continues the discussion occasioned by the prize competition. Indeed, the format of the piece follows closely the order prescribed in the prize question. For Koelreuter the production of plant bastards constitutes a decisive argument for sexual duality. Here, and only here, his analysis parallels that of Linnæus; elsewhere Koelreuter has taken pains to distance himself from Linnæus' wild claims and speculative flights.

THE COMBINATORY OF PARENTAL CHARACTERS

Koelreuter's conception of hybrid production is inseparable from his account of normal generation. He rejected ovist and spermist theories of preformation; he also rejected Linnæus' theory of cortical and medullar layers—views he considered more clever than correct.

According to Koelreuter, two homogeneous fluids of different kinds determined by the Creator for union with one another join to produce organization. The male agent in flowering plants is a product of the pollen grain; the female agent, Koelreuter believed initially, is a sticky secretion of the stigma.[16] These fluids, male and female, differ essentially; that is, "the force of one must be different from the force of the other."[17] From the union of these two fluids results another mean fluid, with a mean force compounded from the two simple forces. This purposive agent which emerges from the organization of the reproductive fluid as a whole is the source of organization for the future plant. For each class of organized beings, a specific compound force produces a determinate structure and specific nature.

16. Experimentally corrected, Koelreuter later regarded the ovary as the true producer of the female matter, and the oily stigmatic fluid as "merely a means of transportions." *II. Forts.*, 124–9.

17. *I. Forts.*, 42.

Although Koelreuter made many observations and experiments concerning the process of fertilization, the mass of quantitative evidence he accumulated to support theories of epigenesis, equality of parental contributions, and sterility of interspecific offspring constitutes the novelty of his contribution to late 18th-century generational theory.[18]

Koelreuter's work contained an important ambiguity concerning the source of organization in the living body. It is impossible to determine unequivocally whether the source of organization is material or nonmaterial.[19] Koelreuter draws an analogy between the union of two seed materials and the production of salt crystals. When acid and alkaline substances unite, a third, intermediate salt results. In the same way, he argues, the intermediate fluid resulting from the union of male and female seed materials either constitutes the origin or the firm foundation of the vital machine, or produces this vital machine out of itself.[20] Neither the male nor the female seed matter suffices to produce this result by itself, no more than an acid or alkaline substance can in and of itself produce the intermediate salt or form crystals. The formation of the plant requires both the compound of two specific seed materials and the composite active and purposive force resulting from that compound.[21]

Nature works in the same way to produce a cross between two species. Once the male and female seed matters unite, formation proceeds rapidly from the nucleus to the flower, and in the process

18. Ernst Mayr, "Koelreuter's contribution," 144–51. Koelreuter's conception of an active and purposive force emerging from a compound of two specific seed materials became the dominant conception of generation among German epigenesists. Variations of the theory can be found in the work of naturalists otherwise as widely divergent in their views as J.F. Blumenbach and C.F. Wolff. Cf. Johann Friedrich Blumenbach, *Über den Bildungstrieb und das Zeugungsgeschäfte* (Göttingen: Dieterich, 1781); Caspar Friedrich Wolff, *Theoria generationis* (Halle: Hendel, 1759); Shirley A. Roe, "The rise of teleological epigenesis," in *Matter, life, and generation* (Cambridge: Cambridge University Press, 1981), 150–6.

19. The same ambiguity can be found in the work of Mendel. Cf. Robert Olby, "Mendel no Mendelian?" *History of science, 17:1* (March 1979), 53–72; Hans Kalmus, "The scholastic origins of Mendel's concepts," ibid., *21:1* (March 1983), 61–83.

20. *I. Forts.*, 42.

21. Ibid., 42–3.

the sharpest eye can find no more imperfection than in the natural plant. The resulting bastard, composed from the seed matter of two separate species, reflects twofold nature in both its intermediate form and its absolute or relative infertility.

When Koelreuter first worked to produce a cross between *Nicotiana paniculata* ♂ and *N. rustica* ♀, he experimented with many flowers. Each time his fertilization succeeded. The result was perfect, somewhat divergent seeds. From 110 seeds he produced 78 plants; of these, he kept 21 over the winter, and in March 1761 they flowered. In the spread of their branches, situation and color of the flower, and individual floral parts, each plant exhibited a mean between the two natural parental species.[22] In repetitions of these experiments, the first-generation bastards consistently exhibited characteristics in "almost geometrical proportion" between differences in the parental species. Koelreuter tabulated his measurements:[23]

Character	N. rustica	Hybrid	N. paniculata
Length:			
entire flower	7‴	9 2/3‴	1″ 1 1/2‴
longest sepals	5 1/2‴	5 1/2‴	3 2/3‴
flowers above sepals	1 1/2‴	4‴	9 1/2‴
corolla tube	6 2/3‴	9‴	1″ 2/3‴
basal corolla tube	1 1/2‴	2 1/4‴	3‴
filaments		4‴	5 1/2‴
style	4 1/2‴	7‴	11 1/4‴
ovary	1 1/2‴	1 1/2‴	1 1/2‴
Diameter:			
corolla at mouth	2 3/4‴	2 1/4‴	1 2/3‴
corolla at throat	3 1/2‴	3‴	2 1/4‴
ovary	1 1/3‴	1‴	3/4‴
Maximum width:			
flower	7 1/3‴	5 1/3‴	4 1/3‴
flower when wilted	2 1/3‴	2‴	1 1/4‴

22. *Vorl. Nach.*, 30–1.
23. *I. Forts.*, 69–70.

Experiments with seven other genera supported the inference of intermediate size in first-generation hybrids. Koelreuter found that even the time of flowering and the odor of hybrid offspring were intermediate between the characteristics of the parents.

The intermediate character of first-generation hybrids, Koelreuter argued, supported the Aristotelian doctrine of reproduction by means of two seed matters. Koelreuter also used it to argue against the contemporary doctrine of generation, that is, "the doctrine of animalcula, or of original embryos and nuclei in the ovaries of animals and plants activated by male seed."[24]

All offspring of the first cross, *Nicotiana paniculata* ♂ × *N. rustica* ♀, were identical. When Koelreuter reversed the direction of the cross, *N. rustica* ♂ × *N. paniculata* ♀, the offspring "agreed all together with the plants of the first experiment, and reacted in the same way to the experiments performed upon them."[25] This phenomenon, now called the "identity of reciprocal crosses," also spoke to another contention about the contributions of parents to offspring. The hybrids with which naturalists were best acquainted, plant and animal, show a greater resemblance to the mother than to the father, leading naturalists to assume that this is a universal character of hybrids. Linnæus used the idea in his two-layer theory of generation: "a bastard offspring is with respect to its inner medullar [essential] substance the exact image of the mother, but in leaves and other outer [nonessential] parts [the image of] the father."[26]

Koelreuter's reciprocal crosses and detailed measurements established, however, that parental contributions could not be distinguished so readily. Since Koelreuter did not come across any examples of sex-linked characters in his experimental plants, his inference that each parent contributed equally to an intermediate result was to an extent justifiable.[27]

24. *Vorl. Nach.*, 31.
25. *I. Forts.*, 53–4.
26. Linnæus, "Nya bevis för sexualitet hos växterna," 127.
27. Olby, "Joseph Koelreuter," 53.

A second essential character of bastard offspring, fully as important as their intermediate form, is their absolute or relative infertility. Although the formation of the bastard tobacco parallels that of the natural species and its flowers are brilliant, the bastard plant is deficient in the most important character of all: fertility, the final cause of all formation. Koelreuter found the pollen containers of the *Nicotiana* bastard to be markedly smaller than in the natural species. What pollen they contained was white and dry; the grains did not cohere with one another as in the natural species. When examined under a microscope, the pollen grains proved to be irregular and shrunken. They contained scarcely any fluid, and most were empty husks.[28] These observations led Koelreuter to doubt the fertility of the bastard plant; his experimental results reinforced this suspicion. Of the many flowers on the bastard plants, not one succeeded in bearing a single seed, even after dusting with a large quantity of the plant's own pollen: "instead of 50,000 [they] contained not a single one, and more than a thousand flowers, one after the other, fell without leaving a single capsule behind."[29] In every sense of the word the tobacco proved to be "a true, and as far as I know, the first botanical mule produced by art."[30]

Koelreuter thought infertility only a relative imperfection, which, from the point of view of ultimate consequences, proved to be a positive good. Nature wields sterility to preserve the order established at the Creation. "What an astonishing confusion would the peculiar and unchanged hybrid characters, and consistent fertility of such plants give rise to in nature? What a monstrous swarm of imperfections would they bear, and what evil and inevitable consequences would ensue?"[31]

Koelreuter's pollination experiments, however, quickly convinced him that the infertility of the bastard was only relative. When pollinated with either of the parent species, the *Nicotiana* hybrid produced some ripe seed; much more was produced when plants of the

28. *Vorl. Nach.*, 31.
29. *I. Forts.*, 44.
30. *Vorl. Nach.*, 31.
31. *I. Forts.*, 44.

original natural species were self-pollinated.[32] For each cross, Koelreuter raised ten plants from the bastard. The results were no longer intermediate between the two parent species, but measurably resembled the pollen parent. In Koelreuter's terminology, the plants were metamorphosed or transformed (his verb is *verwandeln*): they again approached the fundamental nature (*Grundwesen*) of which the initial crossing had deprived them.[33] Again his measurements told the tale of transformation—in the size of plants; the spread of branches and flowers; and the form, size, and number of flowers.

Koelreuter's evidence for the relative infertility of hybrids and for the progressive reversion of hybrid offspring to the parental species parallels Georges Louis Leclerc Buffon's conjectures concerning animals published three years later. Under certain circumstances, Buffon maintained, a male mule can engender progeny and a female mule can conceive and give birth. Buffon also found a wide range in the productivity of different species of animals. This showed him that fertility is variable. Just as species vary in productivity, so, too, must hybrids—a conjecture that squared with Buffon's transformist views.[34]

Koelreuter was less comfortable with the transformist position. Thus, when his experiments began to produce hybrids with varying degrees of fertility, he attempted to read the results as so many potential reversions to the parental species. He carefully constructed a table of his experimental results with respect to fertility. This was, he emphasized, not just another "useless, hasty, and absurd list of chimeric bastards,"[35] but the first systematic catalogue according to a theory of generation certified by experiment.[36] Koelreuter's class of "perfect" bastards includes offspring of two or three natural species of a single genus; normally they are infertile in the highest degree, although some products of only two species prove fertile on the female side when pollinated by either parent, or were fertile to a

32. *Vorl. Nach.*, 32.
33. *I. Forts.*, 46–50.
34. Buffon, "De la dégénération des animaux," in Jean Pivoteau, ed., *Oeuvres philosophiques* (Paris: PUF, 1954), 401–5.
35. *I. Forts.*, 72.
36. Ibid., 73.

diminished degree on both sides when crossed with parental species. "Imperfect" bastards are offspring of two natural species produced when a tincture of pollen from the female supplemented pollen from the male. These offspring are characterized by a diminished degree of fertility on both sides.[37] Finally, what Koelreuter called "varietal" bastards are completely fertile. Koelreuter denied that the parents were different species. He drew a forceful conclusion:[38]

> Such a bastard in the true sense is either wholly infertile, or at most in a very limited and unequally diminished degree, by comparison with the true natural species from which it was produced is fertile. On the other hand, a mere bastard variety retains the degree of fertility of its parents, or at least loses nothing observable of this. Thus, I regard the experiment of crossing [species] in every respect as the only true, certain, and infallible touchstone of all separate species and varieties.

Koelreuter also distinguished between first- and second-generation hybrids. At an early stage in the experimental series, for example, he noticed a marked contrast in uniformity and stability. First-generation bastards for any single cross are all alike—intermediate in form between the parental species and either wholly or relatively sterile. Second-generation hybrids, even when produced from a single ovary, tend to be less like the parental bastard and more like the grandparents. After he fertilized perfect bastards of the first generation with pollen from *Nicotiana paniculata* or from *N. rustica*, he classified the results of such back crosses as descending (to the natural maternal species) or ascending (to the natural paternal species).[39] Later he succeeded in producing second-generation hybrids from the self-pollination of the tobacco hybrid. Thus, contrary to expectation, pollen from the hybrid tobacco finally proved to have a slight degree of fertility. In *Dianthus* and *Mirabilis* bastards he found a greater degree of fertility on both sides.[40]

37. Linnæus's *Tragopogon pratense* × *T. porrifolium* was in Koelreuter's opinion just such an imperfect hybrid; seed from Linnæus's specimen grown at St. Petersburg more closely resembled *T. pratense*. See *Vorl. Nach.*, 31–2.

38. *I. Forts.*, 76.

39. Ibid., 72.

40. According to Olby, Koelreuter's numbers of second-generation and back-crossed bastards are "fairly" representative. Cf. Olby, *Origins of Mendelism*, 27–8.

Koelreuter used a quantitative argument to account for these results. Crosses in which the two seed matters are in equal proportion produce second-generation bastards resembling the first-generation parent; crosses of seed matters in unequal proportions produce second-generation bastards resembling more or less closely one or the other original natural species, depending on which seed matter is dominant and to what degree. These three main groups correspond to the three segregating classes of second-generation hybrids in Mendel's experiments with plants differing in one essential character.[41] In other words, Koelreuter had found roughly the three segregating classes for second-generation hybrids. Koelreuter found puzzling, however, the exception to this neat scheme. Among the characters of second-generation *Mirabilis* bastards, for instance, he saw reversions, completely fresh characters, and colors too varied to be classified.

The contrasting results between first- and second-generation bastards required only a simple adaptation of the theory of normal reproduction. In any given plant the process of formation liberates the compound matter that gave rise to the form of the plant, and divides this compound once again into the two original matters, concentrated in the ovules and the pollen grains. In any natural fertilization the two seed matters and the simple forces inherent in them unite in equal proportions to form the intermediate product with its corresponding compound force, hypostatized during the process of generation in the specific characters of external form.[42] In some original plant species the outer characters are very different, in others very similar; likewise, the seed matters and simple forces of natural species display differing degrees of affinity. Between species with little affinity—where characters and simple forces differ significantly—no crosses can take place. In cases of close affinity, where there is "no slight resemblance between its parents and a suitable agreement of their natures,"[43] crosses can occur, although the Creator had not intended the two ground matters to unite.

41. Ibid., 30.
42. *I. Forts.*, 42–3.
43. Ibid., 74.

Both vital functions in first-generation plant bastards differ from the norm: enhanced vegetative function and completely or partially curtailed reproductive function. Where infertility is not absolute, the next generation of offspring shows an unnaturally wide range of variability. The measurable intermediacy that characterizes all perfect bastards of the first generation characterizes only a small number of individuals in the next generation, the product of seed matters united in equal proportions. More often, however, mixture and union of seed matters do not proceed with the regularity typical of natural products and simple bastards. The principle of equality is broken; the seed matters combine in different proportions, and "All kinds of wrong paths result."

Two routes are open to nature according to Koelreuter. On one, "where she has the laws of close affinity as a guide, she again approaches the high road with something like a straight line; on the other path, where she lacks this guide, she strays...ever more from the high road."[44] Here again Koelreuter's speculation parallels the best-informed contemporary opinion in zoology. In a discussion of sheep, goats, dogs, and domestic fowl the St. Petersburg academician Peter Simon Pallas noted that inconsistency of form, once introduced, increased from one generation to the next. Pallas supposed it to result from a vice introduced into the generative faculties of the original species by way of crosses. This vice, he said, parallels in effect the alteration of fluids and solids in a living body under the influence of a miasma.[45]

The presence of essential characters resembling the original natural species was Koelreuter's key to the contrast between first- and second-generation hybrids. He concluded that although new varieties might arise through selfing, plant bastards do not establish the new and constant species Linnæus had predicted. Eventually hybrid races will revert to one or the other of the original natural species.

Offspring of *Nicotiana rustica* ♂ × *N. paniculata* ♀ are intermediate in form, produce sterile pollen, and are only somewhat fertile on

44. Ibid., 78.

45. Pallas, "Memoire sur la variation des animaux," Academy of Sciences, St. Petersburg, *Acta*, 1783:2, 82–3.

the female side. After pollination with *N. paniculata*, the plants grown from the seed resemble the pollen parent far more than the first generation. Koelreuter measured the essential likenesses with his usual care: situation, form, and substance of the leaves; number of leafless slender branches; shape and size of the corolla; form, color, and breadth of the flower; and form, size, and external perfection of the capsules.[46]

The next summer (1762) Koelreuter pollinated this second generation with *N. paniculata*, and in 1763 he sowed 128 apparently fertile seeds, most of which germinated. Of this crop he retained 12 plants; when they bloomed they so resembled *paniculata* "that one could only have differentiated them with difficulty had they not been labeled with separate numbers."[47] Moreover, the fertility of both pollen and seed had increased noticeably, although this was subject to fluctuations. *Nicotiana paniculata* was well on the way toward dominating *rustica*, and Koelreuter foresaw that this series of experiments would end by producing true *N. paniculata*. "In a word," he wrote, "I no longer have the least doubt of the possibility of transforming one natural species into another."[48]

In the summer of 1765, four years after he had initiated this experimental series, Koelreuter achieved the first complete transformation of *N. rustica* into *Nicotiana paniculata* in the fifth generation, or, as he put it, in "the fourth ascending degree."[49] It was as if "he had seen a cat emerge in the shape of a lion."[50] A comparable reversion in the direction of the original mother plant, which he had followed in a parallel series of experiments, would, he predicted, require a different although proportional number of generations. He was also able to announce the successful transformation of *Dianthus* and *Mirabilis*. From a cross AB indistinguishable from the natural species B,

46. *II. Forts.*, 129–32.

47. Ibid., 132.

48. Ibid., 135.

49. Koelreuter uses two terms, *Oberherrschaft* and *Uebergewicht*, which could be translated as "dominance." But the term is reserved for Mendel's experiments with single characters; it is, in fact, regarded as his most original concept. Cf. Kalmus, "The scholastic origins of Mendel's concepts," 132.

50. *I. Forts.*, 66.

Koelreuter had thus produced offspring B. Since the maternal plant of AB was A, he reasoned that he had thereby transformed species A into species B. Moreover, he showed that the number of generations required for a successful transformation varied from species to species. He concluded that the transformation of "one plant into another [occurs according] to the greater or lesser degree of fertility [of] the bastards produced from their equality."[51]

Transmutation of metals served as an analogy for this experimental series. In theory one metal could be transformed into another by taking from it particular properties, conceived as so many independent substances, and substituting other properties for them. Mercury, for example, could be ennobled by removing the two characters upon which its fluidity and volatility rested and substituting other characters for them. In the process two seed matters were needed: the male agent, of a sulfuric nature, possessed the force to make the fluid mercurial female seed matter capable of resisting fire and forming a stable body. Sulfur, in other words, transformed the nature of the mercurial body. The result was a true metamorphosis in which the male, sulfuric agent asserted its superiority (*Uebergewicht*) over the mercurial agent.

Koelreuter believed he had achieved the alchemist's dream in botany in the course of a few years. In plants the male seed matter is oily and sulfuric; union with the female seed matter produces a stable organic body, "the initial basis of the future plant." In succeeding generations the male agent gradually takes the upper hand, and in the end the nature of the female has been wholly transformed. The transformation depends on "close affinity, fertility not wholly suppressed in the production of bastards, and dominance in a certain degree."[52]

Koelreuter used his success to argue for the possibility of the transformation of metals. Likewise, one species of animal can probably be transformed into another. From a canary might come a linnet. Experiment had already established that the female retains

51. *III. Forts.*, 199.
52. *II. Forts.*, 88.

fertility in "the second descending degree," and it was therefore probable that she would do so in the "second ascending degree."[53]

Central in Koelreuter's reasoning was the contention that a measurable change of characters in the product of any unnatural union reflects the proportions of the two seed matters. Variability increases in the offspring of bastards since the number of possible combinations of the two seed matters is infinite: that thread of Ariadne, natural affinity, is lacking. Koelreuter did see limits to variation, however. By means of crosses a naturalist can transform one affine species into another, or produce one of the infinite number of possible variations between the forms of the two natural species, but cannot produce wholly new species with entirely new characters. The naturalist thus transforms natural bodies by altering or removing particular characters and substituting others.

Transformations of this kind refuted monoparental heredity, and they showed that crossing was a powerful instrument for change in the world of living forms. But how was change itself interpreted? Koelreuter read his experiments against a conceptual tradition that did not permit development away from the original natural forms. He considered bastard plants artificial products, able to survive only in artificial conditions. "In the orderly arrangement and ordinary situation," he wrote, "established by nature in the plant kingdom, bastard plants would be difficult to produce or even to initiate." Even if we suppose the possibility of a true bastard plant in an open field, "the question would remain whether this chance had not taken place in a region where the natural situation as a whole, either mediately or immediately, had been destroyed or changed: true wilderness as it comes from the hand of nature is one thing; a field, free but in respect to a hundred things often very much altered by the hand of man, is another."[54]

Consider *Verbascum*, for which the probability of natural crosses is great. Koelreuter did not believe that unnatural products could transform the natural species, even when, as in *Verbascum*, the

53. Ibid., 89.
54. *III. Forts.*, 192–3.

natural species cross regularly. In an open field, where its own and alien pollen reach the stigmata at approximately the same time, a plant will accept the male agent intended for it by nature, excluding the alien matter from fertilization. Neither ancients nor moderns spoke of bastard *Verbascum* in the field. Linnæus, it is true, mentioned a *Verbascum* bastard, but this was likely a product of the unnatural conditions in the Uppsala botanical garden. Koelreuter continued, "It is to be wished, nonetheless, that Herr von Linné had given us a more careful description, and more according to nature than according to his fantastic theory of generation, which contradicts nature."[55]

As a naturalist, Koelreuter may have sensed the limits and assumptions of his conceptual tradition, but his was a stubborn defense of the traditional essentialist concept of a species. Koelreuter used his experiments to throw light on the nature of the species as a whole, not the inheritance of individual characters. He set out to prove that a cross between two species does not produce new and constant species. We may imagine that his findings were in this respect a relief. He regarded the essence of species as monolithic; the intermediacy that characterized all first generation bastards confirmed this interpretation.[56] However, second-generation and back-crossed hybrids did not lend themselves to the same neat explanations. Here Koelreuter took comfort from the predominance of forms resembling the original natural species; from this he concluded that hybrid offspring will revert sooner or later to one or the other of the original species. They cannot, he thought, form new and constant species.

RECEPTION

Contemporaries were only too happy to accept Koelreuter's conclusions. The rigor and complexity of his experiments were alien to

55. Ibid., 189.

56. "At no time," writes Ernst Mayr, "did he partition the phenotype into individual characters and trace the fate of a given character in different combinations through several generations." Ernst Mayr, *The growth of biological thought. Diversity, evolution, and inheritance* (Cambridge, Mass.: Belknap Press, 1982), 646.

the mentality of many naturalists of the late 18th century, but his results supported very powerfully the theory of limited variability then being elaborated. Nor was Koelreuter's work so obscure or inaccessible that no one read it. Christian Sprengel took up Koelreuter's work on pollination, arguing that floral structure was an adaptation to secure pollination and showing that cross-fertilization was the rule rather than the exception. C.F. von Gärtner, an erudite and industrious plant hybridizer, reaffirmed Koelreuter's entire theory of generation. His evidence consisted of the results of nearly 10,000 separate crossing experiments among 700 species. As Koelreuter's *Vorläufige Nachricht* and the three *Fortsetzungen* were published, a reviewer in the *Göttingsche Anzeigen* astutely picked out the tendency of the argument. Not only had Koelreuter limited severely the possibility of novelty by means of species crosses; he had shown that change was a narrowly defined combinatory of already existing characters.[57]

These summaries, and *mémoires* published by Koelreuter between 1770 and 1796 in the *Commentarii* and *Acta* of St. Petersburg, were read, assimilated, and commented upon by two of the most influential and widely read naturalists of the period, Johann Friedrich Blumenbach and Peter Simon Pallas. In his first important work on the theory of generation, Blumenbach presented Koelreuter's experiments as a refutation of preformation. The very possibility of specific bastards contradicted any conception of preformed germs, while Koelreuter's transformation of one species into another "must undeceive even the most partisan defender of the theory of evolution [i.e., preformationist] from his prejudice."[58]

Peter Simon Pallas claimed that Koelreuter had instituted "experiments on plant crosses the results of which absolutely contradict the opinion of the Chevalier Linné." He continued:[59]

> The great difficulty we have in producing crosses, with all our human industry, between two different species, the impossibility, confirmed by fact, of perpetuating these as crosses or as distinct races, either because of their

57. *Göttingsche Anzeigen von gelehrten Sachen*, St. 102 (1762), 889–91; St. 36 (24 Mar 1764), 281–4; St. 26 (28 Feb 1767), 207–8.

58. Blumenbach, *Uber den Bildungstrieb*, 61.

59. Pallas, "Memoire sur la variation des animaux," 74–5.

> absolute or relative infertility, or because of the regressive degeneration to which their issue is subject, since these crosses require for fertilization the aid of one of their primitive branches, all this opposes facts to a simple probability and forces us to see that all those species that nature affects to render alike as primitive, projected in the first plan of creation, and destined to form that chain of beings that we admire without being able to account for it, any more than [we can account for] the choice, harmony, and combination of colors and ornaments that the same creative force has used to embellish its works.

The experiments Koelreuter undertook to discover the reproductive limits of plant species may seem commonplace. Older natural history had a modest experimental tradition and breeding experiments utilizing plants and animals were common after 1750. Koelreuter was unusual, however, in his diligence and energy: his experiments involved 138 species and over 500 hybridizations, and required him to examine pollen grains from more than 1,000 different species of plants.[60] No other experimental sequence in natural history equals Koelreuter's in extent.

Koelreuter's use of counting and measurement in the analysis of his data was also unusual in 18th-century natural history. Although his contemporaries were masters of impersonal, objective, precise observation, naturalists throughout the 18th century ignored or resisted the use of mathematical methods and symbols in their studies of morphobiology. Natural history, as they understood it, was observational, qualitative, and finalist. The central conceptual tradition was Aristotelian, with its depreciation of mathematics as irreconcilable with the study of final causes.[61] Thus a resistance to measuring and counting came naturally to 18th-century natural history.

Not all naturalists were Aristotelians, however, and even the diehards of the Aristotelian tradition found it impossible to resist the forces toward greater precision that originated in more prestigious spheres of the scientific constellation. Anatomy contributed to the establishment of species characters the procedure of measuring size. Naturalists also counted seeds and pollen grains, observed numerical

60. Olby, *The origins of Mendelism*, 21.
61. As explained in Aristotle, *Metaphysics*, B 2 996a.

ratios, noted numbers of progeny, and proposed mathematical nomenclatures. These efforts toward quantification, however, remained sporadic and uncorrelated; and most naturalists considered the complexity of living forms beyond the power of mathematical analysis.

Koelreuter limited his concern to the boundaries imposed by nature upon crosses between two plant species—"the most confused knot in the entire doctrine of reproduction." He did not intend to untie the knot, but rather to identify its major features. Since only two plant species were involved in any one cross, Koelreuter conceived his problem as analogous to a combinatory. Because specific differentia were susceptible to measure and enumeration, combinations could thus be expressed as a simple sum or difference. Koelreuter applied the instruments of measurement and counting with unusual consistency. The design and scale of his experiments permitted him to determine different forms and sizes of hybrid offspring, arrange them by generation, and estimate their ratios. That his quantitative approach yielded important results was due in large measure to his stubborn concentration on carefully limited problems of plant morphology.

10 Labs in the Woods: The Quantification of Technology During the Late Enlightenment

By Svante Lindqvist

The concept of quantification can be considered an intrinsic component of technology. Two basic characteristics of technology in modern Western civilization are technical efficiency and economy. Practitioners of technology had labored to increase technical efficiency and economy long before the concept of quantification began to be applied to technology. The latter occurred during the 18th century when technology began to borrow its methods from science, that is, when attempts were made to study technical reality by systematic experimentation and quantification in fixed units using precision instruments. These attempts first met with success after 1770.

The phrase "systematic experimentation and quantification in fixed units using precision instruments" describes a laboratory—a table in a quiet, secluded room with sufficient illumination and heat. Such laboratories became more common in universities, academies of science, and affluent homes in the 18th century. But it was quite a different matter to transfer these ideal conditions from the natural philosopher's laboratory to the technical world outside. Where the

For help in revising an earlier version of this chapter, I am deeply indebted to Eugene S. Ferguson and David A. Hounshell. I also gratefully acknowledge suggestions and comments on this version from R. Angus Buchanan, Arthur L. Donovan, Willem D. Hackmann, Roger Hahn, Charles Haines, Richard L. Hills, Karl Hufbauer, Thomas P. Hughes, Melvin Kranzberg, John Law, Edwin T. Layton, Jr., Otto Mayr, Terry S. Reynolds, Sheldon Rothblatt, Richard Sclove, Bruce Sinclair, Martin Trow, and Wolfhard Weber. The *Proceedings* of the Royal Swedish Academy of Sciences (*Kungl. Vetenskapsakademiens Handlingar*) are abbreviated *KVAH*.

laboratory was neat, technology was messy. The central component of the experimental method, controlled experiment, was difficult to achieve in technology, with its comparatively large scale. To apply quantitative methods to technology required a degree of *control* over the material and social world beyond the means of individuals. The late 18th century saw the emergence of institutions with authority and competence to exercise such control.

Three case studies will illustrate general characteristics of this transformation of technology. Our image of 18th-century technology has by and large been shaped by those technologies seen subsequently as spectacular and/or efficient. But the ingenious and complex machines depicted in books of the 18th century—like those depicted by Ramelli, Zonca, Besson, and others in the Renaissance—were often no more than bold, futuristic speculations directed at a restricted audience, irrelevant to the daily reality in which most people lived and worked. We may identify numerous forerunners to 19th-century industrialization, but technological reality in the 18th century lay a long way from the elegant engravings. The average 18th-century man lived by the sweat of his brow, aided by simple wooden implements at a low level of mechanization. Only occasionally was his work made easier by machines powered by his fellow-men or by oxen, horses, water, wind, or steam. As a period in technological history, the 18th century was characterized by the use of wood, water, and work, and this is reflected in the choice of examples for this study.

WATER POWER

Industries dependent on mechanical energy in the 18th century included mines, blast furnaces, tilt hammers, other metal works, saw and paper mills, gunpowder factories, brickworks, oil mills, and glassworks. Such industry required *high power* (work per unit of time) and *continuous operation*; the extent to which these requirements could be met by traditional sources of power varied. Muscle power (animal and human) provided continuous operation at limited output. Wind produced high power, but proved inappropriate for industrial production since continuous operation could never be guaranteed. Only water power fulfilled both requirements. But existing hydraulic resources were limited. In order to utilize fully the available water power, the efficiency of waterwheels had to be increased. For each type of water mill a particular speed gives maximum power. Finding the rules for determining this speed was a problem to which solutions were sought throughout the 18th century. The design of waterwheels was the subject of vigorous debate in learned journals and technical literature in the 18th century, although theoretical considerations had "no direct effect...on the construction and installation of wheels," which were governed by a tradition little changed until the end of the century.[1]

Two approaches characterized 18th-century attempts to establish general rules for the most efficient design of waterwheels.[2] In the *deductive* approach mathematical analysis permitted derivation of general rules from fundamental laws of motion. The *inductive* approach used systematic experiments with parameter variation and optimization to achieve the same end.[3] Great expectations were matched by great investments of work; the rate of return, at least until the early 19th century, was disappointing. The deductive method resulted in formulas too complicated for ready application. The inductive method yielded unmanageable amounts of unreliable quantitative data. A waterwheel turning peacefully in a stream proved far more complicated than the heavenly clockwork.

Around 1700 French scientists in the Academy of Sciences in Paris made the first analysis of waterwheels in dynamical terms. In his comprehensive history of the waterwheel, Terry Reynolds distinguishes five approximate and overlapping periods following the work in Paris.[4] The first was the establishment of theoretical analysis by

1. Maurice Daumas and Paul Gille, "Methods of producing power," in Maurice Daumas, ed., *A history of technology and invention: Progress through the ages*, Eng. transl., 3 vols. (New York: Crown Publishers, Inc., 1969–79), *3*, 24–5.

2. Svante Lindqvist, *Technology on trial: The introduction of steam-power technology into Sweden, 1715–1736* (Uppsala: Almqvist & Wiksell International, 1984), 62–92.

3. "Parameter variation and optimization" was introduced and defined in Walter G. Vincenti, "The air-propeller tests of W.F. Durand and E.P. Lesley: A case study in technological methodology," *Technology and culture, 20* (1979), 712–51.

4. Terry S. Reynolds, *Stronger than a hundred men: A history of the vertical water*

academicians like Parent and his followers between 1700 and 1750. The second was a period of experimental work from 1750 to 1770 by engineers such as De Parcieux, John Smeaton, and Charles Bossut. A third, simultaneous development was the attempt during the same period by Johann Euler and Charles Borda to reconcile the discrepancy between theoretical predictions and experimental findings. This next forty years, 1770–1810, Reynolds labels "the era of theoretical confusion."[5] This confusion abated during the fifth period, 1810–50, when Borda's analysis of 1767 achieved general acceptance.

Facilitating the transition from general theoretical confusion to the reconciliation between theory and experiment were institutions capable of large-scale experiments beyond the means of individuals. The importance of such institutions in water power technology during the late Enlightenment and early 19th century is underscored by comparing a successful Swedish attempt to quantify water power with an earlier, unsuccessful one. From 1701 to 1705 the well-known Swedish inventor Christopher Polhem performed some 25,000 experiments with a hydrodynamic apparatus of his own design. But this heroic experimental effort produced only meager results. A century later, between 1811 and 1815, the Swedish Ironmasters' Association (*Jernkontoret*), a private organization of independent ironworks established in 1747, undertook a major investigation of hydrodynamics.[6] Originally intended as a credit agency for the ironworks, the Association assumed an important role in technological development toward the end of the 18th century. Its officers exercised quality control over the various stages in the process of iron manufacturing, and the Association financed a number of large development projects that were

---

wheel (Baltimore: Johns Hopkins University Press, 1983), 204–7. Cf. Reynolds, "Scientific influences on technology: The case of the overshot waterwheel, 1752–1754," *Technology and culture, 20* (1979), 270–95; Roger Hahn, *L'hydrodynamique au XVIIIe siècle: Aspects scientifiques et sociologiques* (Paris: Palais de la Découverte, 1964), and *The anatomy of a scientific institution: The Paris Academy of Sciences, 1666–1803* (Berkeley: University of California Press, 1971).

5. Reynolds, *Stronger than a hundred men*, 243.

6. Bertil Boethius and Åke Kromnow, *Jernkontorets historia*, 3 vols. (Stockholm: P.A. Norstedt & Söner, 1947–68).

too expensive for any individual ironworks. A large experimental apparatus had been built at the Great Copper Mine in Falun in 1804, which was originally intended for investigating the efficiency of winding gear. However, the metallurgist Eric Thomas Svedenstierna and others supported the idea that the Swedish Ironmasters' Association should finance a lengthy series of experiments in order to establish a general theory of water power.[7] The resulting hydrodynamic investigation of 1811–5 stands in glaring contrast to Polhem's experiments.

Theoretical, instrumental, and, most importantly, institutional factors led to this successful quantification of water power technology during the late Enlightenment. A more profound theoretical framework was available by the end of the 18th century; a heightened appreciation of the applications of mathematics was reinforced by a broader and more critical knowledge of the international literature in the field and an awareness of fundamental principles of experimentation. The institutionalization of science contributed to this conceptual change: technological innovation became verbalized and was documented in journals and monographs produced and distributed under the auspices of new scientific institutions. A higher degree of precision in scientific instruments resulted from a more vigorous market for scientific apparatus and information. In the social organization of technology, responsibility shifted from individuals to institutions. This last was a sine qua non for applying the concept of quantification to the expanded spatial and temporal dimensions of technology. The importance of these three factors in the Swedish case is described in the following subsections.

Theoretical

Although the experimental apparatus used by Lagerhjelm for the investigations of 1811–5 resembled that of Polhem a century earlier, there the resemblance ends. Polhem's experimental apparatus seems to show the influence of the French physicist Edmé Mariotte, who

7. Lindqvist, *Technology on trial*, 86–9.

had carried out experimental studies of water and wind mills.[8] Certainly they adhered to the same empirical tradition. Like Mariotte, Polhem was more concerned with articulating and applying generalizations based on experiments than reducing them to fundamental principles; both relied on common sense to guide their reasoning.[9] Although Polhem's work contains an early example of parameter variation and optimization,[10] he was never able to convert his many experimental results into general rules for the design of waterwheels.[11] Nor did he fully appreciate the merit of mathematical analysis of hydrodynamic phenomena. This is evident in his faint praise for the work of his younger colleague Pehr Elvius, who in 1742 published *A mathematical treatise on the effect of water mills.*[12] Polhem commented in the *Proceedings* of the Royal Swedish Academy of Sciences: "Although [Elvius'] book is really written for the learned, who are already familiar with *the modern mathematics*, which by its discoverer the learned Leibniz is called *calculus differentialis* and by Newton, *fluxio curvarum*, so does yet Mr. Elvius show his profound knowledge of such puzzling matters, that he gives hope of becoming a good *Mechanicus* with time, as well in Practice as now to begin with in Theory."[13] What Polhem had considered "modern mathematics" was a standard tool in the hands of the mine official Pehr Lagerhjelm, whose report on the hydrodynamic experiments of 1811–5 financed by the Swedish Ironmasters' Association was a highly mathematical treatise.[14] Lagerhjelm's report also

8. Sten Lindroth, *Christopher Polhem och Stora Kopparberget: Ett bidrag till bergsmekanikens historia* (Uppsala: Almqvist & Wiksell, 1951), 85. Cf. Reynolds, *Stronger than a hundred men*, 202–3.

9. Michael S. Mahoney, s.v. "Edmé Mariotte," in *Dictionary of scientific biography* (*DSB*), 9, 114–22.

10. Boel Berner, *Teknikens värld: Teknisk förändring och ingenjörsarbete i svensk industri* (Lund, 1981), 249–52. Cf. Vincenti, "The air-propeller tests."

11. Lindqvist, *Technology on trial*, 69–74.

12. Pehr Elvius, *Mathematisk tractat om effecter af vatn-drifter, efter brukliga vatn-värks art och lag* (Stockholm: P.J. Nyström, 1742). Cf. Reynolds, *Stronger than a hundred men*, 235–6.

13. Christopher Polhem, "Fortsättning om theoriens ock practiquens sammanlämpning i mechaniquen," *KVAH*, 1742, 158.

14. Pehr Lagerhjelm, Jacob af Forselles and Georg Samuel Kallstenius, *Hydrauliska försök, anställda vid Fahlu grufva, åren 1811–1815*, 2 vols. (Stockholm, 1818–22).

included a thorough, critical review of relevant international literature. The first fifty pages of the second volume commented on the works of Smeaton, Euler, Borda, Bossut, Banks, Langsdorf, and others.[15]

Lagerhjelm's treatise also evinces a higher level of conceptual awareness. In his preface to the second volume, Lagerhjelm offered an epistemological program to relate theory and experiment for hydrodynamics. His ideas bear a certain resemblance to Kant's theory of knowledge, and the terminology—"phenomenon," "form," and "content"—is similar. For Lagerhjelm, inductive reasoning cannot produce conclusions of universal validity, because "abstractions from a given experience...are only valid under the circumstances and within the boundaries essentially associated with the class of phenomena one experienced."[16] The implication was clear: the inductive method followed by Polhem and others, with their thousands of experiments throughout the 18th century, was epistemologically pointless. So, too, was the deductive method, the "speculative root" of knowledge in which Elvius and others had placed their confidence, inadequate in and of itself. The path to truth required a synthesis between "form" and "content," specifically, theory and experiment.

Instrumental

Inaccurate measurements compromised Polhem's data. Because the pendulum in the clock he used was not of the proper length, he arrived at incorrect values for the speed of the waterwheel and hence incorrect values for the output. Polhem tried in 1710 to reduce all these figures to their proper value by means of a correction coefficient, but found the work "so difficult and tedious, that no amount of patience would have sufficed." More seriously, the protractor he used to measure the inclination of the water trough gave different readings as the waterwheel was placed at different levels.

15. Lagerhjelm, *Hydrauliska försök*, 2, 2–49.
16. Ibid., preface. Cf. Lindqvist, *Technology on trial*, 87–9.

Polhem confided to his assistant: "In fact between ourselves, this work is as useful as a fifth wheel on a carriage."[17]

Lagerhjelm proclaimed explicitly his awareness that calibrated precision instruments were essential if the experiments were to be reproducible and the results of general value. He made linear measurements using "a precisely graduated decimal scale two Swedish feet in length" produced by Johan Gustaf Hasselström, purveyor of mathematical instruments to the Royal Swedish Academy of Sciences. Lagerhjelm also employed "a set of weights calibrated against the Swedish original standard, which is kept in the Archives of the Royal Treasury," and a balance constructed by Gabriel Collin, manufacturer of optical instruments for the Academy, and watched a clock borrowed from the astronomical observatory of the University of Uppsala.[18]

In quality of instruments, Lagerhjelm enjoyed a significant advantage over Polhem. Polhem had used the most accurate instruments he could acquire or construct. Over the course of the 18th century, however, a real market for scientific instruments had developed in Sweden. The market was largely the creation of the Royal Swedish Academy of Sciences, established in 1739. Rivals for this market competed in precision. Instrument-makers like Daniel Ekström, Hasselström, and Collin won the right to call themselves "Purveyors to the Royal Swedish Academy of Sciences";[19] this distinction implied to prospective customers that every instrument produced in their workshops promised the highest possible degree of precision. Market pressures did thus increase the degree of precision, and the market itself was a result of the establishment of the Academy. Before then, no Swedish instrument-maker could acquire such status; hence there had been little or no competition in degree of precision.

17. Axel Liljencrantz, ed., *Christopher Polhems brev* (Uppsala: Almqvist & Wiksell, 1941–6), 37–8.

18. Lagerhjelm, *Hydrauliska försök, 1*, 25–6.

19. Cf. Sten Lindroth, *Kungl. Svenska Vetenskapsakademiens historia 1739–1818*, 2 vols. (Stockholm: Almqvist & Wiksell, 1967), *1*, 789–815, *2*, 519–38, and *Svensk lärdomshistoria: Frihetstiden* (Stockholm: Norstedt, 1978), 373–6; Gunnar Pipping, *The chamber of physics: Instruments in the History of Sciences Collections of the Royal Swedish Academy of Sciences* (Stockholm: Almqvist & Wiksell, 1977).

In this way, the institutionalization of Swedish science contributed to the increased degree of precision in scientific instruments during the late 18th century, a development that contributed to the quantification of technology.

Institutional

The establishment of the Royal Swedish Academy of Sciences contributed, as mentioned above, to a theoretical and instrumental development. But there was also an institutionalization of technology within the largest and most important of Swedish industries, the iron industry: the establishment of the Swedish Ironmasters' Association, which reflected increased interest in general technological problems during the late 18th century.[20] The general importance of this change in the social organization of technology can be described as a shift in responsibility from *individuals* to *institutions.* Technological projects were now being undertaken by and with the *competence* and *authority* of the institution.

The *competence* of an institution, with its hierarchical structure based on academic merit, is apparent in the case of Pehr Lagerhjelm, the leader of the project of 1811–5. He had studied at the University of Uppsala, where he passed the mining examination (*Bergsexamen*) in 1807. This university degree, established in 1750, had become a prerequisite for officials in the service of the Board of Mines, the governmental department that exercised ultimate authority over the mining industry. The degree required examinations in physics, mechanics, geometry, chemistry, and law. After graduation from the University of Uppsala, Lagerhjelm was duly appointed to the Board of Mines in Stockholm. He became a pupil of the chemist Jöns Jacob Berzelius, and assisted him in calculating the percentage composition of nearly 2,000 chemical compounds. This work was published as a supplement to the third volume of Berzelius' textbook, *Lärbok i kemien.*[21] In 1808, Lagerhjelm was appointed under-secretary of the

20. Boethius and Kromnow, *Jerkontorets historia, 3* (1968), is devoted to the technological developments initiated by the Swedish Ironmasters' Association.

21. Jöns Jacob Berzelius, *Tabell som utvisar vigten af större delen vid den oorganiska kemiens studium märkvärdiga enkla och sammansatta kroppars atomer, jemte deras sammansättning, räknad i procent: Bihang till tredje delen av läroboken i kemien*

Swedish Ironmasters' Association. He thus reached his position in 1812 as leader of the hydrodynamic experiments by making a career within the formally organized educational system of the Swedish mining bureaucracy.[22] The merit of the system is proven by his many other contributions to Swedish technology during the early years of the 19th century.

By contrast, Christopher Polhem had gained his position as a favorite of Karl XII in the time of the Absolute Monarchy.[23] Polhem had been appointed director of the *Laboratorium mechanicum*, the section of the Board of Mines with designated responsibility for research and education in mechanical technology. It was here that Polhem's hydrodynamic experiments were undertaken from 1701 to 1705. Although a mechanical genius by any standard, Polhem lacked the broad formal university education Lagerhjelm enjoyed. His investigations were therefore carried out within the context of a more narrow theoretical perspective and were influenced by important personal idiosyncrasies—including his disdain for mathematical analysis.

Comparing these two examples also illustrates the importance of institutional *authority*. The project in 1811–5 involved several persons, all highly qualified. Although they did not cooperate throughout the many years of the project without personal friction, the authority of the Swedish Ironmasters' Association led to the completion of the project and to the publication of the results in two volumes in 1818 and 1822.[24]

On the other hand, Christopher Polhem submitted to the Board in 1705 what he called an interim report; in fact, it was the only report he ever prepared concerning his hydrodynamic experiments. When he resumed work with the experimental apparatus in 1710, he made the distressing discovery that two crucial quantities had been measured inaccurately throughout the whole series of measurements.

(Stockholm: H.A. Nordström, 1818), 10. The calculations by Berzelius and others of the exact amounts of different substances required for chemical processes represented an important stage in the quantification of chemical industry.

22. Lindqvist, *Technology on trial*, 88f.

23. Sten Lindroth, *Svensk lärdomshistoria: Stormaktstiden* (Stockholm: Norstedt, 1975), 534–52.

24. Lagerhjelm, *Hydrauliska försök*.

This rendered the results incommensurable and the whole series of experiments nonreproducible. But Polhem, whose individual reputation exceeded his relatively subordinate position as an official in the Board of Mines, had no difficulty in forbidding his assistant to mention to anyone that the data in the report were useless. An institution stronger in terms of hierarchical authority would have insisted that he submit a final report on the experiments for which he had already received his fees, and, on discovering that the data were useless, demanded that he repeat the experiments.

CHARCOAL-BURNING

We tend to associate the 18th century with the use of coal and iron, and to look back on the 16th and 17th centuries as, in John U. Nef's phrase, "an age of timber."[25] But the growing importance of coal technology, especially in England, should not obscure the dependence on forests of virtually all aspects of material culture during the 18th century.[26] Not only did mining consume large amounts of wood; so, too, did potash plants, tanneries, glassworks, saltpeter works, train-oil works, lime production, and other industries rely on the forests for fuel and raw materials. Domestic demands included fuel for heating houses and drying grain and malt, and timber for houses, fences, ships, carts, barrels, and agricultural implements. This vital natural resource, however, was perceived to be running out in 18th-century Europe. The fear of imminent shortages spurred both legislative actions and interest in technical improvements aimed at reducing the number of trees felled.

In the 18th century, Sweden—a country devoid of fossil fuel resources for all practical purposes—was gripped by general anxiety about a dearth of timber.[27] It was believed that the forests were laid

25. John U. Nef, *The rise of the British coal industry*, 2 vols. (Hamden, Conn.: Archon Books, 1966), *1*, 191.

26. John R. Harris, "Skills, coal and British industry in the eighteenth century," *History, 61* (1976), 167–82, and "The rise of coal technology," *Scientific American, 231:2* (1974), 92–7.

27. Lindqvist, *Technology on trial*, 34–61, and "Natural resources and technology: The debate about energy technology in eighteenth-century Sweden, *Scandinavian*

waste by excessive felling: "many large areas of the realm are in danger of soon becoming desolate because of the shortage of timber and. . .the mines and towns in many parts of the country are likewise threatened with ruin that cannot long be postponed if an early remedy is not found." In the worst-case scenario, "the fatherland will in the course of time be reduced to a miserable condition."[28] Whether or not the fear of timber shortage was well founded does not concern us here. What matters is that the belief in imminent shortage was widespread and influential.

The production of bar iron accounted for about 70 percent of Sweden's exports in the 18th century. About 50,000 tons of bar iron were produced every year, and every stage of production required much timber. Blast furnaces and forges consumed charcoal equivalent to three million cubic meters of timber a year. Charcoal-burning amounted to half of the total industrial consumption of timber.[29] There were two alternative methods of charcoal-burning: stacking the wood either horizontally in piles called *liggmilor* or vertically in *resmilor*. Despite the fear of a forest shortage and the great consumption of wood for charcoal-burning, evaluations of the two methods of extracting charcoal from wood were not easily undertaken; the question was not resolved until the early 19th century.

In 1811 the Swedish Ironmasters' Association financed a series of experiments to assess the relative merits of *resmilor* and *liggmilor* and to determine the pile design yielding the maximum amount of charcoal. The report was published as a monograph three years later by the mining official Carl David af Uhr.[30] It was an impressive volume, comparable in scope and depth to the report on the hydrodynamic experiments undertaken at the same time, also with the support of

---

journal of history, 8 (1983), 83–107.

28. Anonymous, *Swar på Kongl. Wettenskaps Academiens fråga: hwilka författningar äro de bästa, at underhålla tilräckelig tilgång på skog här i landet?* (Stockholm, 1768), 9.

29. Leif Mattsson and Einar Stridsberg, *Det industriinriktade skogsbruket sett ur ett historiskt perspektiv* (Stockholm, 1979), 14. Cf. Lindqvist, *Technology on trial,* 35–6, 318 no. 1.

30. Carl David af Uhr, *Berättelse om kolnings-försök åren 1811, 1812 och 1813: På Bruks-Societetens bekostnad anstälde* (Stockholm, 1814).

the Swedish Ironmasters' Association. Uhr's report described the series of experiments in charcoal burning carried out at Furudahls Ironworks in the province of Dalecarlia during the years 1811–3. No less than forty full-scale piles of different types were tested, with twenty parameters recorded for each pile (fig. 10.1). The systematic study was meticulously planned: for example, a specially designed tool was used to measure the diameter of the billets at both ends, in order to compensate for their taper ("frustra of a cone, as they truly are") when calculating the volume of a pile.[31] The volume of the piles was measured in cubic ells to one or two decimal places. The author discusses the effect of various errors in measurement and ways to compensate for these errors. Output, measured in cubic ells of charcoal, was correlated to the total labor, measured in man-days and horse-days, needed from the day the trees were felled to the day the charcoal arrived at the ironworks. Calorimetric experiments helped determine the quality of the charcoal, prompting Uhr to discuss Lavoisier's opinion of the role of oxygen in combustion. He handled this question with the same facility he showed in computing the number of horse-days needed to build a charcoal pile. Results of this study were summarized in a handbook for charcoal-burners that appeared in three editions—a measure of its success.[32]

One looks in vain for quantitative methods in the literature on charcoal-burning at midcentury. In assessing the relative merits of *resmilor* and *liggmilor* the quantity of wood was not specified; nor was it clear which units were used in measuring the charcoal.[33] Consider Magnus Wallner, who published at his own expense *A brief account of charcoal-burning in Sweden*, a Swedish translation of his dissertation under Celsius at the University of Uppsala.[34] The work

31. Ibid., 81.

32. Carl David af Uhr, *Handbok för kolare* (1st and 2d edn., 1814; 3d edn., Stockholm: H.A. Nordstrom, 1823). Technically, charcoal burning is a chemical process technology. David Hounshell has observed that af Uhr's investigation is in this sense an early example of a successful application of quantitative methods to chemical technology.

33. This lack of clarity throughout the 18th century is reflected in Rinman's well-known mining dictionary; Sven Rinman, *Bergwerks lexicon, 2* (Stockholm: Johan A. Carlbohm, 1789), 26, 411.

34. Magnus Edvardi Wallner, *Kolare konsten uti Swerige, korterligen beskrifwen* (Stockholm, 1746).

contained a brief description of methods and tools, statements by charcoal-burners, a few quotations from foreign literature on the subject, and some of Wallner's own ideas. Entirely lacking was any attempt to give quantities in defined units and to carry out systematic experiments under controlled conditions.

The failure in applying quantitative methods to charcoal-burning may be attributed to the lack of institutions able to recreate a laboratory environment in the forest. No retort on a laboratory bench could reveal the best type of pile for charcoal-burning. "Systematic experimentation" required building many piles of different types and supervising them day and night for several weeks. It was necessary to take into account the species, age, and moisture of the wood; the length and thickness of the billets; the stacking pattern; the total amount of wood; the outer dimensions of the pile; and the ignition method. After the piles were pulled down, the charcoal had to be shoveled into barrels of known size—"quantification in fixed units." All this was far different from laboratory work. It differed first in spatial terms—not only the size of the piles but also the area of the forest required for the tests was large. The temporal requirements were also of a different order. Because building, watching, and pulling down a pile took more than a month, a series of systematic experiments might stretch over several years. Such was the case with the investigations carried out by the Swedish Ironmasters' Association in 1811–3.

The social organization of the work process also argued against a comprehensive study undertaken by an individual and resulting in useful data. Charcoal-burning was a huge, decentralized system of production: peasants and crofters labored under the tenant's obligation to deliver charcoal to the ironworks. Production was the responsibility of individuals, tens of thousands of peasants and crofters, each working independently, deep in the forests.[35] Work in the forest was linked with the changing of the seasons and the tilling of the soil. In autumn, after the harvest, wood was burned to charcoal in the forest

35. Eli F. Heckscher, *Sveriges ekonomiska historia från Gustav Vasa*, 2 vols. (Stockholm: Bonnier, 1935–), *2:1*, 454.

where it had been felled; in winter, when the snow made the pathless forest passable, sledges carried charcoal to the ironworks. The methods of charcoal burning were the product of local conditions and tradition; the variety of circumstances was reflected in the names of different sorts of charcoal piles. To hammer this decentralized variety of methods and measures into a standard form amenable to quantitative comparisons seemed impossible. Hence Wallner contented himself in 1746 with publishing the views he had elicited from charcoal-burners he had met and a few quotations from foreign literature, together with his own individual, casual observations.

That charcoal-burning—like most technologies at the time—was a nonverbal technology complicated the problem of quantification.[36] Witness the failure of the encyclopedists in their attempt to gather and record technical know-how of existing practices like charcoal-burning. John R. Harris has argued that "the difficulties of imparting craft skills by literary and graphic means" limited the "technological gain" possible from the 18th-century encyclopedias.[37] The difficulties of improving a nonverbal technology were akin to the problems of describing it. A process could not easily be formulated as an intellectual problem and reduced to quantitative terms outside the realm of practical experience.

Science first met technology at the edge of the woods, when the sledges loaded with charcoal approached the ironworks to deliver their products. Only here could the individual ironmaster exercise some control over production by making sure that the peasants and crofters delivered the agreed amount of charcoal. Here the ironmaster might enlist the aid of geometry. Two articles in the *Proceedings* of the Royal Swedish Academy of Sciences addressed his concerns.

36. "Nonverbal technology" was introduced by Eugene S. Ferguson, "The mind's eye: Nonverbal thought in technology," *Science, 197* (1977), 827–36.

37. Harris, "Skills, coal and British industry," 167 no. 2. Cf. Roger Hahn, "Science and the arts in France: The limitations of an encyclopedic ideology," *Studies in eighteenth-century culture, 10* (1981), 89: "Because the practical arts have rarely been improved by this essentially polite, but largely literary modification, it is not surprising that the ideology was never effectively put into practice." See also Bertrand Gille, ed., *The history of techniques* (New York: Gordon and Breach, 1986), 596–600, 1150–6.

They were devoted to the mathematical problem of calculating the volume of the rhombically shaped sledges that carried the charcoal (fig. 10.2).[38] Together, however, the ironmasters could marshal resources enough to launch a concerted, large-scale attack on the technical problem of charcoal production.

UTILIZATION OF MANPOWER

It is only too easy for historians to overlook the importance of muscle power during the 18th century. Horse whims and treadwheels, despite their widespread use and importance in their day, have been overshadowed by more efficient and spectacular technologies like waterwheels, windmills, and steam engines. But an innovation does not immediately cause the abandonment of earlier, inferior technologies as obsolete. On the contrary, the total use of old technologies declines slowly and asymptotically. Treadwheels, for example, were used in Swedish mines as late as the 1880s, and one is even reported to have been in use as late as 1896.[39] Muscle power was the dominant power technology during the 18th century: the total work produced by men, horses, and oxen in fields, roads, forests, mines, mills, and harbors probably exceeded the combined power of all steam engines, waterwheels, and windmills.[40]

Industry needed sources of mechanical power capable of high output and continuous operation. The muscle power of men and animals

38. Zacharias Z. Plantin, "Mathematiskt theorem," *KVAH,* 1778, 292–302, and "Construction och uträkning af en kol-ryss eller kol-stig öfver en gifven botten," *KVAH,* 1784, 309–15.

39. Harald Carlborg, "Om tramphjul och andra motorer i äldre tid vid svenska malmgruvor," *Med hammare och fackla, 25* (1967), 77.

40. The textile industry was based almost entirely on manual labor until 1769 when Arkwright invented his spinning machine; see Richard L. Hills, *Power in the industrial revolution* (Manchester: Manchester University Press, 1970), 15. The qualitative importance of this invention must not lead us to forget that the majority of people during the 18th century (and before 1770, *all* people) were dressed in textiles that had been spun and woven by human energy. For a quantitative evaluation of the importance of steam power during the 18th century, see Georg N. von Tunzelmann, *Steam power and British industrialization to 1860* (Oxford: Clarendon Press, 1978).

could provide continuous operation, but output was relatively low. A man working a ten-hour day produced approximately 0.1 horsepower, a horse in a good harness roughly six times as much. Horse whims used at the mines, driven by two or four horses, could thus develop approximately 1.2 or 2.4 horsepower.[41] Continuous operation in three shifts required as many as a dozen horses. Perhaps eighty or even forty men could have produced an equivalent amount of work, but animals offered certain advantages when continuous power of this magnitude was required. In other cases, however, the power supplied by a few persons was not only sufficient but preferable—turning a crank or a windlass, pulling at ropes, carrying burdens, tramping in treadwheels. Compared to horses and oxen, men were relatively small and movable; their power output could be regulated with a word or a glance. The factor of control was significant when it came to loading or unloading ships, turning lathes, grinding and polishing, operating textile machines, or building.

Early attempts to determine how much physical labor a man could be expected to do in a day were made around 1700 at the Académie royale des sciences.[42] Inquiries continued through the 18th century. Bernard Forest de Bélidor, Charles Augustin Coulomb, John T. Desaguliers, Johann Euler, and Philippe de La Hire were among those who addressed related questions of a fair day's work and the comparative strength of men and horses.[43] No consensus was reached, but a large body of data was generated.[44]

41. Cf. Frank Atkinson, "The horse as a source of rotary power," Newcomen Society, *Transactions, 33* (1960–1), 31–55; Jennifer Tann, "Horse power, 1780–1880," in Francis M.L. Thompson, ed., *Horses in European economic history: A preliminary canter* (London: British Agricultural Historical Society, 1983), 21–30.

42. Ferguson, "The measurement of the 'man-day'," *Scientific American, 225:4* (1971), 96–103. These studies were identical with the earliest attempts to analyze the waterwheel in dynamical terms.

43. In Sweden the problem was discussed in 1800 by Erik Nordwall. He said that a horse was usually considered able to do the work of seven men, but that this was not universally accepted, and he referred to the results of Desaguliers, Bernoulli, La Hire, Bélidor and others. See Erik Nordwall, *Afhandling rörande mechaniquen, med tillämpning i synnerhet till bruk och bergverk, 1* (Stockholm: F.A. Carlbohm, 1800), 95–100.

44. The data was quoted in engineering handbooks throughout the 19th and early 20th centuries. Ferguson, "The measurement of the 'man-day'," 101.

In these 18th-century studies, Eugene S. Ferguson has written, "the most casual and fragmentary data were being worked up, with the help of algebraic operations, into definite and precise conclusions."[45] A case in point, described in more detail by Ferguson in an earlier paper, is a study published by Coulomb in 1798. On the basis of two single observations of physical labor, Coulomb wrote an equation for the useful work done while carrying one load of firewood upstairs. This equation was then differentiated and set equal to zero. Coulomb claimed to have obtained thereby the optimum load that would lead to the maximum day's work.[46]

In an article published in 1744, the Swedish mathematician Pehr Elvius compared the efficiency of four treadmills then operating in Stockholm: one at the new Royal Palace under construction, one at a glass factory, and two at the construction of Stockholm's lock. For each treadmill, Elvius measured the rate at which a weight was raised and calculated that "the output of every single fellow is so great that 4 2/5 lispounds was hoisted at a rate of one foot every second."[47] After discussing the differences in design and output for the four treadmills, he attempted to draw general conclusions concerning the ideal design for various applications.

Though Elvius recognized that quantitative methods could be used to improve an existing technology, his method suffered from two interdependent weaknesses, one mathematical and the other social. He used a single observation for each type of treadwheel ("a time of 4 minutes exactly or 240 seconds"),[48] not an average value. Furthermore, his study was based on the types of treadwheels that could be seen in action in the course of a leisurely stroll through the

45. Ibid., 96.

46. Ferguson, "The science of muscle power, 1700–1970" (unpublished paper read at a Smithsonian Institution seminar in 1970). Cf. Ferguson, "The measurement of the 'man-day'," 99; Charles Augustin Coulomb, "Résultat de plusieurs expériences destinées à déterminer la quantité d'action que les hommes peuvent fournir par leur travail journalier, suivant les differentes manières dont ils emploient leurs forces," Institut national des sciences et arts, *Mémoires*, Sciences mathématiques et physiques, 2 (1798–9), 380–428; reprinted in Coulomb, *Théorie des machines simples* (Paris: Bachelier, 1821), 225–97.

47. Pehr Elvius, "Rön vid trampkvarnar," *KVAH*, 1744, 198–208, on 204.

48. Ibid., 199.

capital. But the large majority of treadwheels were at work in the mines of the countryside.[49]

The opportunity for a systematic study of manpower did not arrive in Sweden until the early 1770s.[50] The place was the naval dockyard of Karlskrona in the southeast, where manual labor was used to operate the pumps. To dry-dock a man-of-war, ninety sailors worked in three shifts of ten to thirteen hours depending on the displacement of the ship. The engineer Johan Eric Norberg studied the efficiency of manual labor during dry-docking on ten different occasions during 1772 and 1773, and published his report in the *Proceedings* of the Royal Swedish Academy of Sciences. His primary aim was to increase the efficiency of the pumping operation, but he also had in mind to determine in general the output that could be expected from human muscle power. This, Norberg wrote, had not been attempted before, save in a very small number of tests of limited scope. The results had fallen short in both reliability and extent.[51] Norberg thus demonstrated his awareness that many tests under varying conditions were required to obtain results of general validity. Norberg took hundreds of measurements of the work of the ninety men. Figure 10.3, which shows one of his tables, illustrates the systematic nature of his approach.

Norberg's study is the earliest example in Swedish technology of a systematic, full-scale investigation intended to yield a result of general

49. The same social limitation is to be found in the early attempts to quantify waterwheels. Parent and his followers in the European scientific academies claimed to have reached results of general validity when, in reality, they had only considered undershot wheels. Reynolds has suggested that a reason for this erroneous assumption was the topographical location of most European academies: in regions of low waterfalls and relatively large water volumes where undershot wheels were preferred. The academicians who usually resided in large cities, situated along big but tranquil rivers, were therefore more likely to consider undershot wheels they could easily observe. Reynolds, *Stronger than a hundred men*, 212.

50. Johan Eric Norberg, "Rön öfver den effect, som af manskap kan användas medelst handkraft, å machiner, som sättas i rörelse genom hvef," *KVAH*, 1799, 51–69. The delay in publication, some twenty-five years, may have been caused either by a lack of interest on the part of the Academy or because Norberg only later found it useful for his career or prestige to publish in the *Proceedings*. In neither case is the point made here menaced.

51. Ibid., 52.

applicability. It is scarcely surprising that Norberg chose the navy as the locus for his study. The military was the only social institution sufficient in size and authority to assemble and command the large work force needed for so ambitious a study of the efficiency of manual labor.[52]

In his study in the 1740s Elvius had discussed the performance of only a few individuals. Even the physical characteristics of the individual—in particular, the length of his stride—was an important parameter in Elvius' calculation (dust jacket illustration). But in Norberg's study of the dockyard, individuals are reduced to figures in a table, and the combined product of ninety men's work expressed as a single quantity. Norberg was certainly aware of influential external factors: "harmony, the ration of spirits, the gentle persuasion of the officers, fair or foul weather, etc., have the most important influence on the work, and make a great difference as to output, which should otherwise be nearly the same for equal weights of water." These more human aspects of work at the pumps were not reproduced in Norberg's table, however, which expressed output as a product of measurable physical quantities. Norberg distinguished "the outputs of three kinds of people in various corps." In the sixth column of the table the letter *V* denotes volunteers (*Volontairer*), *M*, marines (*Marinierer*) and *B*, tenement seamen (*Rote-Båtsmän*). The attempt to obtain average rather than individual values necessitated suppression of individual characteristics in favor of general traits.[53]

THE NEED FOR CONTROL

In their aspiration to apply quantitative methods to technology in the hope of improving its efficiency and economy, the engineers of the 18th century found themselves hampered by their own boun-

52. A later Swedish example is the construction of the Göta Canal connecting the Baltic Sea with the North Sea, which was dug in 1810–32 by a total of 60,000 soldiers who during this time produced seven million man-days of work. Samuel E. Bring, ed., *Göta Kanals historia*, 2 vols. (Uppsala: Almqvist & Wiksell, 1922–30), *2:1*, 454–60.

53. Norberg, "Rön öfver den effect," 54–5.

daries as individuals. They had the ambition to reproduce the conditions of the laboratory in the field: to carry out systematic experiments under controlled and reproducible conditions. But this demanded a much larger spatial and temporal dimension than the laboratory—with its desktop apparatus and time scale of days or weeks. It was beyond the means of individuals to embrace the technical reality in time and space. In terms of *space*, their ambition required wooden structures weighing perhaps up to a ton (whether they were treadmills, charcoal piles, or waterwheels). It made even large spatial demands on the immediate surroundings, since it required such assets as large areas of forest, hundreds of men, or a substantial water supply. In terms of *time*, it required that several people work together on the experiments for several years. In short, considerable resources were needed to build these unwieldy pieces of experimental apparatus and to operate them for a period of years, and these resources were beyond the means of any individual.

This argument can be expressed in another way. To say that the spatial and temporal dimensions of technology demanded considerable resources if the conditions of the laboratory were to be reproduced in the field is the same as to say that *one must be able to control the necessary physical and social realm in terms of time and space.* This chapter began with the claim that efficiency and economy are the two quantities that characterize modern technology, and it will end by asserting that its qualitative characteristic is *control.*[54] Technology is an activity that aims at changes of the material world, and this always involves control.

It is therefore not by chance that the first successful attempt to apply quantitative methods to technology discoverable in Swedish history of technology was achieved at the naval dockyard in Karlskrona in the early 1770s. The case of the ninety soldiers, working in shifts at the pumps to dry-dock a frigate while Johan Eric Norberg stood over them with his watch and notebook, symbolizes a turning-point. It marked the beginning of the successful application of quantitative

54. The idea is not novel since technology as control has been a recurrent theme in modern history and philosophy of technology. It has, however, evolved in the present study from an examination of empirical data in a specific historical context.

methods in technology during the late Enlightenment. At that time, only the military possessed the necessary control over the physical and social realm to reproduce laboratory conditions in the field. It would be some time before civilian institutions grew strong enough to be able to exercise the same degree of control, for example, in the federally supported investigation of the Franklin Institute in the 1830s into the causes of steamboat boiler explosions.[55] These institutions were then strong enough in both *authority* and *competence*.

Although this study has been limited to a specific aspect of the question of the relationship between science and technology during the 18th century—the application of quantitative methods to technology—it may have some relevance for the more general question.[56] Roger Hahn has argued that one way to grasp the relationship between science and technology during the 18th century would be to examine the institutional development of technology, and in particular the many societies of arts that flourished in Europe on the model of the scientific academies. Science made an impact, Hahn wrote, "not only by lending its ideology, personnel and theories to technology, but also by offering its social organization as a model to be copied."[57] By copying the social organization of the scientific enterprise, these institutions "also accepted the presuppositions of science itself: rationality, objectivity and publicity."[58] Studies of the history of

55. John G. Burke, "Bursting boilers and federal power," *Technology and culture, 7* (1966), 1–23; Bruce Sinclair, *Early research at the Franklin Institute: The investigation into the causes of steam boiler explosions: 1830–1837* (Philadelphia: Franklin Institute, 1966). See also a review by Burke of Sinclair, *Technology and culture, 9* (1968), 230–2.

56. Cf. Peter Mathias, *The transformation of England: Essays in the economic and social history of England in the eighteenth century* (London: Methuen, 1979), 45–71, 72–87.

57. Hahn, "The application of science to society: The societies of arts," *Studies on Voltaire and the eighteenth century, 25* (1963), 829–36, on 831.

58. Ibid., 836. Edwin T. Layton has shown how the American engineering community in the 19th century borrowed and adapted the social organization, rather than the results, of the scientific community, and with the effect that the communities of science and technology became "mirror-image twins." See Layton, "Mirror-image twins: The communities of science and technology in 19th century America," *Technology and culture, 12* (1971), 562–80, and "Millwrights and engineers, science and social roles, and the evolution of the turbine in America," in Wolfgang Krohn, Edwin T. Layton, and Peter Weingart, eds., *The dynamics of science and technology: Social values, technical norms and scientific criteria in the development of knowledge*

individual societies, such as Robert E. Schofield's *The Lunar Society of Birmingham*,[59] have demonstrated and elucidated this institutional development, but the question of their effectiveness in achieving technological change continues to puzzle historians. A. Rupert Hall has, for example, remarked that it requires "a degree of faith" to find a causal relationship between the popularity of science on the one hand and innovation in technology on the other.[60] But if the benefit of a union between science and technology was conceived only on a cultural level during the 18th century, why was it not put into effect on a political and economic level?[61]

This study suggests that we should continue our search for institutional developments, but look for projects that were undertaken by institutions rather than individual efforts. The characteristics to look for in these technological institutions are not only the rationality, objectivity, and publicity of their scientific counterparts but also the power to control the necessary physical and social realm in order to reproduce laboratory conditions in the field.

More specifically, we should look for institutions that exercised *authority* according to rank in hierarchies based on scientific and technical *competence*. The quest will lead us to that historical borderland that lies between the inquisitive academies of the 18th century and the efficient industries of the 19th century, to the period between 1790 and 1825, which has been recognized by Cardwell as one of the definite periods of decisive change in the course of technological history.[62] Not only were many of the major technologies of

(Dordrecht: Reidel, 1978), 61–87.

59. Robert E. Schofield, *The Lunar Society of Birmingham* (Oxford: Clarendon Press, 1963), and "The industrial orientation of science in the Lunar Society of Birmingham," *Isis, 48* (1957), 408–15, reprinted in A.E. Musson, ed., *Science, technology, and economic growth in the eighteenth century* (London: Methuen, 1972), 136–47.

60. A. Rupert Hall, "What did the industrial revolution in Britain owe to science?," in Neil McKendrick, ed., *Historical perspectives: Studies in English thought and society in honour of J.H. Plumb* (London: Europa, 1974), 143.

61. Cf. Hahn, "Science and the arts in France," 89.

62. D.S.L. Cardwell, *Turning points in western technology: A study of technology, science and history* (New York: Science History Publications, 1972), 111–2; published in Europe under the title, *Technology, science and history* (London: Heineman, 1972).

the 19th and 20th centuries founded then, but "at the same time social changes took place in the organisation of technology and science setting them on the courses that led to modern technological society."[63]

One of the major social changes was the emergence of institutions that could exercise the necessary control for the successful application of quantitative methods to technology; first within the established structure of the military (military academies, arsenals and dockyards) and later in large industries of national-military importance (mining and chemical industry), major civil-engineering projects (canals), and new civilian institutions. These institutions are probably identical with those that Peter Mathias has called "focal points for developing new skills and educational programs" and that were sponsored by the demands of the state for deploying scientific technology for military or official purposes.[64]

63. Ibid., 112. Cf. Mikulàs Teich, "Born's amalgamation process and the International Metallurgic Gathering at Skleno in 1786," *Annals of science, 32* (1975), 305–40, esp. 338.

64. Mathias, *The transformation of England*, 84.

11 The Calculating Forester: Quantification, Cameral Science, and the Emergence of Scientific Forestry Management in Germany

By Henry E. Lowood

In the second half of the 18th century, few occupational groups rivaled government officials in their attention to numbers. Government officials employed in the duchies, kingdoms, and free cities of German-speaking Central Europe pored over the data on population, imports, and taxes that a growing fiscal apparatus produced in unprecedented volume. Those concerned with the prosperity of the prince and his subjects, from low-level tax assessors to ministers of state, developed an attachment to the quantitative spirit proportionate to the expansion of the state's economic agencies.

Reasons of state and forces of social change brought on the bureaucratization of the state financial apparatus in the 18th century. Rather than dulling their initiative, this bureaucratization created new opportunities for officials, professors, and instructors. In the Age of Enlightenment, the improvement of fiscal administration and resource management was seen as requiring a *science* of state finances, while the proliferation of economic facts and figures raised issues of numeracy and appropriate training for office-holders charged with applying the principles of this new science, which became known as the "cameral sciences" in Germany.[1] The term derived from the *Kammer* (chamber) in which the prince's advisors tradition-

1. "Germany" is understood to mean German-speaking Central Europe, including the German *Kleinstaaten* of the Holy Roman Empire, the Hapsburg dominions, and most of the Swiss cantons.

ally deliberated.[2] The subject matter ranged from economics, finance, and *Polizei* to mining, agriculture, and trade.

First introduced in Prussia at the universities of Halle and Frankfurt an der Oder in 1727, the *Kameral-* or *Staatswissenschaften* were firmly established in the university curriculum throughout Germany by the last third of the century. The call for professional training in cameral science and its gradual emancipation from the faculties of law led to the creation of new professorial chairs and schools for teaching a body of theory and techniques needed for the administration of the state and its domains. It has been argued that these cameral sciences represented a mixed bag of professional training, empirical rules, and warmed-over economic theory.[3] This unflattering characterization overlooks the seminal importance of *Kameralwissenschaft* in subjecting a variety of economic, administrative, and social practices to rational or "scientific" scrutiny.

Forest management was one aspect of state administration thus scrutinized, in order to fit "scattered pieces of knowledge...into systems" and to transform "all sorts of activities previously left to habit...into a science."[4] The glue that held these new systems together was economic rationalization. The forest displayed the size of the task of managing the resources from which the prince of the late 18th century ultimately derived his wealth. Discharging the task forged new links between administration and science. The result was quantification and rationalization as applied to both the description of nature and the regulation of economic practice.

German writers on forestry science in the 19th century were struck by the achievements of their compatriots, which "since the middle of the last century can hardly be sufficiently admired." They

2. For the most recent discussion of the derivation and definition of *Kameralwissenschaft*, see Keith Tribe, "Cameralism and the science of government," *Journal of modern history, 9* (1980), 172–96.

3. See Erhard Dittrich, *Die deutschen und österreichischen Kameralisten* (Darmstadt: Wissenchaftliche Buchgesellschaft, 1974), 1–24, which summarizes the critical literature on cameralism, of which cameral *science* is a subset.

4. [J.M. Bechstein], "Anzeige von der Herzoglich-Sächsisch-Gothaischen und Altenburgischen Societät der Forst- und Jagdkunde zu Waltershausen nebst den vorläufigen Statuten derselben," *Diana, 1* (1797), 424–9, on 485.

touted this example of German cameral science in distinctly national terms: "Compare our literature and the number of our educated foresters to what there is abroad! *The beginnings of forestry science are entirely German.*"[5] Beginning around 1765, dozens of books and articles published in Germany had established principles and practices of sound forest management; few kindred publications appeared in languages other than German for nearly a century.[6] Theories, practices, and instructional models from Germany provided the starting point for every other national effort in forestry science and management until the end of the nineteenth century.

The work of the classical writers of German forestry science, such as Georg Hartig, Johann Heinrich Cotta, Johann Hundeshagen, and Friedrich Pfeil, built upon an established tradition of quantitative approaches to the measurement and regulation of the forest. This debt has generally been neglected in the writings of historians of forestry. The origins of rational forest management in the quantitative "forest mathematics" of the last half of the 18th century constitute the subject of this chapter. It will demonstrate that the first advocates of forestry science quantified in spirit in order to bring profits in practice; in the process, they established a tradition of quantitative resource management.

5. J.Ch. Hundeshagen, *Encyclopädie der Forstwissenschaft, systematisch abgefasst*, 3 vols., 4th ed. (Tübingen: Laupp, 1842), *1*, 6. So, too, would university professors and government officials from Prussia, Saxony, southwestern Germany, and Bavaria dominate the theoretical and technical literature of forest management through the middle of the nineteenth century.

6. The "Bradley Bibliography" lists no non-German title in the literature of forestry mathematics before Adolfo Berenger's "Rudimenti di matematica applicata specialmente alla tassazione ed assestamento delle foreste," published in 1865 and 1866. In the same bibliography, Berenger's article is preceded by forty-four German publications, many of them books reprinted in multiple editions. Alfred Rehder, *The Bradley Bibliography; a guide to the literature of the woody plants of the world published before the beginning of the twentieth century* (Cambridge: Riverside Press, 1911–8), *4*, 57–8.

BETTER MANAGEMENT

As a substantial portion of the prince's domain, forests constituted one of the largest sectors of the state economy in central Europe.[7] Other forested lands in Germany belonged to the cities and the landed nobility and provided indispensable products for the local and regional economies under their control. Wood in one form or another was essential for home heating and construction, iron manufacture, glassmaking, shipbuilding, and other crafts and trades, while secondary products of the forest found applications in myriad occupations, such as tanning and agriculture. Before the age of coal, which would not begin in many parts of Germany until the middle of the nineteenth century, wood was king.

After the acute and widespread devastation and neglect that resulted from the Seven Years' War (1756–63), the state fixed its gaze on economic recovery. The specter of shortages of wood fuel caught the attention of a small group of conscientious foresters and enlightened bureaucrats, who saw evidence that the deterioration of the woodlands, reported here and there since the Middle Ages, had dramatically accelerated. In the Palatinate, for example, a survey of the forests carried out between 1767 and 1776 spoke of "woods in places so ruined that...hardly a single bird can fly from tree to tree."[8] The state of Germany's forests reached its nadir just when rulers like Frederick the Great sought to encourage population growth and force the expansion of industry and trade, measures bound to increase the pressure of demand for wood and other forest products. The fear of impending crisis in the supply of wood lodged in the minds of government officials throughout the remainder of the century, and was periodically intensified by reports of rapidly rising prices.[9]

7. With the prominent exception of the Hapsburg lands and Switzerland, in which much of the best forest land was privately owned.

8. Quoted in Anneliese Sturm, *Die Wälder des östlichen Nordpfälzer Berglands. Die Entwicklung der heutigen Forstwirtschaftsformation aus den Waldwirtschaftsformationen während der letzten 300 Jahre* (Speyer: Verl. d. Pfälz. Gesellschaft zur Förderung d. Wissenschaften, 1959), 182.

9. Ibid., 192. In Baden, for example, the price of lumber nearly doubled between 1750 and 1790. "Beytrag zur Kenntniss der Badenschen Forstanstalten,"

Officials vigorously pursued economy in the use of wood. But redesigning fireplaces, door-frames, and spoons offered help only on a limited scale; to expand that scale would be a tedious undertaking. Better understanding of the nature of combustion and material properties of wood offered some hope for greater efficiency in wood burning, and scattered experimental reports on these matters of forest physics appeared before 1800.[10] The alternative of expanding the wood *supply* promised larger gains. Here a bold innovation might succeed in increasing the amount of firewood and lumber available to an entire town, city, or region. Almost in proportion to the potential payoff, however, the complex problem of proper forest management exceeded the meager qualifications of the vast majority of foresters. As a rule their primary appointments as caretakers, game wardens, and masters of the hunt required neither practical nor theoretical training in forestry. In Prussia, for example—even under Frederick the Great—posts in the forest administration, which carried the revealing title of *Jäger*, served as sinecures for military retirees. In the absence of qualified personnel, how could a new approach to forest management arise?

After the middle of the century, the establishment of private forestry schools and publication of books and even journals devoted to forestry began to raise expectations for the training and competence of future foresters and forestry officials. The last year of the Seven Years' War saw the foundation of the first forestry school (by H.D. van Zanthier, in the Harz Forest), the appearance of the first book to use "forestry science" in its title (Johann Beckmann's *Beyträge zur Verbesserung der Forstwissenschaft)*, and the first journal devoted exclusively to forestry (J.F. Stahl's *Allgemeines oekonomisches*

Journal von und für Deutschland, 8 (1791), 816. Cf. Heinrich Rubner, *Forstgeschichte im Zeitalter der industriellen Revolution* (Berlin: Duncker & Humblot, 1967), 59, for the rapid take-off of wood-fuel prices in Württemberg between 1760 and 1780.

10. Johann Wilhelm Hossfeld, "Etwas über die Heizkraft der Hölzer," *Diana, 4* (1816), 213–44; Johann Leonhard Späth, *Anleitung, die Mathematik und physikalische Chemie auf das Forstwesen und forstliche Camerale nützlich anzuwenden* (Nuremberg, 1797), 454–72; Georg Ludwig Hartig, *Physikalische Versuche über das Verhältnis der Brennbarkeit der meisten deutschen Wald-Baum-Hölzer. Ein Beytrag zur höheren Forstwissenschaft* (Marburg, 1794), 454–72.

Forstmagazin).[11] One of the first points to settle was the very definition of forests. Traditional privileges and the continued use of the forest for such agricultural purposes as grazing or mast (windfall nuts) had long discouraged a conceptually precise demarcation of the forest. Beginning in the 1760s, however, better-trained officials, equipped with publications for the exchange of ideas, promoted the notion that the forest could be defined precisely and studied objectively.

The first writers on forestry science were led by men trained in the cameral sciences—financial officials and chief foresters who expected economic disaster if the condition of the forests continued its downward slide. As these officers of the local prince consolidated their control over state-managed economies throughout Germany, they attended to the forests in their jurisdiction. Where bureaucratization and centralization of political authority extended the official's sphere of action, as in Prussia, forestry science flourished.[12] The year 1757 marked the appearance of the first of many books on forestry geared specifically to cameral officials: Wilhelm Gottfried von Moser's *Principles of forest-economy.*[13] Like other cameral officials, the head forester came to his post after considerable study. Every cameralist learned about forest administration, a subject of acknowledged importance: "First, because they are a considerable source of revenue for the state, and second, because they constitute a vital necessity for the sustenance of its citizens, without which these lands—especially in the north—would hardly be habitable."[14] Cameralist writers such as

11. See Johann Gottlieb Beckmann, *Beyträge zur Verbesserung der Forstwissenschaft, als einen dritten Theil der Versuche von der Holzsaat zum allgemeinen Besten herausgegeben* (Chemnitz, 1763). The entire field of forestry management was represented by a mere handful of titles before 1760. The first book in German devoted to this subject, Carlowitz' *Sylvicultura oeconomica*, appeared in 1713.

12. See Hans Rosenberg, *Bureaucracy, aristocracy, and autocracy. The Prussian experience, 1660–1815* (Cambridge, Mass.: Harvard University Press, 1958).

13. W. Gottfried von Moser, *Grundsätze der Forstökonomie* (Frankfurt: H.L. Brönner; Leipzig, 1757). A few decades later, it was claimed that every "servant of the state" could learn what he needed to know about forestry science at the university. Cf. Friedrich Casimir Medicus, "Ueber Forstschulen in jedem einzelnen Forste," *Unächter Acacienbaum, 1:6* (1796), 588–94, on 589.

14. J.H.G. von Justi, "Von der Aufmerksamkeit eines Cameralisten auf die Waldungen und den Holzanbau," in *Gesammelte politische und Finanzschriften* (Copenhagen and Leipzig: Rothen, 1761), *1*, 439–64, on 439–40.

Georg Ludwig Hartig placed the new forestry alongside the "state sciences," since the two "make up a complete whole."[15]

The new breed of officials trained in cameral science described the living forest quantitatively before subjecting it to economic reason. They brought to the task a familiarity with mathematics. Mathematics figured prominently among the required subjects, especially in the first year or two of coursework, in the university curriculum in the cameral sciences and also in special forestry schools. Published curricula and schedules of lectures consistently featured mathematics as a *Hilfswissenschaft*, both for the work of the future government official and as exercise for his mind. At the Cameral College in Kaiserslautern, for example, mathematics was one of the subjects required of every student, and "empiricists" wishing to proceed straight to practical studies without this preparation were not welcome.[16] Heinrich Cotta's Forest Institute at Zillbach, which originated as a site for private instruction in mathematics during the idle Saxon winters, featured the same progression from theoretical to practical.[17] Forestry had become a "complicated science," and it fell to "patriotic men" to ensure that foresters entrusted with the resources of the state were adequately prepared in this new science.[18]

The program won over skeptics. An anonymous reviewer of one book on mathematics for the forester had questioned whether forestry required its own mathematical literature. Careful reading removed his doubts: forest management presented a set of problems worthy of special attention, which they surely would not receive within the body of mathematical literature. Moreover, the reviewer pointed out,

15. Georg Ludwig Hartig, *Grundsätze der Forst-Direction* (Hadamar: In der neuen Gelehrten-Buchhandlung, 1803), 7.

16. Johann Daniel Suckow, "Zweyter Brief über die hohe Kameralschule zu Lautern," *Der Teutsche Merkur* (1776), 56–67, on 66–7.

17. See Heinrich Cotta, "Nachricht an das Publicum von einer im Herzogl. S. Weimar- und Eisenachischen Forstamte Zillbach seit mehreren Jahren bestehenden und nun erweiterten Anstalt zur Bildung angehender Forstmänner und Jäger," *Kaiserlich-privilegirter Reichs-Anzeiger* (1795), 1617–23, on 1620. On the origins of the Forest Institute, see Kurt Mantel and J. Pacher, *Forstliche Biographie vom 14. Jahrhundert bis zur Gegenwart*, vol. 1. *Forstliche Persönlichkeiten und ihre Schriften vom Mittelalter bis zum 19. Jahrhundert* (Hannover: Schaper, 1976), 242.

18. Hartig, *Grundsätze der Forst-Direction*, 7.

new sciences need to stand on their own feet, and specialized textbooks help to disseminate new rules and procedures and to establish new sciences as independent disciplines.[19]

Writers on forestry presented problems and applications of special techniques, not elementary mathematical instruction.[20] Their goal was to demonstrate how the forester should proceed mathematically, not to produce a new mathematics. With the exception of solutions to a few obscure problems of stereometry and xylometry (measurement of volume and specific gravity of wood), mathematical virtuosity was not necessary. Cotta argued that the "practiced algebraist," to whom calculating the value of a forest was a trivial exercise, would not be the least bit interested in applying his art to it. Cotta also knew that most foresters, unencumbered by such mathematical sophistication, were likely to faint at the slightest scent of a mathematical problem.[21] A reviewer of another early book on the mathematics of forestry concurred: "[the author] demands from the forester planimetry, stereometry, trigonometry, levelling, transformation of figures, third-order and second-order equations. Terrible demands for most foresters!"[22] A prominent advocate of forestry schools argued that one cannot make "great scholars out of uneducated people."[23] But one could turn trees into thalers by replacing the time-worn "routines" of the old *Jäger* with *Forstwissenschaft*,[24] it was generally agreed.

This approach was decidedly German. Reforms under Louis XIV had resulted in *plans de forêts* for state-owned forests and promoted the concept of dividing the forest into annual cutting areas. Jean-Baptiste Colbert's ambitious plan for improving France's forests in 1669 had prompted new statutes, administrative reorganizations, and inventories throughout the 18th century. But a scientific forest management did not take root in France until it was imported from

19. *Allgemeine Deutsche Bibliothek, 12:1* (1770), 320.

20. Späth, *Anleitung*, v.

21. Heinrich Cotta, *Systematische Anleitung zur Taxation der Waldungen*, 2 vols. (Berlin, 1804), v.

22. *Allgemeine Deutsche Bibliothek, 2:2* (1766), 280.

23. Medicus, "Ueber Forstchulen in jedem einzelnen Forste," 589.

24. "Wissenschaftliche Staatsdiener;" Hartig, *Grundsätze der Forst-Direction*, 17.

Germany in the 1820s.[25] English authorities, ignoring such expressions of concern as John Evelyn's *Sylva* (1664), did not even inventory the remaining forests until the founding of the Board of Agriculture in 1793. As late as 1885, select committees in Parliament debated the merits of emulating the German model of forestry schools and forest science.[26] In Switzerland and Austria, government officials exerted control over a lesser proportion of the forests than did their counterparts in Prussia and Saxony. Moreover, the physiocratic doctrine fashionable in late 18th-century Vienna and Bern offered a rationale for avoiding the problem by selling off woodland and converting it to farmland.[27]

Doing the Work

In central Germany, particularly in Hesse and Saxony, a few foresters had applied the same enthusiasm to managing the forest as to directing the hunt. These conscientious *holzgerechte Jäger* of the mid-century set annual cuttings according to easy rules based on areal divisions of the forest. After demarcating and measuring the acreage covered by the woods under their supervision, foresters estimated the number of years that the dominant types of trees should be allowed to grow between clearings or cuttings. They then partitioned the forest into a number of divisions equal to the number of years in this growth cycle, from which they proposed to derive equal annual yields, assuming that equal areas yield equal amounts of wood for harvest each year. This straightforward method worked reasonably

25. Rubner, *Forstgeschichte im Zeitalter der industriellen Revolution*, 44–9.

26. N.D.G. James, *A history of English forestry* (Oxford: Blackwell, 1981), 167–72, 194: until 1900, "English forestry had chiefly consisted of allowing trees to grow and then felling them, and forest management, as it is known today, was non-existent." Cf. John Croumbie Brown, *Schools of forestry in Germany, with addenda relative to a desiderated British National School of Forestry* (Edinburgh: Oliver and Boyd, 1887). On the conflict between economic forestry and the English romantic affection for trees, see Keith Thomas, *Man and the natural world: A history of the modern sensibility* (New York: Pantheon Books, 1983), 212–23.

27. On Switzerland, see Heinrich Grossmann, *Der Einfluss der ökonomischen Gesellschaften auf die Entstehung einer eigentlichen Forstwirtschaft in der Schweiz* (Bern: Bücher, 1932), 14–9; on Austria, see Rubner, *Forstgeschichte im Zeitalter der industriellen Revolution*, 71–5.

well for relatively short growth periods typical of coppice farming and the periodic clearing of underwood. It permitted limited variations, such as shelterwood (*Schirmschlag*) or relative cutting (*Proportionalschlag*), in which the harvest from a given section of the forest or the size of individual sections could be adjusted according to soil quality and other contingencies.

These methods may have sufficed for a minimally trained huntsman, but not for the fiscal or forest official imbued with *Wissenschaft*. The crude assumptions underlying the traditional areal division of the forest proved wholly unsatisfactory for the cash crop of forestry—the long-lived high timber, or *Hochwald*; the older the trees, the greater the variation in the timber produced by each of the divisions of the forest. Furthermore, the irregular topography and uneven distribution of German woodlands confounded ocular estimation of area without the aid of instruments. Only in the 1780s did Johann Peter Kling, chief administrator of forests in the Palatinate and Bavaria under Elector Karl Theodor, systematize forest mensuration and cartography into instructions for making forest maps of unprecedented detail.[28]

28. On Kling's work, see Erich Bauer, "An der Wiege der deutschen Forstwissenschaft: Forstgeschichtliches aus Kaiserslautern und der Kurpfalz im letzten Drittel des 18. Jahrhunderts," *Jahrbuch zur Geschichte von Stadt und Kreis Kaiserlautern, 3* (1965), 101–17, on 107–13; Franz Tichy, "Die kurpfälzische Waldstandortskartierung von 1783," *Berichte zur deutschen Landeskunde, 20* (1958), 320–6; Johann Keiper, "Hofkammerrat Johann Peter Kling, kurfürstlicher Forstkommissar zu Mannheim," *Mannheimer Geschichtsblätter, 30* (1924), 138–47. The first text on forest surveying was J.F. Penther's *Praxis geometriae* (1729), which was reprinted as late as 1788, but a new wave of manuals and books on this subject appeared after 1770, stimulated by Karl Christoph Oettelt, *Practischer Beweis, dass die Mathesis bey dem Forstwesen unentbehrliche Dienste thue* (Eisenach, 1765; also 1786, 1798, 1803); and Johann Ehrenfried Vierenklee, *Mathematische Anfangsgründe der Arithmetik und Geometrie, in so fern solche denjenigen, die sich dem höchstnötigen Forstwesen auf eine vernünftige und gründliche Weise widmen wollen, zu wissen nöthig sind* (Leipzig, 1767; 1797, 1822). An example of the new wave is *Instruction für die Land-Messer, welche zu denen, von Sr. Kgl. Majestet allerhöchst verordneten speziellen Vermessungen derer sämmtlicher Herrschaftlicher Heyden und Holtzungen in Ost-Preussen und Lithauen, sollen gebraucht werden* (Berlin, 1771). Cf. "Entwicklung der forstlichen Literatur in Deutschland. Zugleich als Überblick über die Entwicklung der Forstwissenschaft vom Ende des Mittelalters bis zur klassischen Zeit," in Kurt Mantel, ed., *Deutsche forstliche Bibliographie* (Freiburg i. Br.: Universität, Forstgeschichtliches Institut, 1970), iii–xlviii, on xxxiii.

Other fundamental problems also plagued area-based forest management. First, a division of the forest into equal cutting areas did not provide the most useful information to those responsible for fiscal planning and management. They needed to know the *amount* of firewood or lumber. Correlation of acreage with actual distribution of lumber and firewood required principles not formulated and measurements not routinely executed under the old forestry. Second, the prudent forester could not easily respond to inevitable quirks of nature over the many decades in a single forest cycle, because the area-based system did not provide a flexible method for directly adjusting the harvest from year to year, let alone predicting annual yields over the long cycle from the outset. The most meticulous forest management under these methods, while an improvement over neglect, fell short of the high principles of *Kameralwissenschaft*.

After midcentury, an approach to forest economy based on the mass or volume of wood gradually displaced area-based systems. The first prominent advocate of wood-mass as the quantitative basis for sound forestry emerged from the *holzgerechte Jäger*. Johann Gottlieb Beckmann, a forest inspector in Saxony, gave the forest priority over the hunt; his knowledge of forestry derived from experience, not education. Beckmann's deep concern for preserving the wood supply led him to construct a system of forest economy that rested on a practical technique for measuring the quantity of standing wood in the forest. Beckmann instructed his team of assistants, whom he supplied with birch nails of various colors, to walk side by side through the forest at intervals of a few yards. Each member of the formation fixed his gaze to the same side and noted every tree he passed. He made a quick estimate of the size category in which the tree fell and marked it with a nail of the appropriate color. At the end of the day, unused nails were counted and subtracted from the original supply to indicate the number of trees in each category. The forester and his assistants knew from experience the approximate yield of wood from trees in each size category; with multipliers thus assigned, the number of nails used could be converted through a simple calculation into the quantity of standing wood in the forest. Beckmann's case suggests that the clever quantifier need not be a calculator or mathematician nor carry out detailed measurements or stereometric calculations in order to determine the mass of wood. A vigorous and productive author, Beckmann began around 1760 to campaign for the method of forest economy based on wood mass. Soon *Beckmannianer* sprang up throughout Germany to propagate his ideas.[29]

Within a few years, a group of mathematically adept foresters followed along the trail cleared by Beckmann. Carl Christoph Oettelt, Johann Vierenklee, and Johann Hossfeld assigned the task of measuring the area of the forest to the *Forstgeometer*, a surveyor hired to demarcate the borders of the forest, prepare maps, and carry out other prescribed tasks for a set fee.[30] The geometer, along with the army of marching assistants, gathered the data. *Forsttaxation*, or forestry assessment—a mix of calculation, analysis, and planning—fell to the chief forester and his superiors. Forest mathematicians like Oettelt and Vierenklee were moved by a new confidence in the power of mathematics to solve problems associated with the conversion of the forest into an equivalent quantity of wood mass. Assessment, the scientific component in *Forstwissenschaft*, required general principles and techniques based on them. Without them the unrelated numbers and observations reported by foresters and surveyors would overwhelm planners and administrators. Forestry science supplied the necessary organizing principle: "evaluation, or the ascertaining of the mass of wood, which is to be found for a given place at a given time."[31] Identifying wood mass as the crucial variable of forestry set the stage for quantitative forest management.

29. For Beckmann's method, see August Bernhardt, *Geschichte des Waldeigenthums, der Waldwirthschaft und Forstwissenschaft in Deutschland*, 3 vols. (Berlin: J. Springer, 1872–5), 2, 88–90. On his life and publications, see Richard Alexander Hess, *Lebensbilder hervorragender Forstmänner und um das Forstwesen verdienter Mathematiker, Naturforscher und Nationalökonomen* (Berlin: P. Parey, 1885), 12–3; Mantel and Pacher, *Forstliche Persönlichkeiten*, 42–4. For a similar method featuring a "Linie von Jägern," see Johann Gleditsch, *Systematische Einleitung in die neuere aus ihren eigenthümlichen physikalisch-ökonomischen Gründen hergeleitete Forstwissenschaft*, 2 vols. (Berlin: A. Wever, 1774–5), 2, 1142–3.

30. See Georg Ludwig Hartig, *Neue Instructionen für die Königlich-Preussischen Forst-Geometer und Forst-Taxatoren, durch Beispiele erklärt* (Berlin: In Commission bey der Kummerischen Buchhandlung, 1819), Table E, "Taxe für die Forst-Geometer."

31. Späth, *Anleitung*, 195.

Counts to Calculation

Theoretical computations of tree volume began to appear in the 1760s. In the first definitive work of scientific tree measurement (*Holzmesskunde*), Carl Christoph Oettelt's *Practial proof that mathematics performs indispensable services for forestry*, the problem of estimating the quantity of wood on a tree without felling it figured prominently. Oettelt was an experienced surveyor and had held the title of "Forest-Geometer" in the civil service of Saxony-Gotha before taking over the forest department in Ilmenau, where he would later serve under Goethe. In the *Practical proof*, Oettelt criticized the crude techniques commonly used to estimate the quantity of wood.[32] Most foresters used the so-called *Brustständke*, or a stack of wood piled to chest height, to veil their wild guesses as to how many boards a tree had delivered. Estimating in this way, they commonly made the value of a tree proportional to its diameter. Heinrich Wilhelm Döbel, one of the most conscientious writers on forestry around 1750, exemplified the problem. In his influential *Gamekeeper's practicum*, Döbel struggled to find a simple computation for the problem—in fact, relatively easy—of estimating the volume of a felled trunk.[33] Oettelt invoked geometry: "A tree is the same as a cone with a circular base." With the appropriate formula for the volume of a cone, calculating the volume and mass of trees was not so troublesome.[34]

Oettelt's treatment of wood mass as a mathematical quantity was a radical departure. The *holzgerechte Jäger* had shown little potential for forest geometry. Döbel argued vehemently that exact calculations of wood mass were unnecessary, "since you don't measure wood like

32. Oettelt, *Practischer Beweis*, 79.

33. Heinrich Wilhelm Döbel, [*Neu*] *Eröffnete Jäger-Practica, Oder der wohlgeübte und erfahrene Jäger* (1913 reprint of Leipzig: J.S. Heinsius, 1754; other editions in 1754, 1783, 1828), 679–701. Oettelt's example is described in Adam Schwappach, *Handbuch der Forst- und Jagdgeschichte Deutschlands*, 2 vols. (Berlin: J. Springer, 1886–8), *1*, 559–60.

34. Oettelt, *Practischer Beweis*, 82. His *Abschilderung eines redlichen und geschickten Försters zum allgemeinen Besten* (Eisenach, 1768) continued in the same spirit. Oettelt was not the first to apply geometrical and stereometrical analysis to trees. Abraham Gotthelf Kästner worked out simple formulas for the volume of logs as early as 1758; cf. *Anweisung zu der Messkunst der Höhe und Dicke des Holzes* (Frankfurt a/M, 1758), reviewed in *Allgemeine Deutsche Bibliothek, 2:2* (1766), 278.

you do gold." He preferred the simple "farmer's calculation" to disputations and proofs.[35] The mathematically oriented foresters, among them Johann Vierenklee and Carl Wilhelm Hennert, joined Oettelt's cause. They corrected and improved his geometric calculations in a series of books that culminated in 1812 in the definitive work on forest stereometry by Johann Hossfeld.[36] As abstract, mathematics-based forestry gained sway during the 1780s and 1790s, compilations of tables based on controlled measurements replaced the older crude techniques described by Oettelt.[37]

Those who compiled such tables had to bridge the gap between tree conics and precise measurement. Consider the problem of converting from cubic measures of wood mass to *Klafter*, the unit of stacked cordwood familiar to the forester, and back again. The interstices and warping of real wood might defeat the most exact geometrical analysis of its volume. Since mass or volume constituted the central quantity of the new forestry science, small errors due to branches, warped stocks, and imperfections of nature multiplied rapidly as one reasoned from the tree to the forest. Equating the economic measure—volume of stacks of hardwood—and the computed volume did not work out.

35. Quoted in Schwappach, *Handbuch, 1*, 560, from Heinrich Wilhelm Döbel, *Eröffnete Jäger-Practica* (1754).

36. Johann Wilhelm Hossfeld, *Niedere und höhere praktische Stereometrie; oder kurze und leichte Messung und Berechnung aller regel- und unregelmässigen Körper und selbst der Bäume in Walde* (Leipzig: Weidmann, 1812). Cf. variants in F.W. Herzberg, "Versuch aus dem Umfange der Bäume scharfkantige Bauhölzer leicht und richtig zu berechnen," Patriotische Gesellschaft in Schlesien, *Neue oekonomische Nachrichten, 2* (1781), 157–9; and G.W. Schäfer, "Praktischer Erfund des wahren körperlichen Inhalts der Nadelbäume und gründlichen Verhältniss der Stammklafter zur Scheitklafter," *Diana, 3* (1805), 333–62.

37. Vierenklee, *Mathematische Anfangsgründe*, 2d ed. (1797), 670; Carl [Karl] Wilhelm Hennert, *Anweisung zur Taxation der Forsten nach den hierüber ergangenen und bereits bey vielen Forsten in Ausübung gebrachten Königl. Preuss. Verordnungen*, 2 parts (Berlin and Stettin, 1791–5; 2d ed., 1803; 4th ed., 1819), *1*, 205. Cf. Späth, *Anleitung*, 196. *Kubiktabellen* were published in Giessen in 1787. Another important measurement was the height of a standing tree. In order to make quick and reasonably accurate measurements of the height, as well as the thickness or cross section of the trunk, specially designed calipers were invented and improved toward the end of the 18th century. Cf. Schwappach, *Handbuch, 1*, 561.

The quantifiers, beginning with Oettelt and Hennert, searched for scientific sandpaper to achieve a greater semblance of precision. Oettelt measured as accurately as possible the volume of the cord, then ordered the wood chopped into small pieces. The volume of each piece could be measured with greater accuracy. He summed these individual measurements, and compared the sum to the original cord. After repeated tests he determined that a typical span of cordwood measuring approximately 110 cubic feet contained 14 to 18 cubic feet of empty space, about 15 percent of its volume.[38] Hennert borrowed Diogenes' barrel: he poured water into a box filled with wood; the volume of the box less the volume of the water yielded the solid content of wood (*Derbgehalt*).[39] By 1812 Hossfeld, in his *Lower and higher practical stereometry* (1812), had replaced Hennert's water with sand and contrived even more accurate xylometers. Such innovations made feasible "measurement and calculation of all regular and irregular bodies, and especially trees in the forest."[40]

In the German tradition, the mathematician's forest was populated not by the creations of undisciplined nature, but by the *Normalbaum*. Forest scientists planted, grew, and harvested this construct of tables, geometry, and measurements in their treatises and on it based their calculations of inventory, growth, and yield. Writers and instructors gave foresters in the field the tools for reckoning the dimensions of the standard tree. Most treatises contained instructions for averaging measurements made on a test plot, but foresters were happier to use the *Normalbaum*. Tables of numbers representing measurements and calculations, or *Erfahrungstabellen*, provided data organized by classes of trees under specified conditions. A small number of variables governed the forester's choice of one or another of these tables. For example, the wood mass of the typical sixty-year-old pine on good soil was given as a function of its height and circumference. These tables, which appeared in every complete manual of rational forestry

38. Oettelt, *Practischer Beweis*, 96.

39. Hennert, *Anweisung zur Taxation der Forsten, 1*, 214.

40. From the title of Hossfeld, *Niedere und höhere praktische Stereometrie*. Cf. Franz Adolf Gregor von Baur, *Die Holzmesskunde*, 3d ed. (Berlin: P. Parey, 1891), 100–30.

practice, generally did not bother with regional variation, the bugaboo of 18th-century agricultural treatises.

By the end of the 18th century, German writers on forest management had worked out steps for determining, predicting, and controlling wood mass. Heinrich Cotta presented the clearest and most widely read exposition of these steps in his *Systematic instruction for the assessment of woods*, published in 1804; they were elaborated in his *Directions for the organization and assessment of the forest*, which appeared sixteen years later. Cotta's first book, which consisted of lectures originally prepared for students attending the forestry school under his direction, was an example of systematization induced by the necessity of teaching. In his method, the "geometric survey" of the woods supplied the *Taxator* with information about the extent of his forests. The next step required calculations of wood mass of individual trees, then of stands, and finally of the forest as a whole; growth rates were computed for each level of organization.[41] Finally, Cotta's forester qua cameralist linked the forest balance sheet to the monetary budget by determining the value of the yield.[42]

If the standing forest is capital and its yield is interest, the forester can complete the chain of conversions from wood to numbers to units of currency: an estimate for the worth of the forest can thus be used to predict income, calculate taxes, assess the worth of the forest, or determine damage to it resulting from a natural disaster. For Cotta, the fundamental problem of forestry management was determining the "standing value" of a forest, given uninterrupted maintenance costs and full harvest some 100 to 150 years hence.[43] Cotta's forestry science thus consisted of sound methods for inventory and prediction: "From summary investigations based entirely on verified

41. Heinrich Cotta, *Anweisung zur Forst-Einrichtung und Abschätzung durch ein ausgeführtes Beispiel* (Dresden, 1820), iv.

42. Cotta, "Regulierung eines Waldes zu forstwiss. Etat," in *Systematische Anleitung*, 5–7.

43. Ibid., esp. v–vii, 3–6. Cf. Cotta, *Anweisung zur Forst-Einrichtung.* These uses for the taxation were also emphasized in Pfeil's text in 1843, among others, even after the emphasis had long since passed to privately held forests, as in Pfeil's system of forest assessment. Cf. Friedrich Wilhelm Pfeil, *Die Forsttaxation in ihrem ganzen Umfange*, 2d ed. (Berlin: Veit, 1843), the same as his *Neue vollständige Anleitung zur Behandlung, Benutzung, und Schätzung der Forsten,* 3d ed., part 5.

judgment, we go through various stages to more exact investigations, first of individual trees, then of the supply, growth, and yield-determination of individual stands, and finally of whole forests."[44] Similar procedures, from the forest to the tree and back again, also appeared in practical manuals such as Georg Hartig's *New instructions for the Royal Prussian forest-geometers and forest-assessors.*[45]

In one respect, Cotta differed from Oettelt's line. He preferred careful ocular estimates based on tables to geometrical deduction, which he not only considered impractical in the field, but also inaccurate, since branches and other irregularities confound the comparison of trees to cylinders and cones. For Cotta, the only absolutely sure method was to chop up a tree and measure its volume (or mass) in the same unit of measure to be adopted in the taxation itself.[46] This view did not weaken his allegiance to mathematical forestry. He was skeptical only of geometrical estimates, not of quantification.

The *Forstwissenschaftler*, and particularly Cotta, championed use of "experience tables."[47] Their use reinforced the notion of a forest filled with standard trees. The forester was to instruct his assistants in the use of these tables so that a mental picture of a tree encountered in a forest corresponded to an entry in the tables. With sufficient repetition, a good forester could make an instant association from the mental picture triggered by the tree to the value of the wood mass contained in the table. The next step was to generalize: every tree of the same height has the same mass (or volume). The standard forester was trained to find the standard tree. For Cotta, the "eyeball measure" could displace the "measuring hand" if every forester learned to see the archetypical.[48] The practiced eye could indeed attain this mechanical perfection, "as subsequent measurements and calculations prove[d]."[49]

44. Cotta, *Anweisung zur Forst-Einrichtung*, iv.
45. Hartig, *Neue Instructionen.*
46. Cotta, *Systematische Anleitung*, 118–9.
47. Ibid., 123. See also the chart on 124.
48. Cotta, *Anweisung zur Forst-Einrichtung*, 84.
49. Gleditsch, *Systematische Einleitung, 2*, 1146–7.

The head forester thus trained his assistants to internalize *Erfahrungstabellen* and become computers of wood mass. He remained at his desk manipulating the *Normalbaum* and numerical data based on local measurements. He could produce his own tables if necessary; according to Hartig, the *Taxator* was responsible for all "mathematical preliminaries" of forest assessment—determining growth rates, preparing maps and calculating tables—before delegating to his staff routine measurements and the mechanical application of tables.[50] The assistants marched, tallied, catalogued, and marked under the watchful gaze of their supervisor, who—according to Hartig's directions—never counted with them. Instead, his duty was to "dictate principles, record the results in the Assessment-Register, and make sure that there are no mistakes."[51]

By 1800, the forest assessor trained in the cameral sciences specialized in theoretical principles, mathematical preliminaries, and the cumulation and analysis of data, a far cry from Beckmann with his colored nails and squad of assistants. An array of numbers stood for the quantity of wood in the forest. The forester or cameralist trained in forestry science felt no need to step off every acre with the exactness given to the test plot, the geometrical abstraction, or exact measurements of the volume of cordwood. Instead, he could sample and generalize. The work of the assessment and management of the forest thus required only standard trees and *Erfahrungstabellen*. As Cotta argued, the crucial quantities of his science were "determined mathematically" from the "premises" of forestry science, not through "*direct real measurement*."[52] The scientific forester had abandoned Beckmann's empiricism in favor of "sure mathematical deductions, experiments and experiences in the given and understood units of measure."[53] Under the banner of *Wissenschaft*, the new breed of qualified forester breathed the quantitative spirit into administrative practice.

50. Hartig, *Neue Instructionen*, v.
51. Ibid., viii.
52. Cotta, *Systematische Anleitung*, 5–7, 117.
53. Ibid., 153.

PRINCIPLES

By the end of the 18th century, the new breed of foresters in Germany, those with diplomas from forestry schools or degrees from the university, adopted the methods popularized in the clear prose of writers like Hartig and Cotta. Their appetite for a rational synthesis of calculation and cameralism was whetted by identification of mass and yield as suitable quantities to measure. As in Lavoisier's chemistry, new fundamental measures required new terms of analysis. By 1800, the ideal of the "regulated forest" proclaimed the preservation of the forest's maximum yield under a sound system of forest economy. Three *regulae silvarum* found throughout the writings of the *Forstwissenschaftler* linked the desideratum of the regulated forest and the methodological focus on measurement and calculation: "minimum diversity," "the balance sheet," and "sustained yield."

Minimum Diversity

Direct measurements of wood mass or volume would have provided the forester with the data he needed for determining fellings or predicting monetary yield, but such numbers were hidden in the diversity and complexity of vegetation in the forest. New units of analysis gave categories better suited to forest computation than the vast, green sea of individually appreciated trees: the "standard tree" (*Normalbaum*), the "size class" (*Stärkeklasse*), the "sample plot" (*Probemorgen*), and the "age class" (*Periode*, *Altersklasse*), as used in textbooks and instructions.

Johann Wilhelm Hossfeld typifies the *Forstmathematiker* as leveler. His precocious fondness for mathematics, combined with an argumentative temperament, made him unpopular with his teachers; he turned the tables and became an instructor of mathematics at schools specializing in commerce and forestry at Eisenach, Zillbach (under Cotta), and Dreissigacker, where he finally settled in 1801 with the title of *Forstkommissar*. Here he moved his mathematical skills from the lectern to the forest. Hossfeld made his name among foresters as a leading proponent of stereometrical and geometrical methods in the determination of wood volume and as the inventor of methods to calculate the value of the forest. His writings are a train of mathematical exercises, with solutions. Hossfeld worked his way from the volumes of cubic forms representing ideal tree trunks through growth, yield, expected demand, costs, and the budget to an all-encompassing "integral of all results pertaining to the value of a forest."[54] He defended his mathematical approach on the basis of economy of effort: a purely empirical assessment counting every tree in a forest might take one observer several years, whereas a mathematician could produce a useful formula after a dozen or so careful observations. Nature "makes no leaps," he claimed, so that a series of multiplied averages based on one or two easily observed characteristics, such as the height of a stand of trees, is as good as an exact and painstaking summation of all the individual cases.[55] The mathematician need not fear hidden pockets of diversity.

Minimizing nature's diversity and reconstructing the forest to make life easier for foresters and assessors were typical of the authoritative writings of the *Forstklassiker*. Hartig advocated strict adherence to results drawn from a few sample plots. He recommended that the forester keep things simple by following a small number of general rules and reliable methods. With characteristic dogmatism, Hartig ruled that one should always cut out "arbitrary" details of nature that might distract from the systematic *Taxation*.[56] Cotta agreed with Hartig on the need to ignore disparate details and concentrate on useful numbers derived from a sample plot. Cotta argued that selective measurements generate acceptable values for quantities like typical yield or growth, which then become the characteristics of ideal types presented in tables and other summations and

54. Johann Wilhelm Hossfeld, "Vollständiges System zur Taxation der Hölzer und Regulierung der Forste," *Diana, 3* (1805), 91–226, on 96 and passim; "Integral aller Resultate über den Werth des Waldes," 113. For Hossfeld's biography, see Hess, *Lebensbilder*, 162.

55. Hossfeld, "Beantwortung der Frage: wie viel Mathematik auf einer Forstakademie gelehrt werden müsse?" *Diana, 4* (1816), 260–78, on 260. Hossfeld developed the Reductionszahl to facilitate such calculations: the simple ratio of the volume of wood he calculated for a model tree of given height and crown to a rectangle of lumber with the same height and base. This constant and the size and constitution of the stands in a forest permitted the calculation of the volume of wood in an entire forest. Ibid., 267–8.

56. Hartig, *Grundsätze der Forst-Direction*, 47.

multiplications of data from test plots: "the assumed quantum of growth is really abstracted from many trees of the same kind; the sum of the whole is always the basic measure."[57] These sums cannot be directly measured by any practical method; they can only be determined "mathematically, that is, with the aid of an inference based on single values that are known."[58] The source of the values did not really matter; measurements in sample plots, geometrical deductions, or experience-tables were equally acceptable if the method produced a standard—the "single value." One need not worry about the cumulation of errors; individual differences cancel out in the aggregate.[59] This assumption brought freedom from the need to poll every tree, without increasing the risk of error.[60] The new science rewarded the forester who did not see the trees for the forest.

The Balance Sheet

Although cameralists had in common with forest scientists a faith in numbers as worth a thousand words of old forestry, their underlying assumptions differed. Oettelt, Hossfeld, and Cotta saw management as dependent on mathematics, not the reverse: "the workings of nature and mathematical truths do not subjugate themselves to words of authority."[61] Even kings and ministers had to bow to this ruler of the kingdom of reason.

Officials in the fiscal bureaucracy with broader responsibilities than the forester's showed less enthusiasm for the ultimate rule of mathematics in forestry science. They clearly appreciated numbers as the rudimentary facts of accurate inventory and accounting. Sophisticated forest management provided efficient tools for monitoring the quantities that the state bureaucracy sought to control from year to year. If expressed coherently in numbers, represented clearly in charts

57. Cotta, *Systematische Anleitung*, 166.
58. Ibid., 116.
59. Cotta, *Anweisung zur Forst-Einrichtung*, 58.
60. Ibid., 79–80; Cotta, *Systematische Anleitung*, 123.
61. F.A.L. von Burgsdorf, "Abhandlung von den eigentlichen Theilen und Gränzen der systematischen, aus ihren wahren Quellen hergeleiteten, Experimental- und höhern Forstwissenschaft," Berlinische Gesellschaft naturforschender Freunde, *Schriften, 4* (1783), 123.

and tables, and placed in the hands of the cognizant minister at court, these vital signs eased the task of keeping the body economic healthy, much as a thermometer aids the physician. To the cameralist, the role of quantification in forestry science was descriptive, not prescriptive.

A common denominator nonetheless related the disparate values that scientists and cameralists attached to quantitative information. The annual accounting of the bureaucrat had to be linked with a long-term plan of resource management based on scientific principles. One prominent *Forstwissenschaftler*, Friedrich von Burgsdorf, called the common problem "keeping the forest's books," and defined procedures to follow in terms of the quantities of interest to forestry science.[62] The bond between forestry science and cameralism was the conversion from an amount of wood to its value. From that point, the practitioners could go their separate ways, the cameral official to the preparation of the *Geld-Etat*, or monetary budget, and the forestry scientist to the *Forst-Etat*, the budget that compared the yield to what the forest could bear over time.

Hartig described the task of creating the *Forst-Etat* as seeking an equilibrium, as opposed to the bottom line in a fiscal budget. "Where a sure balance sheet of forest use, based on mathematics and natural philosophy, is lacking, wood will always be over- or underutilized."[63] In the former case, balance would have to be restored through conservation, raising more land for the forests, or abandoning a less vital productive arm of the economy; in the latter, by exporting lumber or founding new industries. Hartig used terms like "forest use budget" and "natural forest budget" to describe the related components of planning and biological growth that concerned the forester in his effort to balance supply and demand.[64] Hossfeld likewise spoke of budgets and balances. He explicitly identified forestry assessment with the process of evaluating disturbances to the equilibrium of the forest, whether natural (fires and pests) or artificial (management). After calculating the magnitude of these disturbances, the forester

62. Ibid., 113.
63. Hartig, *Grundsätze der Forst-Direction*, 64.
64. Ibid.: e.g., "Forstbenutzungs-Etat," 65, 86; "Natural-Forst-Etat," 144.

could prescribe means for restoring the equilibrium of growth and yield over time.[65] The image of the budget, whether of nature or gold, linked forestry, cameralism, and quantification, as foresters learned to manage both the *Forst-Etat* and the *Geld-Etat* according to the books.

As we have seen, the books themselves consisted largely of numbers. Hartig wrote hundreds of pages on the gathering of data, calculations, and organization of charts and tables necessary for the production of ledgers; the charts mimicked the columnar arrangement of the accountant's books.[66] Hartig and Cotta both offered book-length examples of their methods of forest bookkeeping, complete with templates for the tables they had used.[67] In general, journals and records kept by low-level foresters were to be turned in quarterly to the supervising forester in each district, who compiled and summarized. A *Forst-Rentmeister* would calculate the monetary budget from these and parallel records according to prescribed forms, while the Forest Commission, consisting of higher financial officials, would review, analyze, and summarize. According to Ernst Friedrich Hartig, Georg's younger brother and colleague, the results concerning consumption, production, and distribution of wood could thereby be arranged so that "the balance in every forest, district, administrative region, and province can be easily reviewed at a glance."[68] The recurring themes of equilibrium and the balance sheet harmonized with those of administrative convenience and scientific resource management.

Sustained Yield

The third quantitative principle in German forestry science was sustained yield (*Nachhaltigkeit*). Chopping down enough trees to meet immediate needs satisfies the balance sheet. The bureaucratic

65. Johann Wilhelm Hossfeld, "Vollständiges System zur Taxation der Hölzer und Regulierung der Forste," *Diana, 3* (1805), 91–226, on 108.

66. See Hartig, *Anleitung zur Berechnung des Geld-Werthes eines in Betreff seines Natural-Ertrages schon taxirten Forstes* (Berlin, 1812), and *Neue Instructionen.*

67. See Hartig, *Neue Instructionen*; Cotta, *Anweisung zur Forst-Einrichtung.*

68. Ernst Friedrich Hartig, *Die Forstbetriebs-Einrichtung nach staatswirthschaftlichen Grundsätzen* (Kassel: In der kriegerschen Buchhandlung, 1825), 80.

annual cycle and associated methods in forestry management deal with immediate and short-term record-keeping and assessment. Year after year, cuttings reduce the wood mass according to ephemeral prices, needs, and the conditions of nature. All can be precisely measured and monitored. But the life of individual trees, let alone the forest as a whole, contains dozens and dozens of annual cycles. Long after the incompetent forester is gone, his mismanagement and irresponsibility survive. As Johann Matthäus Bechstein proclaimed in 1801 to students entering his forestry school, the forester must be capable of calculating "more than one or two generations into the future."[69] Planning the growth, cutting, and replenishment of a forest over the *longue durée* requires an idea more powerful than the balance sheet. Foresters found it in sustained yield.

The rudimentary concept of sustained yield appeared in one of the earliest texts on forestry mathematics, Johann Ehrenfried Vierenklee's *Mathematical first principles of arithmetic and geometry, to the extent they are needed by those who wish to devote themselves to the most necessary subject of forestry,* which appeared in the first of three editions in 1767. Vierenklee judged that the forester must know "how to divide up a forest into a definite number of annual cutting areas, from which he obtains a definite amount of wood each year."[70] Vierenklee relied on mathematics for the formulas to achieve this division, and based his work on growth calculations for high timber.

A full generation of *Forstwissenschaftler* later, sustained yield figured as the cornerstone of Hartig's dogmatic system of forestry management: "always deliver the greatest possible constant volume of wood."[71] The grail of sustained yield has guided the quest for rational forest economy ever since. With this concept, time entered forestry science. How much wood can the forest deliver over a century or two? How should this yield be harvested in one year so as to ensure that the same yield will still be available 100 years hence?

69. [Johann Matthäus Bechstein], "Feyerliche Eröffnung der Sachsen-Coburg-Meiningischen öffentlichen Lehranstalt der Forst- und Jagdkunde in Dreyssigacker den 12ten May 1801," *Diana, 3* (1805), 502–21, on 512.

70. Quoted in Mantel and Pacher, *Forstliche Persönlichkeiten*, 128.

71. Hartig, *Grundsätze der Forst-Direction*, 64.

Questions like these redefined the forester's task as curator of the forest, not simply its measurer. As Hartig put it, "no lasting forest economy is conceivable if the output of wood from the forests is not calculated according to sustained yield."[72]

The proper way to ensure the "permanence, certainty, and relative equality of the yield"[73] is not immediately obvious. Yield, unlike wood mass or forest area, is not a "quantity determined by nature"; it cannot be measured, save for the year at hand.[74] A system of forestry based on sustained yield requires prediction and planning. Some relevant factors, like the present mass of wood in the forest, can be measured; others, such as growth rates, must be extrapolated from the performance of sample plots, and the assumption of "good," "average," and "bad" soil. From this blend of quantities and qualifiers, the scientific forester can determine a schedule of cuttings for the forest of standard-trees under the "particular aspects of each system of culture," such as timber forest, coppice, or a mixed form.[75] Conditions such as the present state of the forest and expected growth rates must then be factored in; these, as Cotta pointed out, cannot be calculated according to "algebraic formulae." Inconsistencies in soil, weather, and natural devastations complicate the application of the method. Moreover, equating annual yield to the expected biological growth is a risky proposition. "How can man presume to determine such events of the future in advance, when they are dependent on a thousand accidental events?"[76]

Undaunted by the obstacles to accurate prediction, the *Forstklassiker* specified procedures for "forest regulation" (*Forsteinrichtung*); before long, many foresters throughout Germany adopted these methods. Unlike descriptive assessment, forest regulation was predictive and prescriptive. It offered a framework of long-term seeding and cutting based on the mathematical forest and standard practices

72. Quoted in G. Baader, "Die Theorie des Nachhalts und Normalwalds, ihre geschichtliche Wandlung und Bedeutung für die Gegenwart," *Allgemeine Forst- und Jagd-Zeitung, 109* (1933), 309–23, on 310.

73. Cotta, *Anweisung zur Forst-Einrichtung*, 55.

74. Ibid., 2.

75. Cotta, *Systematische Anleitung*, 100–1.

76. Cotta, *Anweisung zur Forst-Einrichtung*, 102, 104.

for application in the wooded forest. Scientific forest regulation also exercised many aspects of the forester's art, from cartography, description, and techniques for regeneration to silviculture and assessment. The role and authority of the vigilant chief forester who oversaw and adjusted the plan to circumstances, were reinforced by the scientific principles of forestry management.

Approaches to forest regulation multiplied quickly and differed considerably. Hossfeld and Cotta used geometry and arithmetic to construct flexible systems based primarily on wood mass and areal divisions of the forest; plans derived from their methods could be adjusted as local conditions dictated.[77] Heinrich Christoph Moser, Commissar of Forests in Bayreuth, published a method of determining the "periodic yield" based on "proportion constants" and sample plots.[78] Johann Leonhard Späth, Professor of Mathematics and Physics at Altdorf, proposed a detailed algebraic method.[79] The result of these investigations in almost all cases was a visual arrangement of age-classes and plots, linked with the quantities of wood and cuttings over time. Fold-out tables were common; Hartig used one to extend his plan into the 21st century (fig. 11.1). Like the business plans of a later day, attention to graphic clarity propped even the most chimerical of schemes against the firm oak of faith in numbers.

EPILOGUE

During the 19th century, the tradition of German forestry science persisted as cameralism gave way to economic liberalism. It produced the monocultural, even-age forests that eventually transformed the *Normalbaum* from abstraction to reality. The German forest became an archetype for imposing on disorderly nature the neatly arranged

77. See Hossfeld, "Vollständige System." Cotta's system, known as *Flächenfachwerk*, was built on the methods of converting area and mass described in Cotta, *Systematische Anleitung*. For the system itself, see Cotta, *Anweisung zur Forst-Einrichtung*.

78. Heinrich Christoph Moser, "Ideen zur Verbesserung der Taxations-Methode in Fichten-Waldungen," *Diana, 2* (1801), 71–117.

79. Späth, *Anleitung*, 206–35.

constructs of science. Witness the forest Cotta chose as an example of his new science: over the decades, his plan transformed a ragged patchwork into a neat chessboard (fig. 11.2).[80] Practical goals had encouraged mathematical utilitarianism, which seemed, in turn, to promote geometric perfection as the outward sign of the well-managed forest; in turn, the rationally ordered arrangement of trees offered new possibilities for controlling nature. For example, the technique of periodic area allotment (*Flächenfachwerk*) favored by Cotta generated the now familiar checkerboard scheme of growth periods. The mathematical exercise that generated the pattern could be modified to order the sequence of cutting so that older stands protected younger trees against prevailing winds.[81] In the hands of a suitably trained forester, mathematical order and practical utility became one enterprise.

During the 19th century, *Forstwissenschaft* advanced along the lines established by the early forest mathematicians: sustained yield, regulation according to age-classes and wood mass, and construction of the "normal forest" as an artifact of mathematical reasoning applied to quantitative data.[82] By the end of the 19th century, reformers of forestry in other natiøns—France, England (via the Indian Forest Service under Sir Dietrich Brandis), and the United States—had also discovered the need for conservation and forest management based on professional training and scientific principles. In each country, beginning with France during the 1820s and culminating with the American conservation movement, the inspiration and example was German *Forstwissenschaft*.[83]

80. Franz Heske, *German forestry* (New Haven: Yale University Press, 1938), 33–5.

81. Ibid., 33.

82. The key names associated with the development of forestry science in the 19th century include J.C. Hundeshagen, C.G. Heyer, Gottlob König, Max Pressler, and Martin Faustmann. For biographical sketches, see Mantel, *Forstliche Biographie* and Pacher, *Forstliche Persönlichkeiten*; for an assessment from abroad, see Brown, *Schools of forestry in Germany*.

83. For France, see Rubner, *Forstgeschichte im Zeitalter der industriellen Revolution*. In the United States, the leading lights of the conservation movement included both emigrés from Germany and Americans who had been influenced by the German examples of professional training and academic forestry science. These included Carl Schurz, Bernhard Fernow, Gifford Pinchot, and Carl Schenck. Schenck and

In Germany, however, resistance to the mathematically formulated forest economy began to grow, spurred by natural devastations caused largely by strict reliance on monocultures. By the end of the 19th century, foresters such as Karl Gayer, inspired by new-found loyalty to the natural diversity of species, called for turning "back to nature." Careful consideration of the forest as a multi-faceted biological ecosystem came into vogue.[84] Even in the face of this opposition, quantitative techniques elaborated by the *Forstklassiker* survived in practice. Above all, the doctrine of sustained yield remained sacrosanct.[85] Franz Heske, writing in 1938 for American foresters, reaffirmed the legacy of 19th-century *Forstwissenschaft* based on the work of the classical writers:[86]

> For all time, this century [the 19th] of systematic forest management in Germany, during which the depleted, abused woods were transformed into well-managed forests with steadily increasing yields, will be a shining example for forestry in all the world. German experience over a century makes it considerably easier for the rest of the world to pursue a similar course, because the attainable goal is now known, at least in principle. The sponsors of sustained-yield management in countries where forestry is still new can find in the results of this large-scale German experiment a strong support in their battle with those who know nothing, who believe nothing, and who wish to do nothing. This experiment and its outcome have rendered inestimable service in the cause of a regulated, planned development and use of the earth's raw materials, which will be an essential feature of the coming organic world economy.

Fernow, both from Germany, represented forestry science in the United States during the early years, and Fernow headed the first academic forestry school, founded at Cornell University in 1898. German-born Dietrich Brandis, later Sir Dietrich Brandis, introduced German forestry science to the British Empire and inspired the American forestry movement. See Carl Alwin Schenck, *The birth of forestry in America: Biltmore Forestry School, 1898–1913*, ed. Ovid Butler (Santa Cruz, Calif.: Forest History Society and Appalachian Consortium, 1974). On Britain, see Brown, *Schools of forestry in Germany*; James, *A history of English forestry*.

84. See Karl Hasel, *Forstgeschichte: ein Gundriss für Studium und Praxis* (Hamburg: Parey, 1985), 210–2; Heske, *German forestry*, 40–2.

85. *History of sustained-yield forestry: A symposium, Western Forestry Center, Portland, Oregon, October 18–19, 1983*, ed. Harold K. Steen ([Santa Cruz, Calif.]: Forest History Society, 1984).

86. Heske, *German forestry*, 81.

12 Society in Numbers: The Debate over Quantification in 18th-Century Political Economy

By Karin Johannisson

In his *Modest proposal* of 1729, Jonathan Swift took aim at political arithmetic—at its number mania, its intimate links with state power, its impudence in measuring human worth in money. A half century later, Adam Smith, himself no stranger to issues of quantification, joined in Swift's attack: "I have no great faith in political arithmetic." In a few more years Robert Malthus would express similar sentiments.[1]

Karl Marx, on the other hand, would accept political arithmetic as just the all-embracing social science its founders had envisioned, and would cite William Petty as one of the most inspired and original of economists.[2] In modern surveys of the history of the social sciences, political arithmetic has likewise been assessed in conspicuously favorable terms. Quantitative methods hold prestige in the context of social science; numerical data are seen as a mark of scientific neutrality toward the phenomena under study. Only in recent years has this objectivity come under scrutiny, with the recognition that statistics offer no special guarantee of freedom from ideological influences. Numerical data are not collected, they are selected and sorted according to criteria shaped by ideology and politics.[3]

1. Adam Smith, *An inquiry into the nature and causes of the wealth of nations* (1776; reprint Chicago: Encyclopædia Britannica, 1952), 230. Cf. Jonathan Swift, *A modest proposal, for preventing the children of poor people in Ireland from being a burden to their parents or country, and for making them beneficial to the public* (Dublin: S. Harding, 1729); T.R. Malthus, *An essay on the principle of population* (London: J. Johnson, 1798).

2. Karl Marx, *Das Kapital*, 3 vols. (Hamburg: O. Meissner, 1867–94), *1*, 3.

3. See J. Irvine et al., eds., *Demystifying social statistics* (London: Pluto Press, 1979), "Introduction"; R.M. Young, "Why are figures so significant? The role and the critique of quantification," ibid., 63–74.

The 18th-century debates over the merits of political arithmetic reveal sober recognition of the ideological content of the subject, as well as shining faith in order, systematics, and reason. The optimism that attended the application of political arithmetic was countered by a reluctance to claim much for its methods and utility. What emerged was not the vision of a measurable society, but rather a new view of social statistics as a useful instrument for describing aspects of society and economy.

STAATENKUNDE AND STATISTICS

By the middle of the 18th century, "statistics" or *Staatenkunde* had been a subject of study in German-speaking countries of over fifty years.[4] Its object was comparative description of the resources of different states; its aim was to assess their political strength. The purely verbal descriptions neither employed numbers nor aspired to generalization or to the formulation of general laws. This early *Staatenkunde*, which lacked both a quantitative method and a connection with the natural sciences, grew into a university discipline of great prestige equipped with an increasingly refined methodology. The descriptions it generated served as a bank of knowledge from which facts could be drawn by government officials as they drafted domestic and foreign policy. University statistics thus came to be known as "the right eye of the politician,"[5] whose duty it was to watch out for the nation's resources and prosperity.

4. The term "statistics" was used for the first time in England in 1770, and more systematically by Sir John Sinclair, *The statistical account of Scotland drawn up from the communications of the ministers of the different parishes*, 10 vols. (Edinburgh, 1791–8). Cf. M.J. Cullen, *The statistical movement in early Victorian Britain: The foundations of empirical social research* (Hassocks: Harvester Press, 1975), 10; A.L. von Schlözer, *Theorie der Statistik* (Göttingen: Vandenhoek und Ruprecht, 1804), 16–7.

5. On early German statistics, see Vincenz John, *Geschichte der Statistik* (Stuttgart: F. Enke, 1884); August Meitzen, *Geschichte, Theorie und Technik der Statistik* (Berlin: W. Hertz, 1886; English transl. 1891); P.F. Lazarsfeld, "Notes on the history of quantification in sociology: Trends, sources, and problems," *Isis, 52* (1961), esp. 283–94; M. Rassem and J. Stagl, eds., *Statistik und Staatsbeschreibung in der Neuzeit, vornehmlich im 16.–18. Jahrhundert* (Paderborn: Schoningh, 1980).

To the German statisticians, schooled in Aristotelian philosophy, the welfare of the state was not merely a question of quantities and materials; their concerns also encompassed intangibles like national character, satisfaction of the citizenry, and realization of the aims of the state. From an enormous mass of information, the statistician had to extract the facts pertaining to the happiness of the masses.[6] In his *Theorie der Statistik*, August von Schlözer identified the task of his discipline: "to measure the happiness of peoples, and whether this is increasing or declining." This meant "power and strength, to be sure! But these are only part of the happiness of a people. And not always these: for are there no states which are outwardly all-powerful but whose citizens live in wretchedness?"[7] Von Schlözer's words capture a basic ideology. A country's strength is not to be measured only in the superficial and the visible. Such assessments miss crucial factors like character, quality, and depth, which serve to distinguish the nations of the earth.

The statistical net hauled in unmanageable quantities of information. Toward the end of the 18th century some statisticians began systematically to use the *table* as a means of organizing this information. The tabular form, with its columns of countries and rows of categories, facilitated comparative analysis and offered new perspectives. At first, the tables mixed verbal and numerical information, but the use of columns soon favored facts in the form of figures. Numerical language, uniform and efficient, produced compact tables. The instrument of the table in turn created a corps of advocates, with clear preference for just those categories that could be most easily quantified. Such "tabular statistics," as its detractors dubbed it, differed markedly from the qualitative discipline of university statistics.[8]

Although they wielded a numerical sword, the advocates of tabular statistics did not fight under the banner of quantitative social

6. Gottfried Achenwall, *Staatsverfassung der heutigen vornehmsten europäischen Reiche*, 4th ed. (Göttingen: Witwe Vandenhoek, 1762), 4–5.

7. Schlözer, *Theorie der Statistik*, 35–36, 15.

8. Cf. Lazarsfeld, "Notes on the history of quantification," 292–3; John, *Geschichte der Statistik*, 88–90.

analysis. The purpose of their tables remained description, albeit with the aid of numbers. Such descriptive quantification should be distinguished from analysis of a society believed to be the real product of individual, quantifiable constituents. Nonetheless, tabular statistics was seen as a real threat by the proponents of the older, university statistics. As supposed materialists and social mechanics bent on dismantling a beautiful and intricate reality, the tabular statisticians came under steadily heavier fire.[9]

To the university statisticians, the *word* was the medium of statistics. Numbers might occasionally prove useful as an auxiliary instrument to give concrete form to particular descriptions or to express relationships. But numbers can never pierce the surface, they argued, or explain more than material circumstance. The influential review close to the professoriate of the University of Göttingen dismissed the new "table hacks" and "table fabricators" as common journeymen, whose reliance on the instrument rendered their work both shallow and coarse and reduced a beautiful art to a soulless technology. Only the nobility of university statistics grasped the idealistic factors inherent in the state. "The tabular method [seeks to] reduce everything to figures....If one has a few columns giving the figures for square miles, revenue, population and our dear livestock, one has a summary of the strength of the state; for national spirit, love of freedom, genius and character...there are no columns...and yet it is much less the body than the spirit that determines the strength of the state." The metaphor of the body recurred: "Has not...the whole science of statistics—one of the noblest—been debased to a skeleton, to a veritable corpse, on which one cannot look without loathing?....The state is something nobler than a machine...it forms a moral body."[10]

The note of desperation in these words reflects a profound transformation in the nature of the dispute. A quarrel that appeared to concern the form in which statistical data should appear came to represent rival philosophies. The quantitative approach, regarded as

9. John, *Geschichte der Statistik*, 128–33.
10. *Göttingische gelehrte Anzeigen*, 1806, no. 84; 1807, no. 131.

reducing reality to the material and excising the spiritual, stood as a challenge to the basic ideology of Romanticism, with its idealism, organic concept of the state, and emphasis on individuality.[11] How naive to see the state as machine! How could anything so multifaceted as a state aspiring to fulfillment be expressed in mere numbers? The university statisticians instead saw the state as a "being" (*Wesen*), people as *Volk*, imbued with *Volkgeist*.[12] The collective presupposed a social code based on spiritual and traditional values. People should not, could not be reduced to a factor of production; they were not the means to prosperity but its purpose.

As this colorful rhetoric may suggest, the university statisticians felt their position slipping out from under them. Deprived of more and more content as new specialties (political economy, geography, ethnology, and so on) broke away, university statistics began to wither away. As a political science it would be ruthlessly discredited by events during and after the French Revolution. University statistics, in failing to identify the popular discontent that found its voice and program in the Revolution, or to foresee that mighty Prussia would be trampled like a sand castle under the feet of Napoleon's troops, sounded its own death knell: "Nothing, nothing at all was achieved by the higher [university] statisticians."[13]

POLITICAL ARITHMETIC

The tables to which the German university statisticians objected were, or could be, instruments of political arithmetic. In England and Sweden, debate about the use of statistics centered explicitly on the program of quantitative social and economic analysis that descended from the work of William Petty and his contemporaries at the end of the 17th century.

11. This issue can be related to the struggle between the ideas of simplicity and of complexity; see Arthur O. Lovejoy, *The great chain of being* (Cambridge, Mass.: Harvard University Press, 1936).

12. See *Göttingische gelehrte Anzeigen*, 1807, no. 131, 1299.

13. August Ferdinand Lueder, *Kritik der Statistik und Politik* (Göttingen: Vandenhoek und Ruprecht, 1812), 48.

In Britain

William Petty was held in high esteem in circles populated by the likes of Robert Boyle, John Wallis, Samuel Hartlib, and Thomas Hobbes. Considered by Samuel Pepys "the most rational man that I ever heard speak with a tongue,"[14] Petty fit in well with the successful men of science who met in the Invisible College and the Royal Society. He dreamed of a perfectly rational society, a social edifice as stable and unassailable as a mathematical theorem. Each element had to be weighed, measured, and evaluated before it could be incorporated into this rational system. Petty made the point explicitly: "The Method I take...is not very usual; for instead of using only comparative and superlative Words, and intellectual Arguments, I have taken the course...to express myself in Terms of *Number, Weight* or *Measure*; to use only Arguments of Sense, and to consider only such Causes, as have visible Foundations in Nature; leaving those that depend upon the mutable Minds, Opinions, Appetites, and Passions of particular Men."[15]

Petty aimed at a science that used quantitative methods (counting and measuring) to isolate, describe, and analyze the elements that created a society's prosperity. Available observations of population, land, and resources—all in numerical form—were used in the calculation of values of other, as yet unknown, resources.[16]

14. Quoted in C.H. Hull, ed., *The economic writings of Sir William Petty* (Cambridge: The University Press, 1899), *1*, xxxiii.

15. Ibid., *1*, 244. Cf. E. Fitzmaurice, *The life of Sir William Petty* (London: J. Murray, 1895); E. Strauss, *Sir William Petty* (London: Bodley Head, 1954); W. Letwin, *The origins of scientific economics: English economic thought 1660–1776* (London: Methuen, 1963), chap. 5; *The Petty papers: Some unpublished writings of Sir William Petty*, ed. the Marquess of Lansdowne, 2 vols. (London: Constable & Co., 1927). On political arithmetic, see Lazarsfeld, "Notes of the history of quantification"; Peter Buck, "Seventeenth-century political arithmetic: Civil strife and vital statistics," *Isis, 68* (1977), and "People who counted: Political arithmetic in the eighteenth century," ibid., *73* (1982); Robert Kargon, "John Graunt, Francis Bacon and the Royal Society: The reception of statistics," *Journal of the history of medicine and allied sciences, 18* (1963).

16. Cf. Petty, "Verbum sapienti" (1664), "The political anatomy of Ireland" (1672), "Political arithmetick" (1671–6), "Two essays in political arithmetick" (1687), and "Five essays in political arithmetick" (1687), all in his *Economic writings.*

To reduce Petty himself to arithmetic, or rather geometry, his program lay at the intersection of inspiration of Baconian empiricism and Newtonian natural philosophy.[17] He called for the assembly of individual measurements (of, for example, birth and mortality rates, work capacity, output, consumption, and fertility) to form a valid picture of a general reality. In practice, however, he often relied on estimates and averages; and his columns of numbers necessarily remained isolated from one another since correcting principles analogous to the law of gravity eluded him. Petty and his successors Gregory King and Charles D'Avenant framed a general program of social analysis based on computations and systematic collection of facts: "He who will pretend to Compute, must draw his Conclusions from many Premises; he must not argue from single Instances, but from a thorough view of many Particulars; and that Body of Political Arithmetick, which is to frame Schemes reduceable to Practise, must be compos'd of a great variety of Members."[18]

The strong political context of political arithmetic may be discovered in a fight in Parliament over a proposal to provide it with one of its basic instruments. In 1753 advocates of a coherent program of social statistics, among them the mathematician James Dodson, called for a general census. But the census bill fell victim to strong opposition fed by fears of an expanding government. Five years later Parliament rejected a similar proposal for mandatory registration of births, marriages, and deaths.[19] Prosperity appeared better served by capital, industry, and steam power than by the calculations of a power-hungry central government. Parliament thereby denied to proponents of political arithmetic both the means of collecting

17. Cf. Robert Schware, *Quantification in the history of political thought: Towards a qualitative approach* (Westport, Conn.: Greenwood Press, 1981), 55–82; J.H. Cassedy, "Medicine and the rise of statistics," in A.G. Debus, ed., *Medicine in seventeenth-century England* (Berkeley: University of California Press, 1974), 283–312.

18. C. D'Avenant, "Of the use of political arithmetick," *Discourses on the public revenues and on the trades of England* (London: Printed for J. Knapton, 1698), *1,* on 29.

19. See Martin Shaw and Ian Miles, "The social roots of statistical knowledge," in Irvine et al., eds., *Demystifying social statistics*, 27–38, on 32 and Buck, "People who counted."

data—the census and the registration of vital statistics—and the function intended by Petty and his contemporaries—that of a political instrument of scientific legitimacy.

That, of course, did not prevent private groups from compiling vital statistics for their political purposes. Republicans and religious dissenters picked up political arithmetic and wielded it to combat the "faithful guardians of the state" and reduce their authority.[20] These advocates of local power counted local conditions as more significant than national aggregations of class, rank, and occupation. A good example is Richard Price's statistical studies of national debt and local prosperity based on data privately collected. Another is development of the work of John Graunt and Edmund Halley on the mathematical analysis of life expectancy. Graunt had demonstrated the utility of vital statistics for establishing the laws of demography; Halley's studies (published in 1693) of mortality lists from the city of Breslau had shown that generally valid calculations of life expectancy could be based on mathematical analysis of available, incomplete data on births and deaths.[21] As a basis for calculating insurance premiums and annuities, quantitative data thus offered real practical value. By the second half of the 18th century, both governments and private investors recognized insurance ventures as a promising prospect. Price saw annuity societies, *if* guided by mathematicians, as a solution to England's economic ills.[22]

In Sweden

While republicans and dissenters in England reserved political arithmetic for their own purposes and rejected its use as a tool of central government policy, Swedish officials sought a socioeconomic

20. Richard Price, quoted in Buck, "People who counted," 38.

21. Edmund Halley, "An estimate of the degrees of the mortality of mankind, drawn from curious tables of the births and funerals at the city of Breslaw" and "Some further considerations on the Breslaw bills of mortality," Royal Society of London, *Philosophical transactions*, 1693. Cf. Kargon, "John Graunt, Francis Bacon, and the Royal Society."

22. See Shaw and Miles, "Social roots of statistical knowledge"; Buck, "People who counted," 40–42; H.E. Raynes, *A history of British insurance*, 2d ed. (London: Pitnam, 1964).

strategy based on social measurement that might replace the capital and manpower the country had lost during long years of war. To remedy these ills, members of the Swedish parliament followed orthodox mercantilist lines based on the conception that population is the best measure of a nation's real riches. "A plenitude of poor people is a country's greatest wealth."[23] In the mercantilist analysis a surplus of people meant a surplus of need, which would inspire economic initiative, industry, and cultivation of land. A Swedish economist of the time wrote, borrowing a line from D'Avenant, "It may be better for the people to suffer a shortage of land than for the land to suffer a shortage of people."[24] The line symbolized a tie: in 18th-century Sweden, the old English idea of political arithmetic would enjoy political support, the notice of the Royal Academy of Sciences, an institutional platform in the parliament, and enthusiastic public backing. The key role of quantitative analysis in the Swedish debate on ways and means to national prosperity can be traced in pamphlets and programmes; in the *Transactions* of the Royal Academy of Sciences; in parish surveys, state memoranda, and confidential parliamentary reports.[25]

In accord with the proposition that people attracted industry, rather than the reverse, colonization projects of all kinds, resettlement projects, and measures to stimulate population growth abounded in Sweden beginning in the 1730s. So, too, did optimism about the rate at which the population of Sweden might increase. Some fortune tellers saw a doubling in the space of twenty years. In the long run, the population of Sweden (which then included Finland) might reach ten, twenty, even thirty million inhabitants. (The actual population figures in the mid-18th century, not disclosed at the time, were closer to two million.) The mainstream of public

23. Anders Nordencrantz, *Arcana oeconomia* (Stockholm: Joh. Laur. Horn, 1730), 145. This statement, with minor variations, can be found in the writings of most Swedish mercantilists.

24. D'Avenant, as quoted by A. Berch, *Sätt at igenom Politisk Aritmetica utröna Länders och Rikens Hushåldning* (Stockholm, 1746), 16. Cf. C. D'Avenant, *An essay upon ways and means of supplying the war*, 3d ed. (London: J. Tonson, 1701), 140–3.

25. For details, see Karin Johannisson, *Det mätbara samhället* (Stockholm: Norstedts, 1988), 96–179.

discussion overflowed with optimistic calculations from leading social theorists, and from reports submitted to the Riksdag by the Office of Tables.

How could these calculators arrive at such preposterous figures? In part, the optimism derived from conceptions of Sweden's great natural resources inherited from the patriotic historiography of the preceding century. In the years when Sweden was emerging as a great power, the country of the north was often depicted as the *vagina gentium*, the land chosen by the sons of Noah for its natural riches.

Mercantilism reinforced this patriotic tradition in a land of unexploited natural resources. "No man dares keep any land unusable and unfruitful, now and in the future, once he has seen the new ideas that can spring from ingenuity when spurred by need in a populous country," proclaimed the director of the Land Survey Board in 1758.[26] In Sweden, all agreed, the good Lord had "let abundance drip from His footsteps."[27] He had also arranged for an advantageous climate. The cold protected Sweden's populace from infectious diseases and made them "merry, lively, and manly." Snow on the ground prevented the evaporation of nutritious substances; once it melted, rotting leaves and needles gave way to rich humus soil. The woods teemed with useful game ("if anyone would seriously try to domesticate our elk, they might well become our camels"); lakes and rivers overflowed with salmon and other splendid fish, pearl-filled mollusks, oysters, and lobsters.[28] Nonetheless, God's handiwork could be improved upon. "If wildernesses and wastes are cultivated, a whole new land can be created, even more fruitful, milder in climate, more pleasant in every way, rich and able to support and feed millions more people than today."[29]

26. E.O. Runeberg, "Beskrifning öfver Lajhela socken i Österbotten," Kungl. Vetenskapsakademien *Handlingar* (hereafter abbreviated as *KVAH*), 1758, on 118–9.

27. J. Kraftman, quoted in K. Forsman, *Studier i det svenska 1700-talets ekonomiska litteratur* (Helsingfors, 1947), 147.

28. *Svar på Vetenskapsakademiens fråga, Hvilka äro Svenska Climatets förmoner och olägenheter?* (Stockholm, 1765 and 1766).

29. Ibid.

The language and the imagery were as rich as Sweden herself. But how to measure and master these natural resources? Political arithmetic offered a key to calculating Sweden's potential and devising a new strategy for growth and prosperity. For three decades, from 1740 to 1770, intense public debate would focus on the interpretation and application of political arithmetic. Three varieties of Swedish statistics emerged in these years: utopian, practical, and descriptive.

SWEDISH STATISTICIANS AT WORK

Utopian Statistics

Utopian statisticians regarded political arithmetic chiefly as a means of forecasting and calculating the prosperity they took for granted. Their method went back to Petty: a mixture of exact measures and approximations, the equation of human worth and capital value, and a willingness to mask imprecision with strings of decimal places.

The visionary statisticians put their program to work in quantitative parish surveys. Their *Description of Lajhela Parish in Österbotten* was explicitly intended as a model for similar surveys of all Sweden's parishes. The director of the Land Survey Board, E.O. Runeberg, drew up the plans and published the results in the *Transactions* of the Academy of Sciences for 1758–9. The report breaks down the area of the parish into exact figures for cultivated, cultivable, and uncultivable land; refines the analysis with a series of subdivisions; and specifies watercourses (numbers of lakes, rivers, streams, and springs) and roads (classed by degree of passability). It analyzes woodland ("450 trees on each *tunnland* [approximately 1.2 acres], yielding 36,733,120 trees in the parish") and animals ("there are 590 horses in the parish, 2,124 cows, 236 oxen, 944 young steers, 4,720 sheep, and 474 calves, [which] together total 9,086 head of livestock, fed over 7 months with 6,443 bales of hay at 7/10 bale a head").[30]

30. *KVAH*, 1758, 157–8.

The report includes details on the number of dwellings, barns, water mills, windmills, and so on, but excludes churches from the count as unproductive resources.

The descriptions had the higher purpose: making it possible to calculate the potential resources of the parish. Most striking in this regard was Runeberg's quantitative analysis of population. Averaging yielded a peculiar figure of 1,800 1/4 for the total population of the parish. Runeberg then analyzed the parish mortality and fertility rates. "In Lajhela 3.83 marriages, or for each marriage 3.83 years, are required to produce a child, but since the majority of children die, 9.15 marriages, or 9.15 years of each marriage, are required to increase the populace by one child."[31] Runeberg found these figures all the more dispiriting when he computed from the parish's potential natural resources it could support 28,000 inhabitants.

Runeberg's most significant calculations addressed the parish's capacity for work. Like the political arithmeticians of the preceding century, Runeberg drew a clear distinction between a *person* and a *worker*. He set the highest value on a married workman (2,390.99 dalers), somewhat less on an unmarried workman, and even less on "a person in general," by which he meant an average value over the total population. A woman was assigned a capital value three-fourths that of a man. Then came children, divided by age group. Runeberg judged infants of negligible worth, since they required on average "one-fifth of a person in care and tending." They thus had to be reckoned as a debit equal to one-fifth of the full value of an adult. The reasoning went as follows: "If we assume that by the eighteenth year a youth is equivalent to a full adult workman, and that a child of common people does not begin to be of use until his ninth year, and that not until the eighteenth year has he atoned for all the inconvenience and damage he caused before his ninth year, then a youth can be seen as non-withdrawable capital, increased by the accrual of compound interest, which only begins to yield an annual return through simple interest after the eighteenth year. Thus if a youth is assigned a political value of 1,195 dalers at the eighteenth

31. Ibid., 131–2.

year, he must be valued at 998.8 at the fifteenth, 746.3 at the tenth, 557.6 at the fifth, and 416.7 dalers in the cradle."[32] On this basis, Runeberg reduced the value of the almost 2,000 inhabitants of the parish to that of 800 workers.

The report then presents the calculations in tabular form and compares actual and potential resources. Note the exactitude of the entries for what Runeberg called "political evaluation:"

Political value

	Interest	Capital
One "mantal" (land assessment unit)	637	10,616
One person in general	36 19/52	612 9/13
One married ditto	73 1/2	1,224 12/13
One workman in general	71.73	1,195.29
One married ditto	143.46	2,390.99
One mantal could be	9,165	152,741
The parish is	57,330	955,440
could be	824,915	13,747,720
One square Swedish mile is	8,928	148,524
(100 sq. km.) could be	128,251	2,137,394

Practical Statistics

Instead of gazing at distant vistas of great wealth, practical statisticians focused on the foreground of poverty and wretchedness, a country in crisis. To some observers the cause was obvious: serious imbalance between different sectors of the economy. The solution seemed equally clear: restore the balance, guided by measurement, counting, and calculation. Here quantification acquired an instrumental function. Quantitative data served as the foundation for functional social models, which could then be translated into immediate political action. Like utopian statistics, practical statistics aimed at improving society, not merely describing it.

32. Ibid., 1759, 194.

The basis of practical statistics lay in the doctrine of proportions. As prescribed in the biblical text, all things are ordered by measure and number and weight. A social order called for well-defined harmony and balance among population, land, industry, and so on; and the definition of harmony rested on number. The task at hand was to measure social and economic components, compare them to the ideal, and shift and alter the components to achieve the desired harmony.

The leading advocate of practical statistics was Anders Berch, a professor of economics at Uppsala University. As a representative of the ruling party in the Riksdag (the party known as "the Hats"), Berch was strongly influenced by mercantilism. The title of his *Politisk arithmetica* (1746) shows where he stood. His ambition was to establish political arithmetic as a science; then and only then could a solid, exact economic policy be constructed. The transformation of social analysis into social strategy required four clearly defined stages.

The first stage was to uncover and understand the omniscient Creator's plan for a world in perfect balance. In the ideal state, everything would stand in harmonious proportion: population to area, economic activity to natural resources, production to consumption, men to women, and the duration of human life to "the supply of all human need." Next came measurement and collection of data—quantitative analysis of the present state of society. Anything and everything was to be measured: people, land, natural resources, productivity, efficiency, and consumption. A crucial factor was human productive capacity, which required careful assessment of the results of work in terms of time expended. A sufficiently broad base of calculations and data could overcome individual variations and yield an accurate value for the country's work force as a whole.

The data gathered would inevitably reveal imbalance in respect to the ideal proportions. The third stage was thus to calculate and balance every conceivable factor affecting the nation's capital strength against every other: people, agriculture, industry, and trade. The practical statistician used computations and estimates to settle on the suitable production of offspring or consumption of aquavit for a stated number of cities or of farmhands per farm. He also needed to balance the costs of war against the value of war booty, and the expense of ambassadorial travel against diplomatic advantage.

In his insistence on the fourth and final stage of implementation, Berch showed his perception of the gulf between theory and practice, or calculation and political action. He criticized political arithmetic in England, which had been allowed to remain the preserve of scientific circles and never approached implementation. Graunt, Petty, and D'Avenant also came in for rebuke for their narrowness of vision; in their hands measurement broke down into fragments without consistency or system. The grandeur of the Swedish program lay in the intent to make of quantitative social analysis an instrument for regulating the *whole* economy.

Berch's dream of a fully planned economy was founded on a faith in the state and its officials and a presumption of the loyalty of individuals to a powerful state. Once the data were put in the hands of the authorities and the balances struck, laws and ordinances would oblige individuals to distribute themselves and their resources to conform with the proposed model for prosperity. Slowly but surely a new social edifice would emerge. But what if all attempts to force the data into tabular form failed? How then to derive mathematical formulas for prosperity? Realization of the difficulties inherent in practical statistics prompted some to retreat to a less ambitious enterprise. For them, "statistics" meant the art of compiling and processing numerical information, with no purpose beyond the figures themselves.

Descriptive Statistics

The leading representative of descriptive statistics was the astronomer Pehr Wilhelm Wargentin, long-time secretary of the Royal Swedish Academy of Sciences, who played an instrumental role in securing an institutional base for Swedish statistics. Descriptive statistics had as its objective to reveal, describe, and interpret data, but not to prescribe how the data might be used. In parallel with the limitation of its aims, descriptive statistics came to be confined to a subject where data might be gathered without insuperable difficulty—to population studies. Slowly, sound and methodologically conscious population statistics began to squeeze out the extravagant attempts at precision and the lofty social aspirations of utopian and practical statistics. By the 1770s, population statistics would become the only type of statistical work undertaken in Sweden.

The general outlines of the growth of Swedish population statistics are well known.[33] In 1749 influential Swedish mercantilists and the Academy of Sciences succeeded in their campaign to establish an Office of Tables (which would become the Central Bureau of Statistics in 1858). Parish priests were required annually to complete printed forms reporting the numbers of births (classed by sex), deaths (by sex, age, and cause), and marriages, as well as the total population (by age, sex, estate, and occupation) for the parish. The tables were forwarded through a series of governmental agencies to the Commission of Tables, whose task was to summarize the results and transmit them to the Riksdag and the king. The Office of Tables thereby compiled the first set of population statistics in the world based on regular counts of total population. An efficient parish registration system that did not miss a single soul, a permanent institutional base, and a population unusual for its ethnic and religious homogeneity, disciplined by an established church with ample opportunity to exercise formal and informal control, contributed to the success of the venture.

But the very success of the Office of Tables represented a retreat from larger ambitions. Its reason for existence derived from the central importance of population in the mercantilist program, but population studies alone were only part of the social analysis urged by the utopian and practical statisticians.

Initially, the staff of the Office of Tables shared the optimism of other statisticians. In particular, the Commission of Tables (dominated by statisticians and civil servants) dreamed of a gigantic survey of all components of the economy. Collated, combined, and

33. See August Johannes Hjelt, *Det svenska tabellverkets uppkomst, organisation och tidigare verksamhet* (Helsingsfors: O.W. Backmann, 1900); E. Arosenius, *Bidrag till det svenska tabellverkets historia* (Stockholm: Nord. Bokh., 1928); and "The history and organization of Swedish official statistics," in John Koren, ed., *The history of statistics* (New York: American Statistical Association, 1918).

compared, the numbers and tables would constitute a map of Sweden's resources, strengths, and weaknesses, and provide the political authorities with an effective instrument for governing the country. This ambitious program is evident in the highly secret reports delivered by the Commission to the Riksdag and the king in 1755, 1761, and 1765.[34] Mind-boggling arrays of figures classified Sweden's population under a total of sixty-one headings; virtually all individuals were linked to their work and capital-producing capacities.

In their aim, these reports appear to be a faithful application of political arithmetic. Individuals were assigned categories according to their economic and hence political value to the state. First came providers, then consumers, and finally a category of wholly "superfluous members" (notably tavern staff and servants) numbering 10,336. The figure for emigration—8,059 in 1761—is just as precise; when converted into value using the methods of utopian statistics, it represented an annual capital loss to Sweden of 9 1/2 million dalers. When the potential of the emigrants to produce offspring was figured in, the loss amounted to no less than 19 million dalers.

At first the Swedish parliament showed much interest in the data and their implications. It appointed commissions and ordered certain reforms, especially in the medical field. Soon, however, the initiatives were tabled or defeated. Decisions disappeared mysteriously en route to the king for implementation. The important table of estates and occupations was originally required annually. But already in the 1750s the requirement was changed to reporting once every three years; later this was reduced to once every five years. Quantitative analysis of natural data like births and deaths remained noncontroversial, but attempts to derive social diagnosis or prescribe social therapy from the figures excited objections. Political arithmetic fell out of favor as a political instrument. With the constitution of 1772, parliamentary reponsibility for the Office of Tables formally ceased.

As officials lost enthusiasm for statistics, so too did advocates of descriptive statistics rebel against the use of their subject as an

34. Reproduced in August Johannes Hjelt, *De första officiela relationerna om Svenska tabellverket* (Helsingfors, 1899), and N.V.E. Nordenmark, *Pehr Wilhelm Wargentin* (Uppsala: Almqvist & Wiksell, 1939).

instrument of state power. For fifteen years, population information, broken down by estate, occupation, and age group, had been kept under wraps by the Office on Tables. During the early 1760s this suppression of population figures as a state secret occasioned heated debate. Only in 1764 was the official population of Sweden (2,383,113) first disclosed. Before long the detailed information underlying the estimate was available for study by anyone who wanted it.[35] As the gap between statistics and state widened, statistics had the opportunity to develop independently of power interests or practical applications. A gradual drop in the number of references in the Academy's *Transactions* to practical political aspects of statistics reflects this shift.

CONCLUSION

A quantitative social science was born of high hopes that social phenomena could be studied with the same precision as natural phenomena, yielding exact knowledge applicable in practical and political contexts. Yet the 18th-century conviction that the methods of natural science could be made to apply to all fields of knowledge hesitated at the crossroads of theoretical and practical goals in political economy.

The practical obstacles were daunting. Efficient collection and utilization of data required not only a firm institutional base (such as political arithmetic enjoyed in Sweden), but also workable methods for reducing masses of information to manageable and functional tables. How were consumption, efficiency, or the utility of diplomacy to be assessed and expressed in numerical terms? How were soil quality, popular morale, or unexploited natural resources to be set down in tables? The ambition to embrace society in its entirety overreached the practical limitations of 18th-century quantitative analysis.

35. E. Runeberg, "Om Svea Rikes folk-nummer och naturliga styrka," *KVAH*, 1764, 1765, 1767, 1770.

The relationship between state and statistics in the 18th century was a complicated one. The more closely the quantitative method was linked to the interests of the state and the more obviously its political function was defined, the greater the danger that the method itself would be undermined. If the practical application of a science normally strengthens its empirical character, here the opposite seems to prevail. In England the quantitative method strayed from empiricism as it became more closely identified with a national strategy for prosperity. In the state-directed, accelerated program for progress in Sweden, quantitative social analysis slipped into a rut of utopianism that led nowhere. Only by reducing its field of operation to vital statistics did quantitative social analysis meet with success. In Germany, the development of a quantitative approach was contingent upon freeing statistics from the ideology of the state.

In England and Sweden, mercantilism had fostered a mechanistic view of society that favored quantitative social analysis. As society was broken down into its material components of population, resources, industry, and so on, so the populace was composed of faceless, voiceless atoms. The quantitative program further reduced the individual to an equivalence in work or capital value. In Germany, where cameralism, not mercantilism, held sway, more complex concepts of *Land* and *Leute* argued against the reduction of social well-being to a set of material components or the reduction of human beings to interchangeable particles.

It is noteworthy that social statistics on the quantitative, English model reached a zenith in Sweden around 1750, just when Swedish natural science was flourishing. Thus in 18th-century Sweden, as in 17th-century England, quantitative social science grew in the same soil as a vigorous and prestigious natural science.

13 The Laplacean View of Calculation

By Roger Hahn

Laplace's adherence to the geometrical spirit of the late Enlightenment appears at first glance so conventional that its discussion might serve best as an example or summary of the new attitudes of the age. One is therefore tempted to look at his work as reflecting each aspect of mathematization prevalent in the century. Many of the individual features that mark his predecessors are in evidence: his approval of the concision afforded by the language of equations; his adherence to a rigorous and logical scheme of presentation that echoed the Euclidean model; his insistence on accurate and precise measurements necessary to concretize explanatory theories and to provide them with unambiguous empirical tests; and his repeated efforts to transform the vague uncertainties attached to empirical laws into measurable degrees of certitude by applying the calculus of probabilities wherever appropriate. Laplace is perhaps the most consistent consumer (and certainly the most influential one) of a philosophical attitude that considers mathematization as the key feature of modern science's success and the guarantor of its continued prosperity in all its branches. He epitomizes the movement we are describing.

Every scientific article Laplace wrote reflects this belief. Our task will be to try to understand how he reached this position and to indicate how it fit with his general philosophy. The difficulty lies in Laplace's peculiar aversion to self-reflective discussions. He clearly preferred monographic treatment of scientific issues to general treatises of philosophy. In his vocabulary, the term "philosophy"—used in the sense of systematic statements referring to epistemological or ontological issues—is rarely in evidence, featured prominently only once in the title adopted for the introductory section to his *Traité analytique des probabilités*, the "Essai philosophique" of 1813.[1]

1. A different dimension of these issues, with special reference to the *Traité de mécanique céleste*, is treated by Michel Serres in his "Histoire des sciences et histoire

Laplace is no outspoken philosopher; nor does he engage in debate with the prominent philosophers whose works he must have read and studied. Yet he was taken by his peers as an exemplar and commonly recommended in France as the spokesman of the scientific approach. He wrote an immensely popular *Exposition du système du monde*; published the lectures on mathematics he gave to huge audiences at the short-lived Ecole normale in 1795; and these and his other treatises were often awarded as prize-books for school valedictorians. Most of the noted teachers of science of the early 19th century in France—Prony, Lacroix, Biot, Arago, Poisson—and many other practicing scientists, including his colleague Berthollet and his disciples Gay-Lussac, Malus, Alexander von Humboldt, and Quetelet (to name but a few), adopted his techniques and their implied philosophy. Even though the principles were not set down explicitly, Laplace's approach was taught at the Ecole polytechnique, and his philosophy was tested in the examinations given at the Faculty of Sciences and adopted as a general yardstick for measuring accomplishments at the Bureau des longitudes and the Academy of Sciences. His attitudes permeated all of the physical sciences in France in the early 19th century, and may even have been influential in the newly emerging life sciences.

Since Laplace is so reluctant to speak for himself in a systematic fashion, we will attempt to squeeze out his message by a select review of his life's activities. The biographical approach promises to offer insights into the development of his "geometrical spirit" that cannot easily be extracted from textual analysis.[2] There were three important stages in his life when the issue of mathematization was prominent.

de la philosophie au XIXe siècle," *Revue de l'Université de Bruxelles*, 1973, 433–44.

2. For details, see Roger Hahn, *Laplace as a Newtonian scientist* (Los Angeles: William Andrews Clark Library, 1967); Charles C. Gillispie et al., "Laplace, Pierre-Simon," *Dictionary of scientific biography* (*DSB*), *15*, 273–403.

Mobility through Mathematics

The most pivotal of these turns on Laplace's decision to become a scientist. His first appearance on the scientific stage is in September 1769 as a teacher of elementary mathematics at the Ecole royale militaire in Paris, where for seven years he dispensed his knowledge to teenage sons of impoverished nobles. Since he received neither special praise nor blame in this occupation, one presumes he discharged his duties competently. The lectures he presented on elementary mathematics twenty-five years later at the Ecole normale—the only other teaching stint he accepted in his long career—show a solid command of the main features of mathematics and, in at least one instance, creative abilities as a mathematician probing the foundations of calculus.[3] But one must not look to Laplace's career for major contributions to the mathematical discipline, or cast him in the role of a rival to his creative contemporaries Lagrange, Legendre, or Gauss. Lecturing on mathematics was a job he initially accepted as a means to escape the provinces where family plans would undoubtedly have taken him in a totally different direction. For Laplace, the move to Paris was a cathartic declaration of independence from his peasant origins and their social implications. Mathematics was the liberating agent that offered him a chance to begin a new life. The experience was also traumatic because mathematics constituted the philosophical antipode of the career his family had chosen for him in theology.

Laplace came from a modest family in Normandy who sent him to a local school from which he was recruited by the University of Caen for the Church. The standard career path for an intelligent youth born in a rural district in the middle of the 18th century led to administration or the clergy. His uncle, who taught him the "three R's," held a sinecure at a nearby parish that left him in a state of secure indolence. Young Pierre Simon could aspire to a similar position after completing his Master's in Theology, which he was awarded only three months before his move to Paris. In his student

3. Robin E. Rider, *Mathematics in the Enlightenment: A study of algebra, 1685–1800* (Ph.D. dissertation, University of California, Berkeley, 1980; *DAI* 42/01A, 351), 70–2.

days, however, Laplace had already succumbed to the lure of mathematics.

Laplace's teacher of metaphysics and natural philosophy was an unreconstructed Aristotelian who bore the distinctively primitive name of Jean Adam. Judged by the size of his classes, Adam was an enticing popular lecturer who performed divertingly on demonstration equipment he bought from the outlawed Jesuits. But he became a pathetic figure whose exploits were ridiculed in a scathing anonymous play (written by a student) ingeniously entitled *Nostradamus.*[4] Whereas contemporaries were arguing about the merits of Newton over Descartes, our abbé Adam was still championing Aristotle. His printed lectures could have been written centuries earlier.[5] They display the worst features of scholasticism, filled with hair-splitting verbal distinctions, elaborate and useless Latin terminology, and questionable principles. His attempt to modernize his manual with examples from current studies on hydraulics and electricity were so inept that his students found in it cause for mockery. The abbé Adam was also a fierce defender of the faith, particularly those values championed by Jesuits, who had been a dominant force earlier in the century at Caen. Adam took special delight in bringing before the bar of justice a fellow priest and professor at Caen named Christophe Gadbled on the grounds that he had failed in his duties as a teacher and clergyman. In his lectures, Gadbled had criticized Adam for his philosophical ineptness, had raised metaphysical questions about God's omnipotence, and had even ventured to hypothesize about a physical world operating without His immanence. The abbé Gadbled was also accused of offering flimsy excuses for failing to attend Mass, thereby setting an intolerable example for youth.

As an impressionable and curious youth, Laplace naturally took notice of Gadbled. A totally new world opened up to him. His new teacher was thoroughly versed in Newtonian physics. He substituted the succinct language of calculus for the tangled verbiage of Aristotle.

4. M.E.M.B.C.D.S.M. [Etienne Mauger], *Nostradamus, ou le Physicien plaideur,* (Leyde [Caen], 1779).

5. Jean Adam, *Philosophia ad usum scholarum accomodata,* 3 vols. (Caen: J.C. Pyron, 1771–5).

He wove a beautiful tapestry of the system of the world, using mathematical threads of gold that dazzled the youth and challenged his mind. To appreciate the new picture, Laplace quickly absorbed the calculus, probably using the new texts of Euler recommended by Gadbled. There was in the manipulation of mathematics a clarity and sharpness that could not be extracted from Adam's lectures. Gadbled was also in touch with practitioners of modern natural philosophy, incorporating the latest advances in the scientific literature, particularly from the work of Clairaut and d'Alembert. Moreover, Gadbled applied mathematics in a useful fashion to navigation in his lectures on hydrography. As one contemporary document puts it, Gadbled had "made calculation and geometry fashionable in Caen."[6]

The critical moment in Laplace's conversion probably came when the hapless Adam, not content to squabble with Gadbled, decided to take on d'Alembert as well. The only known copy of his combative pamphlet was destroyed during the D-Day invasion. It may be assumed, however, that it accused the Newtonian coauthor of the *Encyclopédie* of criminal disrespect for the teachings of the Church and of Aristotle. While Laplace's Adam may have dazzled some of his students, he did not draw as much as a rejoinder from d'Alembert, who was most likely informed about the attack by his correspondent Le Canu, Gadbled's major assistant. It was this same Le Canu who apprised the high priest of the mathematical sciences about a promising twenty-year-old abbé named Laplace, and provided the young man with a letter of introduction. Pierre Simon set off for Paris armed with this letter and an audacious essay criticizing one of d'Alembert's minor writings on the law of inertia. D'Alembert had missed a mathematical point. Laplace's conversion from theology to science was completed when a properly chastened d'Alembert found him a well-paid position in mathematics at the Ecole militaire.

D'Alembert had additional reasons to empathize with Laplace, since he had in his youth experienced a similar transition from theological studies to the sciences. Recruitment into a mathematical occupation for most of the century often involved painful shifts from

6. Archives nationales, Paris, M196.

parental expectations or initial career paths.[7] La Caille, for example, renounced a life as a priest once he "discovered" Euclid for himself. Condorcet, Laplace's elder by six years, and tired of his traditional Jesuit education, turned his back on his uncle the bishop of Auxerre and a military life, embracing instead the uncertainties of a mathematical career, supported at first only by a gifted teacher at the Collège de Navarre. Others fortunate to be in Paris—like Coulomb, Legendre, and Lacroix—took to mathematics because of inspiring lecturers, while still others including Borda, Monge, and Carnot were drawn to the subject through military schooling. The individual paths may each have been different, but they all signaled a break from the security of established patterns for bright youths. In Paris especially, mathematics was an exciting and burgeoning activity that captivated many risk-takers.[8]

As the central feature of Laplace's new life, mathematics came to be quite naturally the symbol of emancipation from the errors of the past and the agent of personal success. In his first three years in Paris, Laplace wrote sixteen original papers on a variety of topics that immediately established him as a leading contributor to the mathematical sciences of his era.[9] He was twenty-three when he took his seat in the Academy of Sciences on a bench behind d'Alembert, following the same path as his elders Condorcet, Vandermonde, and Cousin.

Once in Paris, Laplace's professional life until the French Revolution was entirely centered on the Academy of Sciences. As an academician, he was called upon to review a host of papers and projects submitted for approval to the learned body. On the average, he sat on over a dozen review committees annually and wrote many of the reports, the texts of which still survive in good numbers. His membership in the Academy's section on mechanics and his mathe-

7. Charles B. Paul, *Science and immortality. The eloges of the Paris Academy of Sciences (1699–1791)* (Berkeley: University of California Press, 1980), 75–9.

8. Roger Hahn, "Scientific research as an occupation in eighteenth-century Paris," *Minerva, 13* (1975), 506–8.

9. Stephen M. Stigler, "Laplace's early work: Chronology and citations," *Isis, 69* (1978), 234–54.

matical abilities brought him the tedious task of reporting on eccentric projects for the making of mechanical devices to douse fires, to float on water, or to erect perpetual-motion wheels. Most of them were rejected, often with scathing comments that reveal the mathematical illiteracy of the projectors. The task consumed so much time that in 1775 the Academy accepted d'Alembert's proposal to refuse outright to consider papers on squaring the circle, trisecting an angle, or perpetual motion schemes. Several years later, Laplace sought to use elementary mathematics tests to screen out artisan crackpots who did not deserve serious attention, much to the chagrin of amateurs like Marat and Brissot. A mathematics test was already in use by examiners of military and naval schools to rank aspiring officers. In effect, Laplace had already been practicing what he preached for artisans since 1783 in his capacity as entrance examiner of artillery and naval engineering schools.[10] Next to proof of noble origin, mathematical competence was the only means adopted for separating the wheat from the chaff in the officer corps of the Old Regime. The extension of this practice to artisans seemed natural to Laplace; it also reveals in a pointed way the great importance he attached to this kind of literacy. There was a threshold of learning beneath which one could not be considered as a serious contributor to science.

Equally instructive were his positive reactions as a referee. Laplace was asked to comment on several projects on demography, life insurance schemes, and tables of amortization, all involving mathematical counts and statistics. In these instances, he was either an enthusiastic supporter or found ways to criticize the projects constructively. Laplace was constantly called upon to examine new scientific devices, particularly scales to record numerical data, techniques of interpolation, or other means to improve thermometers, barometers, microscopes, pyrometers, and other apparatus. The reports invariably focused on the improvements the new instruments provided for measuring. He hailed each device leading to more precise and reliable data as a progressive step.

10. Denis I. Duveen and Roger Hahn, "Laplace's succession to Bézout's post of *Examinateur des élèves de l'artillerie*," *Isis, 48* (1957), 416–27.

Laplace's voice was heard and respected even before he made a name for himself in this line of work. In 1781 he devised a telescopic pyrometer to assist Lavoisier in experiments on thermal expansion of solids.[11] A year later, he invented the ice calorimeter, whose sole purpose was to turn qualitative arguments about exothermic reactions into measurable data that could be used to test hypotheses. Lavoisier had enlisted Laplace in his experimental work as a keen critic of instrument-making and as a scientist endowed with a sharp mind trained to uncover errors and to devise methods to turn them. Despite its design limitations, the calorimeter was evidence that they appreciated the critical role instruments must play in a developing science still encumbered with vague principles. We know that later during the Revolution, Laplace participated with singular attachment in every phase of the establishment of the meter and the gram as national standards, and that he continued to promote the diffusion of the metric system in France and its conquered territories as Minister of Interior and later, as Senator.[12] His public service included calling the first international congress of weights and measures in 1798 and laboring at the implementation of its decisions.[13] His correspondence is replete with details about techniques of measurement, sources of error, and the consistency yielded by repeated measurement.

Laplace was not the first scientist to be so concerned. No doubt he developed an appreciation for instruments from close association with observational astronomers on whom he constantly relied for his work in celestial mechanics. Though not an expert observer himself, Laplace was as keenly concerned with sighting and timing devices as Tycho Brahe had been two centuries earlier. What is particularly

11. Henry Guerlac, "Chemistry as a branch of physics: Laplace's collaboration with Lavoisier," *Historical studies in the physical sciences, 7* (1976), 183–276, on 224.

12. For the connection between the calculations surrounding heat experiments and the metric system, see Guerlac, ibid., 253.

13. Laplace to Delambre, 10 pluviôse an VI, published in *Revue d'histoire des sciences et de leurs applications, 14* (1961), 287–90; Maurice P. Crosland, "The Congress on definitive metric standards, 1798–1799: The first international scientific conference?" *Isis, 60* (1969), 226–31.

noteworthy is Laplace's desire to extend this attitude to all the sciences, particularly newly emerging ones. Two examples are in order here.

The Genevan physicist Deluc was one of Laplace's favorite scientific correspondents, despite Deluc's admittedly weak grasp of mathematical theories. In his mature years, Deluc turned into an irrepressible type quite distasteful to Laplace. Nevertheless, at an earlier time, Laplace vigorously encouraged him to reorganize the field of atmospheric research by elaborating his critical history of the thermometer and barometer, published in 1772. Laplace's stated hope was that Deluc could give meteorology its proper place among the more legitimate and established physical sciences. All that meteorology needed, so he imagined, was a carefully developed theory linking altitude to heat and pressure supported by accurate data taken from Alpine climes. So encouraging did Laplace's support prove to be that Deluc originally wrote his revised treatise on modifications of gases as a series of letters to Laplace.[14] Even though the results did not fully meet Laplace's expectations, except in hypsometry, the goal pursued for almost two decades is testimony to Laplace's commitment to an increasingly accurate quantitative approach.[15]

A more successful sponsorship following the same lines was Laplace's advocacy of the abbé Haüy's research into crystallography. In late 1783, Laplace was on the committee assigned to examine the manuscript of an *Essay on a theory of the structure of crystals.* He wrote a glowing report.[16] Haüy had discovered a theory that had all the makings of a legitimate organizing principle linked directly to observation. Moreover, the theory of lattice structure he proposed was acknowledged to stem from a conscious effort to apply mathematics to natural history, as recommended by the mathematician Bézout. Unlike Bergman and Romé de l'Isle, who had tried to

14. Jean André Deluc, *Idées sur la météorologie,* 2 vols. (London: Spilsbury, 1786–7), *1,* 5.

15. Theodore S. Feldman, "Applied mathematics and the quantification of experimental physics: The example of barometric hypsometry," *Historical studies in the physical sciences, 15* (1985), 127–95.

16. Académie des Sciences, Registre des procès-verbaux, 26 Nov 1783.

organize crystals into classes or to relate their formation to the supposedly analogous processes of the formation of the earth, Haüy sought to link the geometry of standard crystals to the geometry of elementary particles that constituted their fundamental units. He succeeded in explaining various simple, quantitative phenomena.[17]

Laplace considered this approach so promising that he developed a close personal attachment to the abbé, seeking to appropriate his talents for the physical sciences, symbolically capturing a portion of natural history (the mineral realm) for mathematics. The strategy worked so well that Haüy's next project brought him squarely into a newly quantified portion of physics. In 1787 Haüy offered the Academy a theory of electricity and magnetism that won him overwhelming praise for combining the theories of Aepinus and the empirical laws of Coulomb.[18] During the Revolution, Haüy was selected (probably on the recommendation of Laplace) to lecture on physics at the Ecole normale; and in 1803 he was commissioned by the government to prepare the standard textbook on physics to be used in secondary schools. In all of these works, he expressed a philosophy totally in accord with and probably inspired by Laplace. Haüy demonstrated what a well-articulated theory intimately linked to verifiable, quantified data could be. If we take the writings of the abbé as a mirror of Laplace's views, we realize that their devotion to mathematics was not an end in itself, but a powerful tool for taming nature to human understanding. In describing the significance of his work in 1792, Haüy stated:[19]

> The theory of the structure of crystals can only be furthered with the aid of calculation (*calcul analytique*). Analysis has the merit of encompassing in a single formula the solution to a large number of varied problems, and it can

17. John G. Burke, *Origins of the science of crystals* (Berkeley and Los Angeles: University of California Press, 1966); Marika Blondel-Mégrelis, "Le modèle et la théorie: Analyse d'un exemple de la cristallographie de Haüy," *Revue philosophique de la France et de l'Etranger, 171* (1981), 283–302.

18. J.L. Heilbron, *Electricity in the 17th and 18th centuries: A study of early modern physics* (Berkeley: University of California Press, 1979), 426.

19. René Just Haüy, "Exposition abrégée de la théorie sur la structure des cristaux," *Journal d'histoire naturelle, 1* (1792), 159 and *Journal de physique*, Aug 1793, 1.

> alone impress upon theory the mark of rigorous certitude by arriving at results completely in accord with those of observation.

Variations on this theme were repeatedly asserted in scientific literature of the early 19th century, sometimes with direct reference to the inspiration provided by Laplace. The chemist Berthollet (Laplace's neighbor in Arcueil), who shared the same attitudes in his *Essai de statique chimique* (1803), acknowledged that his views were shaped by prolonged exchanges with Laplace. The latter never seems to have wavered from the belief that the advancement of scientific learning depended centrally upon the intelligent and appropriate use of mathematization. The view seemed fully sanctioned by many examples of progress in the physical sciences during the late Enlightenment.

Mathematics and Revelation

Laplace experienced a second major change in his life that affected his career and provided an opportunity to develop his view on the calculating spirit. He suffered through the trauma of the French Revolution. Until the eve of the Revolution, his life had been singularly focused on his career. His social circle was limited to professional colleagues who served to establish or reinforce his academic stature. But marriage at the ripe age of thirty-nine and becoming the father of two children imposed a reorientation of his habits. He took time to be more accessible, mixing with the educated public from both the aristocracy and the urban middle class. Political events eventually forced him to find a refuge for his family outside of the capital. With the surge of populist power came the inevitable criticism of elitist science and its practitioners. Old patterns at the Academy were disrupted and ultimately destroyed; many friends went their separate ways to exile, the provinces, prison, or the guillotine; and his sources of income were seriously threatened. While remaining true to the professional ideals he had espoused earlier, Laplace was led by the new circumstances to add a public dimension to his life.

He cleansed himself of political suspicion by serving on several governmental committees connected with science. He became friendly with men of influence, but not so openly that he was compromised when they happened to fall out of political favor. He wisely adopted the role of the technocrat, an expert on science and its applications, loyal to the nation rather than to any particular political party. This strategy paid off handsomely. As France emerged from the Terror, he was recognized as a leading spokesman of science and consulted on all the important phases of the reorganization of the cultural life of France. Good political instincts led him into Bonaparte's circle, and he wound up temporarily as Minister of Interior after the 1799 coup that brought the General to power. Kicked upstairs to the Senate, he became one of the favorite courtiers of Napoleon. When the Emperor fell, Laplace was so highly respected that the new Bourbon regime had to hold him in esteem despite his earlier close association with Napoleon. By then, he was a fixture in public circles and an integral part of France's cultural elite.

In late 1794, Laplace was selected with the aging Lagrange to lecture on mathematics at the Ecole normale before 1,200 auditors, the flower of French educators, sent to the capital to absorb the quintessence of learning from the foremost scientists of the day.[20] The lectures were recorded by stenographers and immediately published in the media, which represented the event as evidence that the French could continue to absorb themselves in cultural activities despite their new-won reputation as blood-thirsty barbarians. Everything, even the chiding remarks about Leibniz' silly belief in a deity and Newton's aberrant interest in the Apocalypse, was reported in the press and spread abroad. It was the first time Laplace had faced the world so openly.

He and Lagrange did a remarkable job, considering that neither had tried such a thing before. Setting aside the grand manner of philosophizing reminiscent of Condillac, d'Alembert, and the Académie française, Laplace chose a direct style of exposition on a fairly sophisticated level with little moralizing. He employed a minimum of mathematical symbols, mixing common-sense platitudes with profound conclusions about the nature of numbers, lengths, and their

20. Jean Dhombres, "L'enseignement des mathématiques par la méthode révolutionnaire: Les leçons de Laplace à l'Ecole Normale de l'an III," *Revue d'histoire des sciences et de leurs applications, 33* (1980), 315–48.

manipulation. The so-called elementary lectures were, in fact, a superior condensation and synthesis of the best thought of the age about mathematics and its uses in everyday life and in the scientific enterprise. He included a lecture on the metric system and announced the composition of a new work meant for the same audience to cover astronomy, the *Exposition du système du monde.*

It is in this *Exposition*, published first in 1796, rather than in the ten lessons on mathematics, that one sees how Laplace's new role as a popularizer expanded his thoughts on mathematization. Not only was it presented as the central means to effect the progress of science, but that progress itself was offered as an example of the noblest aspects of humanity. In the midst of the Revolution, Laplace became a grand spokesman for his profession, justifying the activity he had chosen to pursue as the most glorious of all secular pursuits.

Challenged to expound his life's work to an educated audience presumed to be mathematically untutored, he eschewed all algebraic equations in the body of the work. That did not make it, however, "a handbook of cosmology."[21] It was in fact a nonmathematical version of his planned *Celestial mechanics*, written for literate readers following the format of standard popularizations. It presents astronomy in a simulated inductive fashion, dealing first with direct observations of apparent motions in the heavens, followed by a description of the real (Copernican) movements of the planets and their satellites. These sections lead to a purely verbal discussion of the laws of motion and the theory of universal gravitation. The final section is a review and summary of the entire work offered in the form of an elementary history of astronomical discoveries. It is this last section that provides the key to its author's central purpose. Laplace wants to present the evolution of astronomy as a model for the finest features of modern science and the current state of celestial science as a reflection of the most elevated characteristics of liberated humanity.

21. Gillispie, "Laplace, Pierre-Simon," 343.

22. Pierre Simon Laplace, *Exposition du système du monde* (Paris: Cercle Social, an IV [1796]), *2*, 310–11.

Here is his sermon-like peroration:[22]

> Taken as a whole, astronomy is the most beautiful monument of the human mind—the noblest voucher of its intelligence. Seduced by the illusion of the senses and of vanity, man considered himself for a long time as the center about which the celestial bodies revolved, and his pride was punished by the vain fears they inspired. The labor of many ages has at length withdrawn the screen that concealed the system of the world. And man now appears [to dwell] upon a small planet, almost imperceptible in the vast extent of the solar system, itself only an insensible point in the immensity of space. The sublime results to which this discovery has led may console him for the [inferior] rank assigned to him in the universe.

In a later edition, he added that thinking beings should especially take pride in their ability to have measured this universe, given the tiny base from which they were operating. "Let us carefully preserve, and even augment the number of these sublime discoveries, which constitute the delight of thinking beings."[23] The loss of anthropomorphic centrality is compensated by the power of the human mind, which despite its translocation, is able to contemplate and provide a proper assessment of the heavens.

The originality of Laplace's treatment lies in his characterization of the historic path to progress by the systematic movement from observation to induced laws of nature, and from laws to their causes. Within this framework, he identifies improved observation and mathematics as the two propellants responsible for progress, while unsupported speculation and religious obsession are its most notorious obstacles. Time and again he proclaims "observation and calculation as the only [solid] grounds for human knowledge."[24] His historical analysis does not rest alone, however, on this trite generalization. By "calculation" he means several things, each of which denotes the spirit of the *géomètre philosophe*.

Most obvious is the way increased precision has led to the discovery of celestial regularities (and hence to the possibility of establishing laws of nature). Laplace points to Hipparchus, Ulugh Beg, Tycho Brahe, Galileo, and Kepler. Improvement in precision, he

23. Ibid., 6th ed. (Brussels: De Vroom, 1827), 531.
24. Ibid., 442, note; ibid., 1st ed., 2, 292.

observes, may come about in a variety of ways: through the construction of better observatories and instruments; with the introduction of simplifying techniques of calculation, such as logarithms; by the more systematic tabular collection of data; or simply through the greater conscious attention to detail by observers.

A second dimension is the critical evaluation of data that comes from juxtaposing expected positions calculated from elements of planetary orbits alongside the results of observation. As the degree of conviction about the validity of the calculations increases over time (especially since Newton), observation is subjected to ever greater criticism. Laplace repeatedly reminds his readers that it is "this analytical connection of particular with general facts that constitutes theory," and that this theoretical outlook distinguishes modern Western astronomy from its predecessors.[25]

A third and more complex notion advanced by Laplace refers to the new potentialities of algebra and the calculus for astronomy. It was an issue close to his heart because in the *Traité de mécanique céleste*, he replaced all of Newton's geometrical demonstrations with what was then called *analyse*. The same term was used to denote a particular epistemological method, and Laplace at times shifts from one usage to another without realizing he is dealing with distinct issues. Thus when praising Newton's methodological approach, he adopts the distinction between induction (*analyse*) and deduction (*synthèse*). In the *Principia*, Newton argued the truth of his system by means of synthesis, using a geometrical form. While admiring Newton's use of induction to discover his principles, however, Laplace criticizes him for having chosen a geometrical form of exposition to establish their truth. To be sure, he recognized that there were extenuating circumstances: "The state of imperfection in which the infinitesimal calculus was in the hands of its inventor did not permit him to resolve completely the difficulties pertaining to the theory of the system of the world; and he was often given to positing uncertain conjectures, until such time as they were to be verified by rigorous analysis."[26] Nonetheless, the further advancement of

25. Ibid., 6th ed., 457, 503.
26. Ibid., 1st ed., 2, 285.

astronomy depended upon abandoning the geometrical approach, of which Newton was the last grand master. Laplace credits calculus not only with bettering the inductive or analytical processes that led from evidence to general principles, but also with advancing the deductive or synthetic phase that allows one to derive particulars from the new theory.

Whatever confusion exists in Laplace's use of terminology, he emerges as a strong advocate of the value of calculus for further progress:[27]

> Geometrical synthesis has the advantage of never allowing us to lose sight of its goal...; whereas algebraic analysis quickly allows us to forget the principal goal in the form of abstract combinations, and it is only at the end [of the operation] that it brings us back to it. But in isolating itself from this goal after having taken what is needed to arrive at the required result; and then by giving ourselves over to the operations of analysis...one is led by the power and generality of this method [and by the inestimable advantage of transfering reasoning into a mechanical process to arrive at] results often inaccessible to synthesis.
>
> No other [mathematical] language lends itself so elegantly...to the long train of interconnected expressions, all flowing from one fundamental equation. Analysis also offers the advantage of always leading us to the simplest methods. One need only make a judicious selection of unknowns using the proper methods and give the results the form most easily reducible to...numerical calculation.

He offers as an illustration of the truth of his observation the solution of lunar inequalities, which "would be impossible to arrive at by synthesis."[28]

Mathematics against Uncertainty

To this inventory of virtues in quantification for astronomy, Laplace was later to add the merits of the calculus of probability. Chronologically, his *Philosophical essay on the theory of probability* falls in a dark period of his life, near what he thought was the end of his career in 1813. This third crisis brought him face to face with the

27. Ibid., 1st ed., *2*, 289–290. For the section given in brackets, 2d ed. (Paris: Crapelet, 1799), 339.

28. Ibid., 6th ed., 506, n. 1.

limits of mortality and the uncertainties of life. His bachelor son, an artillery officer and aide-de-camp of Napoleon in the Russian campaign, nearly lost his life; his only daughter died in childbirth, causing him to sink into a profound depression; and Napoleon, who had come to power as a force of stability, was leading the Empire to dissolution and France to surrender. One might easily expect in this essay a Pascalian cry of despair about the weakness of the human mind or a Stoic sense of fatalistic resignation. But Laplace's intellectual resilience and self-confidence prevailed to turn his essay into a remarkably optimistic piece worthy of comparison with his departed friend Condorcet's *Sketch for an historical picture of the human mind.*

In a tour de force, Laplace stands traditional worries about fortuity on their head and asserts his famous view of determinism at the outset. According to his philosophy, the world operates by immutable laws that we can begin to know by applying the calculus of probabilities to the phenomena we observe. His concern with this calculus came directly from a desire dating back to his youth to find a systematic way of moving from the gathering of observations to the statement of true laws of nature. He had invented (or perfected) a calculus of statistical inference that allowed him to estimate the likelihood that a particular configuration of events would lead to a subsequent arrangement known through observation. This a posteriori calculus, presented in a pioneering paper of 1774 he subsequently developed, promised to become a powerful tool for his natural philosophy. Philosophically, its power stemmed from capturing game theory from the domain of conjecture (chance), and turning it into a method for calculating likelihood (probability). The uncertainties of chance were replaced by the manipulable concept of degrees of likelihood, bringing the operation within the reach of the mathematician. As a technical tool, statistical inference was prized for helping natural philosophers to distinguish between likely and spurious causes, thus preventing them from lapsing into unfounded speculation. Systematic errors of observation attributable to instruments could thus be distinguished from those dependent upon human failings. Statistical inference also offered the possibility for treating problems of civic life—such as the differential birth rates for the two sexes, annuities, sampling techniques, voting methods, and judicial decisions—in a rigorous and rational manner, bringing mathematical light to obscure social problems.

For Laplace, calculation was no mere scientific tool of limited significance. Its scope went far beyond the technical advantages it had already supplied for the progress of natural philosophy. Mathematization offered a central path to an enhanced epistemology that would progressively reduce the errors of human ways and permit the assertion of intellectual powers. Given enough evidence, and using the proper analytic tools provided by mathematicians, humans could conceivably attain the skills necessary to rival the Supreme Being. It is no accident that Laplace proclaimed the possibility of an ideal, supreme, but human calculator (an absolute Intelligence) in the only philosophical essay he devoted to a mathematical subject.[29] For him, mathematization truly embodied the spirit of rationalism, the greatest virtue offered by the Age of Enlightenment.

29. Roger Hahn, "Laplace and the mechanistic universe," in David C. Lindberg and Ronald L. Numbers, eds., *God and Nature* (Berkeley: University of California Press, 1986), 269.

Bibliographical Afterword

By Robin E. Rider

The intent of this afterword is to suggest some sources and directions that might prove helpful in further investigation of what we call the quantifying spirit. To keep the task within bounds, the emphasis is on publications of the 1980s. The essays in or into historiography in *The ferment of knowledge. Studies in the historiography of eighteenth-century science* afford a convenient point of departure. Its editors hoped that a historiographical approach would gather together the "detailed and discrete publications" that chronicle the complexity of 18th-century science—complexity that bedevils attempts to craft synthetic accounts of 18th-century knowledge about nature. To that end, they assembled a dozen detailed, not always discrete responses to the historiographic challenge, a challenge individual contributors defined in quite different ways.[1]

Several essays in *The ferment of knowledge* speak directly to themes of the present volume. Henk Bos, for example, underscores the "direct and intense interaction" of mathematics with its applications in the 18th century.[2] Simon Schaffer discusses, among other topics, Buffon's view of "natural philosophy as the discourse of order," drawing our attention to Foucault's distinction between a "general science of order" and mathematization.[3] In making frequent reference to a study of the *méthode numérique* in 18th-century medicine, W. Bynum alerts us to quantification of medical experience

1. G.S. Rousseau and Roy Porter, eds., *The ferment of knowledge. Studies in the historiography of eighteenth-century science* (Cambridge: Cambridge University Press, 1980), quotation on 5.
2. H.J.M. Bos, "Mathematics and rational mechanics," in ibid., 327–56, on 353.
3. Simon Schaffer, "Natural philosophy," in ibid., 55–92, on 85, footnote 89 on 89 for Foucault.

from the 1760s on.[4] D.S.L. Cardwell highlights the significance of the "quantification of power" in 18th-century technology; for Cardwell, the roots of the Smeatonian method—"rational, systematic and comprehensive," aimed at "testable and preferably quantitative" answers—likely lie in Newtonianism.[5] Roy Porter contends that quantification proved of dubious value "for explicating the environment" of the terraqueous globe, but he sets considerable store by the 18th-century view of nature as economical, "uniform, organized and integrative." He also underscores the challenge posed by "the growth of knowledge"—"the sheer quantity of newly available information" to be managed, lest chaos reign.[6] In Rom Harré's discussion of scientific epistemology, the model of mathematical thought makes several appearances: Locke's views on certainty, seen as grounded in geometry; d'Alembert's adherence to the Cartesian doctrine of "mathematical abstractionism"; Adam Smith's "philosophical economics," aimed at optimizing the quantity of happiness.[7] Maurice Crosland, in his stress on Lavoisier and the chemical revolution, plays familiar themes of classification and measurement.[8]

J.L. Heilbron's essay offers the clearest invocation of the quantifying spirit. Heilbron reasons from the development of 18th-century electricity to a periodization of experimental natural philosophy in which quantification plays a crucial role. In the emphasis on experimental physics in the latter part of the century, in the preference for an instrumentalist approach stripping away metaphysical niceties, we hear tunes sounded in several chapters of the present volume.[9]

The editors of *The ferment of knowledge* also asked their contributors to mark out fruitful directions for further research. From the desiderata listed by the contributors emerge several common concerns relevant to the influence of mathematics and the uses of quantifica-

4. W.F. Bynum, "Health, disease and medical care," in ibid., 211–55, on 230.
5. D.S.L. Cardwell, "Science, technology and industry," in ibid., 449–84, on 466, 470.
6. Roy Porter, "The terraqueous globe," in ibid., 282–324, on 320, 314, 305, 309.
7. Rom Harré, "Knowledge," in ibid., 11–54, on 21, 46, 34.
8. Maurice Crosland, "Chemistry and the chemical revolution," in ibid., 389–416.
9. J.L. Heilbron, "Experimental natural philosophy," in ibid., 357–88.

tion. Many of the authors urge that the disciplinary map be redrawn along lines more faithful to 18th-century categories of knowledge; most request greater attention to social, political, or economic context. Several recommend listening carefully to language, to the forms of discourse. All find plenty for historians to do.

The sheer quantity of scholarship on 18th-century science published in the past decade or so speaks to needs and opportunities in this field, and tempts quantification of the art of bibliography. Beginning with the *Isis Critical bibliography* for 1981, the 18th century broke free, as a chronological classification, from the 17th. Though the total number of entries in the *CB*s has varied considerably for the 1980s (and though the numbers are softened by individual decisions regarding compilations of articles), the market share for the 18th century has held fairly steady at 9 to 10 percent of all *CB* entries, with a peak of 397 entries; annual figures for the 17th century have shown more variation. If we take into account all the entries in the *Cumulative bibliography* for the years 1976–85, we find that entries for 18th-century topics outnumber those for the 17th century by about one-third (2,008 versus 1,572). Anniversaries of late 18th-century events have helped to swell the totals. Just as the 250th anniversary of Newton's death proved the occasion for another round of scholarly attention to his work and influence, the 200th anniversaries of scientific or political events or of the deaths of other 18th-century scientific figures have recently triggered reassessments, new editions of classic works, and bibliographies deserving of attention.[10] In what follows, the stress is put on other manifestations of the quantifying spirit and the ways in which the message of quantification was spread.

10. See, for example, *Dix-huitième siècle, 16* (1984), primarily about d'Alembert; A.T. Grigorian, A.P. Yushkevich, and B.D. Kovalev, "In memoriam Daniel Bernoulli," *NTM, 19:2* (1982), 1–13; Ashot T. Grigorian and Boris D. Kovalev, *Daniil Bernulli, 1700–1782* (Moscow: Nauka, 1981); *Leonhard Euler, 1707–1783: Beiträge zu Leben und Werk* (Basel: Birkhäuser, 1983); Arthur Donovan, ed., *The Chemical Revolution: Essays in reinterpretation*, in *Osiris, 4* (1988); *Ambix, 36:1* (March 1989), primarily about Lavoisier and the chemical revolution. Cf. note 36.

OTHER EXPRESSIONS OF THE SPIRIT

The opinions of Fontenelle are frequently trotted out in characterizations of the Enlightenment. Fontenelle commended the *esprit géométrique* for its promise of certainty and rationality, which might improve polity, morality, literature, and oratory as it had enriched astronomy and mechanics.[11] By the end of the 18th century, a wealth of treatises and textbooks had praised the merits of mathematics and the solidity of reasoning in Euclidean geometry, often in terms borrowed from the previous century. Recent scholarship has pointed to John Arbuthnot's arguments in favor of the utility of mathematics,[12] the appeal of the axiomatic approach and the persistence of Cartesian methods as mixed with mathematics,[13] and the wide-ranging

11. Thomas Hankins, *Science and the Enlightenment* (Cambridge: Cambridge University Press, 1985), 2. Cf. Charles B. Paul, *Science and immortality: The éloges of the Paris Academy of Sciences (1699–1791)* (Berkeley: University of California Press, 1980); Alain Niderst, "Fontenelle et la science de son temps," *Studies on Voltaire and the eighteenth century, 216* (1983), 323–4; Birgit Fenner, "Vernünftige Skepsis—skeptische Vernunft: Fontenelle und die Anfänge der Aufklärung," *Germanisch-Romanische Monatsschrift, 32* (1982), 156–73.

12. Eddie Shoesmith, "The continental controversy over Arbuthnot's argument for divine providence," *Historia mathematica, 14* (1987), 133–46; Richard Olson, "Tory-High Church opposition to science and scientism in the 18th century: The works of John Arbuthnot, Jonathan Swift, and Samuel Johnson," in John G. Burke, ed., *The uses of science in the age of Newton* (Berkeley: University of California Press, 1983), 171–204.

13. Michael Heyd, *Between orthodoxy and the Enlightenment: Jean-Robert Chouet and the introduction of Cartesian science in the Academy of Geneva,* International archives of the history of ideas, 96 (The Hague: Nijhoff; Jerusalem: Magnes Press, 1982); Gereon Wolters, *Basis und Deduktion: Studien zur Entstehung und Bedeutung der Theorie der axiomatischen Methode bei J.H. Lambert (1728–1777),* Quellen und Studien zur Philosophie, 15 (Berlin: de Gruyter, 1980); C. Hakfoort, "Christian Wolff tussen Cartesianen en Newtonianen," *Tijdschrift voor de geschiedenis der geneeskunde, natuurwetenschappen, wiskunde en techniek, 5* (1982), 27–38; Jean Dhombres, "Un style axiomatique dans l'écriture de la physique mathématique au 18ème siècle: Daniel Bernoulli et la composition des forces," *Sciences et techniques en perspective, 11* (1986–7), 1–68; Daniel Klein, "Deductive economic methodology in the French Enlightenment: Condillac and Destutt de Tracy," *History of political economy, 17* (1985), 51–71; Wolfgang Röd, "Descartes dans la philosophie universitaire allemande du XVIIIe siècle," *Etudes philosophiques,* 1985, 161–73; Pierre Costabel, "Euler lecteur de Descartes," *Dix-huitième siècle, 18* (1986), 281–8; Mariafranca Spallanzani, "Descartes dans l'Encyclopédie: La méthode," *Recherches sur le XVIIème siècle, 8* (1986), 103–25; cf. Spallanzani, "Notes sur le cartésianisme dans l'Encyclopédie," *Studies on Voltaire and the eighteenth century, 216* (1983), 326–7.

influence of what one author calls the mathematical method-model.[14]

What was reasonable or certain in the 17th and 18th centuries was intimately bound up with questions of probability and risk, in contexts ranging from law to morality, economics to public health. The richness of the analysis and examples in recent studies on probability will warrant careful attention in subsequent exploration of the *esprit géométrique* and the quantifying spirit.[15] As the 18th-century controversy over inoculation against smallpox makes clear, the stakes could be high in disputes over the validity of quantitative arguments.[16]

14. Hans-Jürgen Engfer, *Philosophie als Analysis: Studien zur Entwicklung philosophischer Analysiskonzeptionen unter dem Einfluss mathematischer Methodenmodelle im 17. um frühen 18. Jahrhundert* (Stuttgart-Bad Cannstatt: Frommann-Holzboog, 1982). For Wolff's influence and the ways in which he built on 17th-century *systèmes*, see Sonia Carboncini, "L'Encyclopédie et Christian Wolff: A propos de quelques articles anonymes," *Etudes philosophiques,* 1987, 489–504; Hakfoort, "Christian Wolff tussen Cartesianen en Newtonianen" (note 13); Werner Schneiders, ed., *Christian Wolff, 1679–1754: Interpretation zu seiner Philosophie und deren Wirkung. Mit einer Bibliographie der Wolff-Literatur,* (Hamburg: Meiner, 1983); Röd, "Descartes dans la philosophie universitaire allemande"; and Fabio Todesco, "Dal 'calcolo logico' alla 'riforma della metafisica': Johann Heinrich Lambert tra Wolff e Locke," *Rivista di storia della filosofia, 77* (1986), 337–58.

15. Lorraine J. Daston, "Probabilistic expectation and rationality in classical probability theory," *Historia mathematica, 7* (1980), 234–60, "Mathematical probability and the reasonable man of the 18th century," in *History and philosophy of science: Selected papers,* ed. Joseph W. Dauben and Virginia Staudt Sexton (New York: New York Academy of Sciences, 1983), 57–62, and *Classical probability in the Enlightenment* (Princeton: Princeton University Press, 1988); Lorenz Krüger, Lorraine J. Daston, and Michael Heidelberger, eds., *The probabilistic revolution,* 2 vols. (Cambridge: MIT Press, 1987), esp. *1: Ideas in history*; Luigi Cataldi Madonna, "Wahrscheinlichkeit und wahrscheinliches Wissen in der Philosophie Christian Wolffs," *Studia Leibnitiana, 19* (1987), 2–40; Douglas Lane Patey, *Probability and literary form: Philosophic theory and literary practice in the Augustan age* (Cambridge: Cambridge University Press, 1984).

16. The many studies of smallpox and its prevention in the 18th century include Andrea Rusnock, "When counting counts: The reception of quantitative arguments in eighteenth-century England and France" (unpublished paper, History of Science Society meeting, 1989); Jean-Claude David, "À la querelle de l'inoculation en 1763: Trois lettres inédites de Suard et du Chevalier d'Eon," *Dix-huitième siècle, 17* (1985), 271–84; Maxine Van de Wetering, "A reconsideration of the inoculation controversy," *New England quarterly, 58* (1985), 46–67; Jean-François de Raymond, *Querelle de l'inoculation, ou, Préhistoire de la vaccination* (Paris: Vrin, 1982); Antoinette S. Emch-Dériaz, "L'inoculation justifiée—or was it?" *Eighteenth century life, 7:2* (1982), 65–72. Patricia Cline Cohen, *A calculating people: The spread of numeracy in early America* (Chicago: University of Chicago Press, 1982) also addresses the issue of smallpox statistics and risks.

Such arguments could still carry weight even when the numbers were soft or when measurement was out of the question, as in discussions of moral arithmetic, meandering rivers, or thermometers for female emotional response, as calibrated from modesty through impudence. Here it is necessary to attend to what quantification promised: useful comparisons whatever the scale, informative models without measurement, precision (clarity, distinctness, intelligibility) rather than a close fit with the real world.[17]

Although moral barometers scarcely belong to the realm of exact science, comparisons of soft and hard quantification may prove instructive. In particular, it is worthwhile measuring the play of the quantifying spirit in the 18th century against conspicuous accomplishments in the mathematization of science in the 17th century and the successes of mathematical physics in the late 18th and early 19th centuries. The articles in *Nature mathematized* explore examples of 17th-century exact science that enlightened thinkers subsequently found so persuasive. A recent issue of *Revue d'histoire des sciences* investigates the "conquest of new territories" by mathematical science between 1780 and 1830. And Jean Dhombres links the achievements of *mathématisation* with the nature of the French scientific community in the half-century between 1775 and 1825 in another recent article.[18]

17. Daston, "The quantification of probability" (unpublished paper, History of Science Society meeting, 1989) and *Classical probability in the Enlightenment* (note 15); Garland P. Brooks and Sergei K. Aalto, "The rise and fall of moral algebra: Francis Hutcheson and the mathematization of psychology," *Journal of the history of the behavioral sciences, 17* (1981), 343–56; Robin E. Rider, *Mathematics in the Enlightenment: A study of algebra, 1685–1800* (Ph.D. dissertation, University of California, Berkeley, 1980; *DAI* 42/01A, 351), esp. chap. 9; Marguerite Carozzi, "From the concept of salient and reentrant angles by Louis Bourguet to Nicolas Desmarest's description of meandering rivers," *Archives des sciences* (Geneva), *39* (1986), 25–51; Terry Castle, "The female thermometer," *Representations, 17* (1987), 1–27. Cf. Mark H. Waymack, *Moral philosophy and Newtonianism in the Scottish Enlightenment: A study of the moral philosophies of Gershom Carmichael, Francis Hutcheson, David Hume, and Adam Smith* (Ph.D. dissertation, Johns Hopkins, 1986; *DAI* 48/02A, 413), and a late 17th-century example in Paul McReynolds and Klaus Ludwig, "Christian Thomasius and the origin of psychological rating scales," *Isis, 75* (1984), 546–53.

18. William R. Shea, ed., *Nature mathematized: Historical and philosophical case studies in classical modern natural philosophy* (Dordrecht: D. Reidel, 1983); *Revue d'histoire des sciences, 42:1* (1989), special issue "La mathématisation 1780–1830,"

INCULCATING THE SPIRIT

Academies

It is a commonplace to see organized science as "entirely recast under the dominion" of the scientific academies and societies of the 18th century.[19] Scientific academies and societies played a significant role in establishing the importance and extending the application of mathematical or quantitative methods. Much recent literature addresses questions of the founding and organization of academies and societies, and may thus enlighten us about the 18th-century disciplinary map. It may also contain clues about the place of mathematics and the mathematically adept in these institutions and about the ways in which academic prestige or authority could encourage the use of mathematics. Recent studies look at academies and societies in general[20] and in particular.[21] Studies of the use of mathematical

---

quotation on 3; Jean Dhombres, "Mathématisation et communauté scientifique française (1775–1825)," *Archives internationales d'histoire des sciences, 36* (1986), 249–93. The papers delivered at the international workshop, "The quantification of scientific concepts in their social context" (Tel Aviv, 1986) also speak to issues of quantification from the 17th century through the late 20th century.

19. James E. McClellan III, *Science reorganized: Scientific societies in the 18th century* (New York: Columbia University Press, 1985), xix.

20. Ibid.; Daniel Roche, *Le siècle des lumières en province: Académies et académiciens provinciaux, 1680–1789*, 2 vols. (Paris: Ecole des hautes etudes en sciences sociales, Mouton, 1978); Henry E. Lowood, *Patriotism, profit, and the promotion of science in the German enlightenment: The economic and scientific societies, 1760–1815* (Ph.D. dissertation, University of California, Berkeley, 1987; *DAI* 49/06A, 1563).

21. David M. Griffiths, "The early years of the Petersburg Academy of sciences as reflected in recent Soviet literature," *Canadian-American Slavic studies, 14* (1980), 436–45; Elizabeth R. Kindleberger, *The Société Royal des Sciences de Montpellier, 1706–1793* (Ph.D. dissertation, Johns Hopkins University, 1979; *DAI* 40/05A, 2816); Roy S. Porter, "Science, provincial culture, and public opinion in Enlightenment England," *British journal for eighteenth-century studies, 3* (1980), 2–46; Roger L. Emerson, "The Philosophical Society of Edinburgh, 1748–1768," *British journal of history of science, 14* (1981), 133–76; Rhoda Rappaport, "The liberties of the Paris Academy of Sciences, 1716–1785," in Harry Woolf, ed., *The analytic spirit: Essays in the history of science in honor of Henry Guerlac* (Ithaca, NY: Cornell University Press, 1981), 225–53; James R. Hansen, *Scientific fellowship in a Swiss community enlightenment: A history of Zurich's Physical Society, 1746–1798* (Ph.D. dissertation, Ohio State University, 1981; *DAI* 42/05A, 2269); Ludwig Hammermayer, *Geschichte der Bayerischen Akademie der Wissenschaften, 1759–1807*, 2 vols. (Munich: Beck, 1983); Vincenzo Ferrone, "Tecnocrati militari e scienziati nel Piemonte dell'Antico

methods in the 18th century might follow the lead of works that discuss the place of particular sciences in specific academic contexts.[22]

Studies that examine the role of individuals in academies and hence in the life of science, especially in conjunction with archival or edited correspondence, promise useful information. Historians have begun, for example, to trace the powerful academic influence of Leonhard Euler.[23] Analysis of the éloges of the Paris Academy of Sciences yields insights both about its perpetual secretaries, including Fontenelle and Condorcet, and about the scientific ideologies and standards they promoted.[24] The publications and prizes of academies and societies served to spread the quantifying spirit, and deserve more attention.

Regime: Alle origini della Reale Accademia delle Scienze di Torino," *Rivista di storia italiana, 96* (1984), 414–509; James E. McClellan III, "The Académie Royale des Sciences, 1699–1793: A statistical portrait," *Isis, 72* (1981), 541–76; Ugo Baldini, "L'attività scientifica nelle accademie lombarde del Settecento," in *Economia, istituzioni, cultura in Lombardia nell'età di Maria Teresa*, 3 vols. (Bologna: Il Mulino, 1982), *2*, 503–32; Valeria Molla Losito, "La Società Patriottica di Milano (1776–1796)," in ibid., *3*, 1039–56; Andreas Kleinert, "Mathematik und anorganische Naturwissenschaften," in Rudolf Vierhaus, ed., *Wissenschaften im Zeitalter der Aufklärung* (Göttingen: Vendenhoeck & Ruprecht, 1985), 218–48, esp. section IV; Michel Taillefer, *Une académie interprète des Lumières: L'Académie des Sciences, Inscriptions, et Belles-Lettres de Toulouse au XVIIIe siècle* (Paris: CNRS, 1984).

22. E.g., E.P. Ozhigova, *Matematika v Peterburgskoy Akademii Nauk v kontse XVIII-pervoy polovine XIX veka* (Leningrad: Nauka, 1980); Jacqueline Giroux, "Genèse de la météorologie scientifique dans les milieux de l'Académie de Dijon au XVIIIe siècle," Académie des sciences, arts, et belles-lettres (Dijon), *Mémoires, 125* (1981–2), 135–55; Dhombres, "Mathématisation" (note 18). Cf. Theodore S. Feldman, *The history of meteorology, 1750–1800: A study in the quantification of experimental physics* (Ph.D. dissertation, University of California, Berkeley, 1983, *DAI* 43/05A, 922); and Karl Hufbauer, *The formation of the German chemical community, 1720–1795* (Berkeley: University of California Press, 1982).

23. Wolfgang Knobloch, ed., *Leonhard Eulers Wirken an der Berliner Akademie der Wissenschaften, 1741–1766: Specialinventar Regesten der Euler-Documente aus dem Zentralen Archiv der Akademie der Wissenschaften der DDR* (Berlin: Akademie-Verlag, 1974); *Leonhard Euler, 1707–1783: Beiträge zu Leben und Werk* (Basel: Birkhäuser, 1983), including Judith Kh. Kopelevic, "L. Euler und die Petersburger Akademie," 373–84; Rüdiger Thiele, *Leonhard Euler* (Leipzig: BSB B.G. Teubner Verlagsgesellschaft, 1982).

24. Paul, *Science and immortality* (note 11).

Education

Educational institutions and strategies also played a role. The significance of science, especially mathematics, in the technical training establishment in France was sketched in older works that remain useful.[25] Such works have recently been supplemented by studies of the royal engineering corps and military school in France;[26] a comparative perspective is afforded by analysis of military and engineering training in old and New Spain.[27] All need to be read against the evidence assembled in the work of Charles Gillispie,[28] in biographical dictionaries of 18th-century engineers,[29] in documentary histories of engineering,[30] and in general histories of education.[31] The roles played by academicians in beefing up the technical curricula bear more scrutiny.[32] Data and studies of the teaching of mathematics and

25. E.g., René Taton, ed., *Enseignement et diffusion des sciences en France au XVIIIe siècle*, recently republished in six parts (Paris: Hermann, 1986); the studies by François de Dainville collected in *L'éducation des jésuites, 16e–18e siècles* (Paris: Minuit, 1978). For England, cf. Geoffrey Howson, *A history of mathematics education in England* (Cambridge: Cambridge University Press, 1982).

26. Anne Blanchard, *Les ingénieurs du "Roy" de Louis XIV à Louis XVI: Etude du Corps des Fortifications*, Collection du Centre d'histoire militaire et d'etudes de defense nationale de Montepellier, 9 (Montpellier, 1979); Jean-Claude David, "Grimm, Lalande et le quart de cercle de l'Ecole Royale Militaire," *Dix-huitième siècle, 14* (1982), 277–87; J.-F. Pernot, "Une important contribution à la connaissance des ingénieurs d'Ancien Régime," *Revue d'histoire des sciences, 35* (1982), 275–80.

27. José Omar Moncada Maya, "Una aproximación al estudio del Cuerpo de Ingenieros Militares en la Nueva España," *Quipu, 3* (1986), 55–66; Antonio Lafuente and José Luis Peset, "Las academias militares y la inversión en ciencia en la España ilustrada (1750–1760)," *Dynamis, 2* (1982), 193–209.

28. C.C. Gillispie, *Science and polity in France at the end of the Old Regime* (Princeton: Princeton University Press, 1980).

29. Anne Blanchard's massive *Dictionnaire des ingénieurs militaires, 1691–1791* (Montpellier: CNRS, 1981); *Los ingenieros militares en España siglo XVIII: Repertorio biográfico e inventario de su labor cientifica y espacial* (Barcelona: University Barcelona, 1983).

30. Paul K. Walker, *Engineers of independence: A documentary history of the Army Engineers in the American Revolution, 1775–1783* (Washington: Historical Division, Office of the Chief of Engineers, 1981).

31. L.W.B. Brockliss, *French higher education in the 17th and 18th centuries: A cultural history* (Oxford: Clarendon Press, 1987), esp. 337–443 on natural sciences.

32. Cf. Ferrone, "Tecnocrati militari e scienziati" (note 21), and the works cited in note 38 below. Cf. Thérèse Charmasson, *L'enseignement technique de la Revolution à nos jours. Textes officiels avec introduction, notes et annexes* (Paris: Institut national de recherche pedagogique, 1987).

science in colleges and universities offer possibilities for comparison.[33]

The mathematical content, rhetoric, and appeal of popular courses and self-help books also helped to secure a place for quantification in popular culture. Studies of public courses and of the marketing of science thus warrant further investigation.[34] So too do mathematical periodicals, textbooks for the self-taught, and the corps of mathematical author/practitioners.[35]

For the end of the century, consideration of education and of the French Revolution raises twin issues of continuity and reform, both

33. Hanspeter Marti, *Philosophische Dissertationen deutscher Universitäten, 1660–1750: Eine Auswahlbibliographie* (Munich: Saur, 1982); Karl A.F. Fischer, "Jesuiten-Mathematiker in der französischen und italienischen Assistenz bis 1762 bzw. 1773," *Archivum historicum Societatis Jesu, 52* (1983), 52–92; Marie-Madeleine Compère and Dominique Julia, eds., *Les Collèges français, 16e–18e siècles* (Paris: INRP and CNRS, 1984); Ugo Baldini, "L'insegnamento fisico-matematico a Pavia alle soglie dell'età teresiana," in *Economia, istituzioni, cultura in Lombardia, 3* (note 21), 863–86; Luigi Belloni, "L'insegnamento delle scienze sperimentali a Milano," in ibid., *2*, 441–9.

34. John S. Reid, "Late 18th-century adult education in the sciences at Aberdeen: The natural philosophy classes of Professor Patrick Copland," in Jennifer J. Carter and Joan H. Pittock, eds., *Aberdeen and the Enlightenment* (Aberdeen: Aberdeen University Press, 1987), 168–79; Gad Freudenthal, "Littérature et sciences de la nature en France au début du XVIIIe siècle: Pierre Polinière, l'introduction de l'enseignement de la physique expérimentale, l'Université de Paris, et l'Arrêt burlesque de Boileau," *Revue de synthèse, 101* (1980), 267–95; John R. Millburn, "The London evening courses of Benjamin Martin and James Ferguson: 18th-century lectures on experimental philosophy," *Annals of science, 40* (1983), 437–55, and *Retailer of the sciences: Benjamin Martin's scientific instrument catalogues, 1756–1782* (London: Vade-Mecum, 1986); Simon Schaffer, "Natural philosophy and public spectacle in the 18th century," *History of science, 21* (1983), 1–43; Alexander Rüger, "Populäre Naturwissenschaft in Nürnberg am Ende des 18. Jahrhunderts: Reisende Experimentatoren, öffentliche Vorlesungen und physikalisches Spielzeug," *Berichte zur Wissenschaftsgeschichte, 5* (1982), 173–91. Cf. J.L. Heilbron, *Physics at the Royal Society during Newton's presidency* (Los Angeles: William Andrews Clark Memorial Library, 1983); and Robin E. Rider, *The show of science* (Berkeley: The Bancroft Library, 1983).

35. Ruth Wallis and Peter Wallis, "Female philomaths," *Historia mathematica, 7* (1980), 57–64; Robert R. Bataille, "Robert Heath, Thomas Cowper, and 18th-century mathematical journalism," Bibliographical Society of America, *Papers, 81* (1987), 339–43; Dhombres, "Mathématisation" (note 18), and his account of scientific books in Robert Darnton and Daniel Roche, eds., *Revolution in print: The press in France, 1775–1800* (Berkeley: University of California Press, 1989); R.V. Wallis and P.J. Wallis, *Biobibliography of British mathematics and its applications*, Part 2: 1701–1760 (Newcastle upon Tyne: Project for Historical Biobibliography, 1986).

pertinent to the place of mathematics in a revolutionary curriculum.[36] Numerous studies unravel the organization and ideology of new institutions and trace their teachers, students, and lessons.[37]

OTHER DIRECTIONS

It is in the nature of bibliographical essays to point to topics or directions of research worth pursuing. Out of the diversity of the chapters in this volume on *The quantifying spirit*, as well as from a

36. For bibliography, see Ronald J. Caldwell, *The era of the French Revolution: A bibliography of the history of western civilization, 1789–1799,* 2 vols. (New York: Garland Pub., 1985) and *Aux livres citoyens!* (Livry Gargan: Sciences Sociales Export, 1989), reprinted from a special issue of *Préfaces*, 1989; cf. Samuel F. Scott and Barry Rothaus, eds., *Historical dictionary of the French Revolution 1789-1799* (Westport, Conn.: Greenwood Press, 1985). On a pre-revolutionary reform proposal, see Arnold de Lesseux, "L'éducation publique d'après Guyton de Morveau," Académie des sciences, arts, et belles-lettres (Dijon), *Mémoires, 123* (1976–8; published 1979), 207–39.

37. In general, Robert R. Palmer, *The improvement of humanity: Education and the French Revolution* (Princeton: Princeton University Press, 1985); John Lough, *The philosophes and post-revolutionary France* (Oxford: Clarendon Press, 1982); Jean Dhombres, "Formation des cadres scientifiques et techniques: La marque des débuts de la Révolution," Institut des recherches marxistes, *Cahiers d'histoire, 32* (1988), 189–200; and Janis Langins, *La République avait besoin de savants: Les débuts de L'Ecole Polytechnique—L'Ecole Centrale des Travaux Publics et les cours révolutionnaires de l'an III* (Paris: Belin, 1987). On the Ecole polytechnique, Terry Shinn, *L'Ecole polytechnique: 1794–1914* (Paris: Presses de la Fondation nationale des sciences politiques, 1980); Ambroise Fourcy, *Histoire de l'Ecole polytechnique*, new ed. (Paris: Belin, 1987); Jean-Pierre Callot with Philippe Journau, *Histoire de l'Ecole polytechnique* (Paris: Charles Lavauzelle, 1982); Janis Langins, *The Ecole Polytechnique (1794–1804): From encyclopaedic school to military institution* (Ph.D. dissertation, University of Toronto, 1979; *DAI* 40/12A, 6396); Ivor Grattan-Guinness, "Euler's mathematics in French science, 1795–1815," in *Leonhard Euler, 1707–1783: Beiträge zu Leben und Werk* (note 23), 395–408, esp. section 6. On the role of mathematicians in educational reform, Langins, "Sur l'enseignement et les examens à l'Ecole Polytechnique sous la Directoire: À propos d'une lettre inédite de Laplace," *Revue d'histoire des sciences, 40* (1987), 145–177; Catherine Kintzler, *Condorcet: L'instruction publique et la naissance du citoyen* (Paris: Le Sycomore, 1984); Langins, "Une lettre inédite de Fourier sur l'enseignement destiné aux ingénieurs en 1797," *Revue d'histoire des sciences, 34* (1981), 193–207; Robin E. Rider, "Poisson and algebra: Against an 18th-century background," Michel Métivier, Pierre Costabel, and Pierre Dugac, eds., *Siméon Denis Poisson et la science de son temps* (Paris: Ecole polytechnique, 1981), 167–76; Eduard Glas, "On the dynamics of mathematical change in the case of Monge and the French Revolution," *Studies in history and philosophy of science, 17* (1986), 249–68.

survey of recent literature on the 18th century, emerge a number of possibilities, which may be collected under headings of economics and polity, discourse, and evidence of opposition.

Economics and Polity

In the discussions (in the "Introductory essay" and chapters 5, 6, 10, 11, and 12) of political arithmetic, economic development and policy, and meteorology, as well as in the work of Keith Baker and Charles Gillispie, we see signposts pointing to economics and polity as loci for the operation of the quantifying spirit. Vital statistics had economic value to tax collectors, conscription agents, and actuaries, and government bureaus and functionaries sought to police health as they managed princely coffers, coal, bread, and trees. The expanding literature on vital statistics and on public medicine in the 18th century thus promises a reasonable return on scholarly investment.[38]

38. Studies of medicine and public welfare include, but should scarcely be limited to, Theodore M. Brown, "J.P. Frank's 'Medical police' and its implications for medicalization in America," in Marten de Vries et al., eds., *The use and abuse of medicine* (New York: Praeger, 1982), 208–18; Othmar Keel, "The politics of health and the institutionalisation of clinical practices in Europe in the second half of the 18th century," in W.F. Bynum and Roy Porter, eds., *William Hunter and the 18th-century medical world* (Cambridge: Cambridge University Press, 1985); Mary E. Lindemann, *Producing policed man: Poor relief, population policies, and medical care in Hamburg, 1750–1806* (Ph.D. dissertation, University of Cincinnati, 1980; *DAI* 41/10A, 4474); Harvey Mitchell, "Politics in the service of knowledge: The debate over the administration of medicine and welfare in late 18th-century France," *Social history, 6* (1981), 185–207; Jean-Pierre Goubert, "The medicalization of French society at the end of the Ancien Régime," in Lloyd G. Stevenson, ed., *A celebration of medical history* (Baltimore: Johns Hopkins University Press, 1982), 157–79. On vital statistics and demography, see J. Hoock, "Sciences camérales et statistique démographique en Allemagne aux XVIIe et XVIIIe siècles," *Annales de démographie historique* (1979), 145–55; Jacqueline Hecht, "Johann Peter Süssmilch. Point alpha ou omega de la science démographique naive," ibid. (1979), 101–34; Peter Buck, "People who counted: Political arithmetic in the 18th century," *Isis, 73* (1982), 28–45; Thomas R. Forbes, "Births and deaths in a London parish: The record from the registers, 1654-1693 and 1729–1743," *Bulletin of the history of medicine, 55* (1981), 371–91; Patricia Cline Cohen, "Death and taxes: The domain of numbers in 18th century popular culture," in Stephen H. Cutcliffe, ed., *Science and technology in the 18th century: Essays of the Lawrence Henry Gipson Institute for 18th Century Studies* (Bethlehem, Penn.: Gipson Institute, 1984) and her *A calculating people* (note 16); *Pehr Wargentin, den svenska statisikens fader: En minnesskriff me sju originaluppsatser ur Kungl. Svenska Vetenskapsakademiens Handlingar för åren 1754, 1755 samt 1766* (Stockholm: Statistika Centralbyrån, 1983); J. Lecuir, "Deux siècles après: Montyon, véritable auteur des 'Recherches et considérations sur la population de la France' de

Cameralism and political economy, as they developed in theory and practice and in local and national settings, should not be ignored,[39] nor studies of Adam Smith, influences on him, and impact of his ideas.[40] The economic writings and policies of Physiocrats and Idéologues, especially as they relate to the *philosophes* and to intellectual movements in the Enlightenment, merit another look with the quantifying spirit in mind.[41] The economic and political import of

---

Moheau," *Annales de démographie historique* (1979), 195–249; William Coleman, "Inventing demography: Montyon on hygiene and the state," in Everett Mendelsohn, ed., *Transformation and tradition in the sciences: Essays in honor of I. Bernard Cohen* (Cambridge: Cambridge University Press, 1984), 215–33. On J.S. Mill's skepticism in the face of vital statistics, see Spencer Davis, "Scottish philosophy and political economy," in Donald C. Mell, Jr., et al., eds., *Man, God, and nature in the Enlightenment* (East Lansing, Mich.: Colleagues Press, 1988).

39. On political economy, Keith Tribe, "University teaching on cameralism in 18th-century Germany," *Studi settecenteschi, 7–8* (1985–6), 55–69, and his *Governing economy: The reformation of German economic discourse, 1750–1840* (Cambridge and New York: Cambridge University Press, 1988); Terence Hutchison, *Before Adam Smith: The emergence of political economy, 1662–1776* (Oxford: Blackwell, 1988); Sergio Cremaschi, *Il sistema della ricchezza: Economia politica e problema del metodo in Adam Smith* (Milan: Angeli, 1984); Girolamo Imbruglia, "Economia e politica nel *De l'esprit* di Helvétius," Naples, University, Facoltà di Lettere, Filosofia e Magistero, *Annali, 20* (1977–8), 237–81; Hans Erich Bödeker, "Der staatswissenschaftliche Fächersystem im 18. Jahrhundert," in *Wissenschaften im Zeitalter der Aufklärung,* 143–62; F. Etner, "L'Ancien Régime et le calcul économique," *Economy and society, 18:3* (1984), special issue on "Aspects de l'économie politique en France au XVIIIe siècle," and his book on "calcul économique" in France. On 18th-century appeals to quantification and classification in social analysis, P.J. Corfield, "Class by name and number in 18th-century Britain," *History, 72* (1987), 38–61; Gianni Vaggi, "Social classes and income distribution in 18th-century economics," *History of European ideas, 9* (1988), 171–82.

40. Among them, Andrew S. Skinner, "Adam Smith: Rhetoric and the communication of ideas," in A.W. Coats, ed., *Methodological controversy in economics: Historical essays in honor of T.W. Hutchison* (Greenwich, Conn.: JAI Press, 1983), 71–88; Donald Winch, "Science and the legislator: Adam Smith and after," *Economics journal, 93* (1983), 501–20; M. Dey, *Adam Smith and Adam Ferguson: Philosophy, economic change, and class limitations in 18th-century Scotland* (Ph.D. dissertation, Aberdeen University, 1985); Cremaschi, *Il sistema della ricchezza* (note 39); Norriss S. Hetherington, "Isaac Newton's influence on Adam Smith's natural laws in economics," *Journal of the history of ideas, 44* (1983), 497–505; Adam Smith, *Essays on philosophical subjects*, ed. W.P.D. Wightman and J.C. Bryce (Oxford: Clarendon Press, 1980), one in a set of recent scholarly editions of Adam Smith's writings; R.H. Campbell and A.S. Skinner, *Adam Smith* (London: Croom Helm, 1982); cf. Martha Bolar Lightwood, *A selected bibliography of significant works about Adam Smith* (Philadelphia: University of Pennsylvania Press, 1984).

41. Gianni Vaggi, *The economics of François Quesnay* (Durham, N. Car.: Duke University Press, 1987); Christian Bordes and J. Morange, eds., *Turgot, économiste et*

scientism and the place of mathematics in scientism have relevance as well.[42]

Discourse

Mathematical metaphors and images frequently found their way into discourse in the 18th century, whatever the topic, and hence illustrate both the cultural appeal of quantification and a means by which the quantification message was reinforced and spread. Studies of language for other scientific fields explore the malleability and power of metaphor, and suggest useful approaches for considering the language of quantification.[43]

---

administrateur (Paris: PUF, [1981]); Marco Minerbi, "I presupposti dell'analisi economica del fisiocratici," in *Lezioni sull'illuminismo* (Milan: Feltrinelli, 1980); Daniel Klein, "Deductive economic methodology in the French Enlightenment: Condillac and Destutt de Tracy," *History of political economy, 17* (1985), 51–71; Stephen F. Gudeman, "Physiocracy: A natural economics," *American ethnologist, 7* (1980), 240–58; Leonora Cohen Rosenfield, "La Mettrie and Quesnay, physician-*philosophes* of the Enlightenment," in Alfred J. Bingham and Virgil W. Topazio, eds., *Enlightenment studies in honour of Lester G. Crocker* (Oxford: Voltaire Foundation, 1979); cf. Hutchison, *Before Adam Smith* (note 39).

42. A few titles are Keith Michael Baker, "Scientism at the end of the Old Regime: Reflections on a theme of Professor Charles Gillispie," *Minerva, 25* (1987), 21–34; Paul Alkon, "Changing the calendar," *Eighteenth-century life, 7:2* (1982), 1–18; Harvey Mitchell, "Politics in the service of knowledge: The debate over the administration of medicine and welfare in late 18th-century France," *Social history, 6* (1981), 185–207.

43. See, for example, Geoffrey Cantor, "Weighing light: The role of metaphor in 18th-century optical discourse," in Andrew E. Benjamin et al., eds., *The figural and the literal: Problems of language in the history of science and philosophy, 1630–1800* (Manchester: Manchester University Press, 1986), 124–46; Wilda C. Anderson, *Between the library and the laboratory: The language of chemistry in eighteenth-century France* (Baltimore and London: Johns Hopkins University Press, 1984); Rhoda Rappaport, "Borrowed words: Problems of vocabulary in 18th-century geology," *British journal for the history of science, 15* (1982), 27–44; Imbruglia, "Economia e politica nel *De l'esprit* di Helvétius" (note 39); Tribe, *Governing economy* (note 39); Martin Rudwick, "The emergence of a visual language of geology, 1760–1840," *History of science, 14* (1976), 149–95; cf. Roger Lewinter, "La quadrature du cercle: Remarques sur Diderot et l'Encyclopédie," *Revue de métaphysique et de morale, 89* (1984), 226–31. Cantor speaks of metaphors as "malleable" (142) and claims that they quietly direct research.

Opposition

If there was a quantifying spirit or *esprit géométrique* in the 18th century, there was surely a critical spirit as well, and the voices raised in favor of rethinking the place and power of mathematics and quantification ought also to be heeded. Buffon's criticism of mathematics and Diderot's quotable doubts about the future of mathematics[44] speak to the prevailing faith in mathematics and mathematicians.[45] Even d'Alembert wrote with some skepticism about the "esprit de calcul," which he saw as the dominant "goût de philosophie" in the 18th century.[46] Tracts that decried the "abuse of mathematics in natural science,"[47] need to be set against 18th-century pæans in praise of mathematics.

The bibliographical burden can be made more enjoyable by seeing how mathematics and quantification fared in 18th-century scientific satires. Jonathan Swift's Gulliver watched savants labor to extract sunbeams from cucumbers; he heard their scheme for establishing a

44. In *On the interpretation of nature*, quoted for example in Ernst Cassirer, *The philosophy of the Enlightenment*, transl. Fritz C.A. Koelln and James P. Pettegrove (Princeton: Princeton University Press, 1951), 74; cf. Hankins, *Science and the Enlightenment* (note 11), 169.

45. On Diderot and mathematics, Jean Dhombres, "Quelques rencontres de Diderot avec les mathématiques," in Anne-Marie Chouillet, ed., *Colloque international Diderot (1713–1784)* (Paris: Aux Amateurs de Livres, 1985); Robert Morin, *Diderot et l'imagination* (Paris: Les Belles Lettres, 1987), esp. the chapter on mathematical imagination; Michael Kessler, "A puzzle concerning Diderot's presentation of Saunderson's *Palpable arithmetic*," *Diderot studies, 20* (1981), 159–73; Merle L. Perkins, *Diderot and the time-space continuum: His philosophy, aesthetics and politics,* Studies on Voltaire and the 18th century, 211 (Oxford: Voltaire Foundation, 1982), esp. chaps. 3–4; Charles Dedeyan, *Diderot et la pensée anglaise* (Florence: Olschki, 1987), part III; Renate Wahsner, "Das Verhältnis von Mathematik und Physik aus der Sicht von Denis Diderot," *NTM, 24* (1987), 13–20; Anne-Marie Chouillet, ed., *1984: L'année Diderot*, supplement to *Dix-huitième siècle, 17* (1985); and Frederick A. Spear, *Bibliographie de Diderot: Repertoire analytique international*, Histoire des idées et critique littéraire, 187 (Geneva: Droz, 1980), and its supplements.

46. *Encyclopédie, 6* (1756), s.v. "Fluide," quoted in Costabel, "Euler lecteur de Descartes" (note 13), 287. Nonetheless, Costabel considers that it was confidence in mathematics that "profoundly united Euler and d'Alembert," whatever their differences (p. 286). Cf. d'Alembert, *Essai d'une nouvelle théorie de la résistance des fluides* (Paris: Chez David l'ainé, 1752).

47. Giuseppe Mosca, *Vita di Giovambattista Morgagni...con due lettere, l'una intorno all'abuso della matematica nella scienza naturale* (Naples: V. Manfredi, 1764).

universal language by abolishing all words; and he talked with an innovative projector in speculative learning. This worthy had secured the services of forty pupils to crank the handles on a frame twenty feet square: it carried a grid of "all of the Words of their language," based on careful computation of the proportions obtaining among the various parts of speech. Commanded first to crank, then to read and record the results, the pupils might thereby generate automatically books on all arts and sciences.[48] That is not the method by which this book or this essay was composed.

48. Daryll M. Anderson, *Satires of science in the 17th and 18th centuries* (Ph.D. dissertation, University of Louisville, 1980; *DAI* 41/07A, 3114); Olson, "Tory-High Church opposition" (note 12); Brooks and Aalto, "The rise and fall of moral algebra" (note 17); Eric A. Weiss, "Jonathan Swift's computing invention," *Annals of the history of computing,* 7 (1985), 164–5, which reproduces a portion of chap. 5 of Gulliver's third voyage.

Index

Academia del Cimento, 146

Academy of Sciences. *See* Paris Academy of Sciences

Academy of Sciences (St. Petersburg), 270

Accuracy. *See* Measurement

Achromatic lens, 193

Acids (theory of), 95

Adam, Jean, 366

Adanson, Michel: criticized Linnæus, 57, 61, 79; encyclopedic efforts of, 69; "natural method" of botanical classification, 79–81

Æsthetics, 38

Aggregaten, 49–50

Agricola, George, 248

Agriculture. *See* Climatology

Algebra, 118, 136–7, 138, 340, 377; the language of mathematics, 115, 118, 124, 135; order of algebraic curves defined, 100–2

Algology, 69

Allgemeine Geschichte der Natur in alfabetischer Ordnung (Martini), 68

Allgemeines oekonomisches Forstmagazin (Stahl), 319

Alphabet: alphabetical order of encyclopedias, 48; "alphabetic chaos," 123; alphabetism, 52; used by Linnæus, 55. *See also* Language

A mathematical treatise on the effect of water mills (Elvius), 296

Analysis, 83, 116; inductive method of, 293, 297, 377

Analytic geometry. *See under* Geometry

Analytic triangle, 101–2, 102n

Anatomy, 107

Ancien Régime, 22, 207

Anfangsgründe der Probierkunst (Cramer), 252

Animalcule theory, 277

Arbuthnot, John, 384

Architecture, 238, 239

Arcs of meridian. *See* Measurements of arcs of meridian; Metric system

Arithmetic mentality, 71. *See also* Mathematics; Quantifying

Aristotelian tradition, 76, 345, 366

Armaments, 18. *See also* Military

Artificial languages. *See* Universal languages

Astronomy, 375–6, 377; aberration of stars, 7; inclination of earth's axis, 7; nutation of the earth, 7; mathematization of, 88

Atemporality: in Linnæan botany, 78; characterless time in meteorology, 152

Atheism, 32, 35, 37, 41

Atomic theory, 263

Aune, 207–8

Bacon, Francis, 114

Baker, Keith, 392

Balance, 248, 252, 254, 298. *See also* Chemical balance; Hydrostatics

Banks, Joseph, 186, 187
Banksians, 185, 186
Barometers, 6, 148–9, 157, 169
Barometric hypsometry, 8
Barometry, 144, 175
Barruel, Etienne, 103–5
Baseline: establishment of, 200–1, 224, 233
"Bastard" offspring (hybrids), 277–8, 279–81, 285
Baumgarten, Alexander Gottlieb, 38
Beccaria, Giovanni Battista, 181
Becher, J.J., 248, 251
Bechstein, Johann Matthäus, 338
Beckmann, Johann, 319
Berch, Anders, 356
Bergman, Tobern, 93–4, 250, 255, 256, 264
Berlin Academy, 43, 132
Berthollet, Claude Louis, Comte, 94
Berzelius, Jons Jacob, 255
Bessel, Friedrich Wilhelm, 204, 205
Beyträge zur Verbesserung der Forstwissenschaft (Beckmann), 319
Biblioteca botanica (Linnæus), 79
Bichat, Xavier, 107
Bilfinger, Georg Bernhard, 37–8
Binomial nomenclature: of chemicals, 93–4, 95–6; of plants, 21, 74, 79, 80; proposed for minerals, 92
Biot, Jean Baptiste, 223
Black, Joseph, 262
Blowpipe, 255–6
Böckmann, Johann, 168–9
Bomare, Valmont de, 68
Bonde, Gustaf, 41
Borda, Charles, 18, 233, 234, 294; role in Paris-Greenwich triangulation, 217, 222
Boscovich, Roger, 181
Bos, Henk, 381
Botanical classification, 79–81
Botanicum, 66
Botany: animalcule theory, 272–3, 277, 287; comparison of Linnean and Adanson approaches, 81–2; systematic model of, 52, 70, 73, 75–8, 108. *See also* Adanson, Michel; Binomial nomenclature; Linnæus, Carolus; Natural history; Plant reproduction
Boyle, Robert, 124–5
Brockhaus, 49
"Broken circle," 70–1. *See also* Great chain of being; Natural history
Broussonet, Auguste, 160
Brustsärke, 327
Buffon, Georges Louis Leclerc, Comte de, 47, 56, 279; criticism of mathematics, 395
Bureaucracies, 3, 22, 23, 315. *See also* Cameralism; Military

Calcul analytique (calculation), 372, 376
Calculus: advocated, 378; in astronomy, 377; decimal, 212; integral, 175
Calon, General E.N. de, 230–1
Calorimeter, 6, 260, 370
Cambry, Jacques de, 125
Cameral College, 321
Cameralism, 315–6, 340, 393; influence on forest management, 320, 336–7
Cardwell, D.S.L., 382
Carte de Cassini, 204
Cartesian methods, 384
Cartography, 14, 179, 231; militarization of, 200–6; and political economy, 201. *See also* Geodesy
Cassini de Thury, César François, 180, 181, 200; role in Paris-Greenwich triangulation, 186, 193, 217

Cassini, Jacques Dominique, 179, 188

Cassirer, Ernst, 45, 83

Cavendish, Henry, 162

Celestial mechanics (Laplace), 375

Celsius, Anders, 38, 39–40

Census, 12, 13, 14; fears over a general, 349; reforms following, 360; secrecy of, 359–60

Central Bureau of Statistics (Sweden), 358

Certainty, 137, 382; and the calculus of probability, 378–9, 385

Chain of identities, 118

Chambers, Ephraim, 47

Charcoal: studies on, 302–3, 305

Chemia sublimior, 265

Chemia vulgaris, 264–5

Chemical balance, 6, 247–8

Chemical revolution: influence of economic interests on, 243, 250–1; influence of experimental physics on, 243, 256–62, 265–6

Chemical theory, 246–7; instrumentalist approach to, 262–6, phlogiston theory modified, 264–5; qualitative nature of early, 248; role of analytic quantifying in, 253–6, 261, 265; theory of definite proportions, 255, 256, 257

Chemistry: effect of systematic model on, 88; applied, 107–8, 249–50, 252; chemical classification of minerals, 90, 91–2; instruments introduced from physics, 258–60; reform of chemical nomenclature, 93–4, 95–6. *See also* Phlogiston theory

Children: value of, 354, 355

Chronology. *See* Chronometers; Clocks; Time

Chronometers, 6

Chymische Untersuchungen (Pott), 249

Circle. *See* Repeating circle

Circle of knowledge, 46

Climate: relation to public health and agriculture, 154

Climatology, 23, 145–6, 171–2, 176–7; agricultural applications of, 155, 160–1, 165; meteorological parallels, 150, 152; origins of, 144, 152–3; projects of the Société royale de médecine, 164–73; the "weather rule," 151, 152. *See also* Meteorological networks

Clocks, 7, 298

Colbert, Jean Baptiste, 180, 322

Colden, Cadwallader, 147

Coleridge, Samuel Taylor, 50n; criticized state of botanical science, 67

Collin, Gabriel, 298

Combinatorics, 139

Combustion, 94

Commission temporaire des poids et mesures républicains, 228–9

Compass: magnetic, 143, 169

Computers, attempt to eliminate, 212

Condillac, Etienne Bonnot de, 182; analysis of language, 115–20; influenced systematic model, 93–4, 99, 109

Condorcet, Marquis de, 115–20, 133, 219

Contradiction (principle of), 33

Conversationslexicon (Brockhaus), 49

Cook expedition, 65, 69

Cortical layers, 274

Cotta, Heinrich, 322, 329–30

Cotte, Louis, 165–7, 174

Coulomb, Charles Augustin de, 18, 145, 307, 308

Cramer, Gabriel, 100–3

Cramer, Johann, 252

Crest, Michel du, 161

Critica botanica (Linnæus), 74, 78

Cronhielm, Gustaf, 41

Cronstedt, Axel Fredrik, 251, 254

Crystal: geometric form of, 89

Crystallography, 89, 90, 371–2; theory of crystalline structure, 92, 93

Cullen, William, 262

Cuvier, Georges Leopold, Baron, 110

Cyclopedia (Chambers), 47, 50

Dalby, Isaac, 187, 191–2, 198

D'Alembert, Jean Le Rond, 48, 118–9; skepticism about mathematics, 395

Dalton, John, 145, 263

D'Avenant, Charles, 349, 351

Decimal calculus, 212

Decimals, 195–6, 215, 227; as "natural arithmetic," 212–3, 221. *See also* Metric system

Decimal scale, 298

Deductive method, 33–4, 293–7, 377–8

Definitive meter, 231–2, 234, 237

De Gerando, Baron Joseph Marie, 136–8

De Gribeauval, Jean Baptiste, 18

Delambre, J.B.J., 199, 213; role in metric project, 226–7, 229, 231, 232, 233

De l'Isle, Romé, 90

Delormel, Jean, 121–2

Deluc, Jean-André, 156, 160, 169, 371

Demography. *See under* Statistics

Denier, 212

Dépôt de la Marine, 234

De re metallica (Agricola), 248

Derham, William, 147, 149, 150

De Saussure, Horace Bénédict, 157, 175

Des caractères extérieurs des minéraux (de l'Isle), 91

Description of Lajhela Parish in Österbotten, 353–4

Dhombres, Jean, 386

Dictionnaire portatif d'histoire naturelle (Linnæus), 68–9

Dictionnaire raisoné universel d'histoire naturelle (Bomare), 68

Dictionnaire des arts et métiers, 17

Diderot, Denis, 20, 395

Directions for the organization and assessment of the forest (Cotta), 330

Discours préliminaire (d'Alembert), 47

Diseases: classification of. *See* Nosology

Diseases: climate as a source of, 154–5

Diurnal variation, 171

Diversions of Purley (Tooke), 113

Döbel, Heinrich Wilhelm, 327

Doctrine of proportions, 356

Earth: controversy over shape of the, 7, 181, 182, 216–7; ellipsoidal model of, 192; geoid model of, 205–6; spheroidal model of, 190–2. *See also* Geodesy; Measurement of arcs of meridian

Ecole polytechnique, 103

Economics, 382

Encyclopedia Brittanica, 49

Encyclopedia metropolitana, 50

Edelcrantz, Abraham Niklas, 129–30

Ekström, Daniel, 298

Electrometer, 6, 8

Element, 20, 263

Elements of Chemistry (Lavoisier), 212

Elvius, Pehr, 296, 310

Encyclopedias, 46, 48, 50, 305; and information, 51, 70

Encyclopédie (Diderot and d'Alembert, eds.), 20, 22, 44–9, 86, 114, 118

Encyclopédie méthodique, 68

Encyclopedism, 20, 45–51, 267

Entomology, 70

Ephemerides, 170, 171, 172

Epigenesis (theories of), 275

Epp, F.X., 172

Esprit géométrique, 1, 2, 83–6, 395. *See also* Mathematics; Quantifying

Esprit de système, systématique, 45, 46. *See also* Systematic model

Etat major, 203–4

Euler, Leonhard: 100, 102, 115, 388

Evolution: hypothesis of Linnæus, 55–6, 57, 78, 287–8

Experimental physics. *See* Physics

Exposition du système du monde (Laplace), 364, 375

Experiments on magnesia alba (Black), 262

Falun Copper Mine: failure of, 251

Familles des plantes (Adanson), 79

Felbiger, Ignaz, 159

Ferguson, Eugene, 308

Ferment of knowledge, The (Rousseau and Porter), 381–2

Ferraris, Joseph Jean, 183

Feuilles d'analyse (Monge), 102

Firmas-Périès, Comte de, 130

Fischer, Johann Carl, 9

Fontenelle, Bernard le Bovier de, 1, 384

Forest management, 15, 316; based on mass of wood, 325–7, 328, 330, 331–2; ecosystem approach to, 342; resistance to mathematization of, 342; schools of, 319–20, 321

Forestry science: tradition of in Germany, 317, 322, 336, 341; mathematics used in, 327–8, 331, 335–7, 340; measuring volume of the cord, 328–9

Forestry. *See* Forest management

Forests: calculating value of, 33, 331; consumption and depletion of, 14–5, 301, 302, 318; economic importance of, 317–8, 320; the standard tree, 15, 329, 332; mensuration and cartography of, 324, 326, 332, 339; regulation and standardization of, 330, 333–5, 339, 340–1; sustained and periodic yield of, 334–5, 337–40. *See also* Timber

Forstwissenshaft. *See* Forest management

Fortsetzungen (Koelreuter), 274

Foucault, Michel, 52n, 108, 151, 175, 201

Fourcroy, Antoine, 94, 95–6, 106

Frederick the Great, 33

Freind, John, 256

French Revolution, 373; encouraged quantifying, 14, 213, 242, 390; and meteorology, 173–4; revolutionary calendar, 213–4; and the telegraph, 125–6, 127–8. *See also* Metric system; Political economy; Politics

Gadbled, Christophe, 366–7

Gadolin, Jacob, 64

Galileo (Galileo Galilei), 28

Gamble, John, 130–1

Gardens, 65, 66

Gartner, C.F. von, 287

Gauss, Carl Friedrich, 204

Genera morborum (Linnæus), 98

Genera plantarum (Jussieu), 80

Genera plantarum (Linnæus), 55

Genus, 271

Geodesy, 179; methods of compared, 188, 189–90; problems of 18th century, 183. *See also* Earth

Geodetic instruments, 192, 193; achromatic lens, 193; repeating circle, 193, 194–5, 200, 217, 224–5; 225; pendulum, 219–22, 224; theodolites, 6, 200

Geodetic surveys: French led in, 183; a resource for centralized government, 201. *See also* Measurement of arcs of meridian

Geoffroy, Etienne François, 261

Geographische Geschichte des Menschen und der allgemein vierfüssigen Thiere (Zimmermann), 58–9

Geography. *See* Cartography; Earth; Geodesy

Geoid model, 205–6

Géomètre philosophe, 376

Geometrical grid, 201, 236

Geometry: analytic, 100–2; Euclidean, 3, 384; geometric forms, 89; shift from geometric to arithmetic mentality, 71; spirit of, 1, 2, 71, 83–6; three dimensional, 102–3. *See also* Measurement of arcs of meridian

Geschichte der Physik (Fischer), 9

Gillispie, Charles, 389, 392

God: existence of, 35–7; plan of, 21, 356; plenitude dilemma, 63

Goethe, Johann Wolfgang von, 69

Gonde, Gustaf, 41–2

Grand nomenclateur, 123

Great chain of being, 59–60, 61, 70; changes in concept of, 62–3

Great Copper Mine, 295

Greenwood, Isaac, 147

Grosses vollständiges Universal Lexicon (Zedler), 20, 47

Gulliver's Travels (Swift), 395–6

Guns, 18. *See also* Military

Hales, Stephen, 258

Haller, 57

Hamberger, G.A., 146

Hammaraeus, Olof, 39

Happiness: quantity of, 382

Harmony: of the world, 356

Hartig, Georg Ludwig, 320, 331

Hasselström, Johan Gustaf, 298

Haüy, René Just, 90–1, 93, 371–2

Hemmer, Johann Jakob, 168, 169, 171

Henckel, Johann Friedrich, 257

Hennert, Carl Wilhelm, 328

Herder, Johann Gottfried, 132

Hermann, Johann, 63

Hierarchical systems, 52

Histoire naturelle (Buffon), 47

History of technology (Singer et al.), 246

Horse whims, 306–7

Horsley, Samuel, 163–4

Hossfeld, Johann, 326, 328, 329, 336

Hourwitz, Zalkind, 125

Humboldt, Alexander von, 177

Humboldtian science, 69–70

Humors (theory of), 154–5

Huygens, Christian, 7

Hybridization: 267–8, 277; proof of plant sexuality, 270–1

Hybrids: characteristics of bastard offspring, 277–8, 279–81, 285; production of, 273–4, 276–81; reversion of parent species of, 282; self pollination of, 278–9; *Verbascum*, 285–6

Hydraulic power, 293, 295

Hydrauliska försök, anställda vid Fahlu grufva, dren 1811–1815 (Lagerhjelm), 296n, 296–7

Hydrodynamics: apparatus, 294, 298; experiments, 296, 300; hydrostatics, 296

Hydrostatic balance, 246–7, 257

Hydrostatics, 144

Hygrometry, 104, 157, 159, 175

Ideas (quantifying of), 137

Idéologie, 133–4

Idéologues, 133–4
Images. *See* Metaphors
Indicule, 122–3
Inductive method, 293, 297, 377
Industry: dependent on mechanical energy, 292–3; interaction with science, 250–1; productivity and rationalization of, 17, 19, 22, 302
Information: and encyclopedias, 51, 70; explosion of, 3, 45, 382, 383; sources of data, 343, 357, 360; verbal versus numeric, 345–6
Institutions, 312; authority of, 300–1, 313
Institut national de France, 133–4, 233. *See also* Paris Academy of Sciences
Instrumentalism, 4, 5, 20, 28, 262–6; key to quantifying and standardization, 3, 8, 382. *See also* Geodetic instruments; Measurement; Meteorological instruments
Instruments, 6. *See also by individual names*
Insurance: calculation of premiums, 350
Interior relations, 175
Introductio in analysin infinitorum (Euler), 100
Introduction à l'analyse des lignes courbes algébriques (Cramer), 100
"Invitio ad observationes meteorologicas," 147
Ironmasters, 19, 294–5, 299–300
Isis Critical Bibliography, 383
Isotherms, 177

Jars, Gabriel, 252
Jefferson, Thomas, 16, 17, 18, 219, 222
Jurin, James, 147. *See also* "Weather watch," 147
Jussieu, Antoine-Laurent de, 80
Justi, Johann Heinrich Gottlob von, 251

Kalmár, György, 120–1
Kameralwissenschaft. *See* Cameralism
Kanold, Johann, 147
Kant, Emmanuel: criticism of Linnæan natural history, 64
Karl Theodor, Elector, 168
Keill, John, 254
Kew Gardens, 66
King, Gregory, 349
Klingenstierna, Samuel, 40
Koelreuter, Joseph Gottlieb: character and work of, 268, 272–3, 288; homogenous fluid concept of, 274–5; on "identity of reciprocal crosses," 277; quantitative method of, 21, 276t, 288–9. *See also* Hybridization

La belle Marianne, 242
Laboratories, 19, 291; in field and forest, 304–5, 311
La Condamine, Charles-Marie de, 208–9, 220
Lagerhjelm, Pehr, 19, 295; background of, 299–300
Lagrange, Joseph Louis, 374
La langue des calculs (Condillac), 119
Lamarck, Jean-Baptiste de Monet de, 68
Lambert, Johann Heinrich, 85, 159, 175
Lancelin, P.F., 134–5, 137
Land Survey Board (Swedish), 353
Land surveys, 15–6, 353–4
Language: analysis of, 20, 115–20; attempts to rationalize and quantify, 120, 121, 131, 132–7, 139–40; difficulties in reforming, 138–9. *See also* Alphabet; Metaphors; Nomenclature; Rhetoric; Universal languages
Laplace, Pierre Simon de: life of, 365–70, 379; theory of molecular physics, 4, 23; role in science, 363–4

Lärbok i kemien (Berzelius), 299

Lavoisier, Antoine Laurent, 212, 259; arrested and released, 228–9; law of conservation of mass of, 259; overthrew phlogiston theory, 94; theory of acids of, 95

Leeuwenhoek, Antony van, 63

Legendre, Adrien-Marie, 187

Leibniz, Gottfried Wilhelm, Baron von, 29, 30

Lenoir, Etienne, 193. *See also* Repeating circle

Le Roy, Jean Baptiste, 219

Lexicography, 51

Lexicons, 20, 49, 50, 51, 77

Librarianship: schools of, 52

Life expectancy, 350

Liggmilor, 302–3

Light (as a property), 104

Lindley, Joseph, 189

Linnæus, Carolus: character of, 98, 109–10, 274; estimated number of species, 54; evolutionary hypothesis of, 55–6, 57, 287–8; extent of influence of, 21, 79; "sexual system" of, 60, 61, 62, 76–8, 79, 270–1. *See also* Botany, Binomial nomenclature

Linnean Society of London, 65

Lower and higher practical stereometry (Hossfeld), 329

Lunar motion and influence, 163, 171, 213–4

Machines, 106–7, 292. *See also* Mechanics

Magnetism, 145

Maimieux, Joseph de, 121, 122–4

Malouin, Paul, 155

Malthus, Robert, 343

Malves, J.P. de Gua de, 101

Mannheim hours, 170

Manuel interprète de correspondance, ou Vocabulaires polyglottes alphabétiques et numériques en tableaux (Cambry), 125

Maps. *See* Cartography

Mariotte, Edmé, 295

Martini, Friedrich Heinrich, 68

Marx, Karl, 343

Maskelyne, Nevil, 187, 188–9

Matériaux primitifs, 105–6

Mathematics: criticism of mathematical method, 393; desire for a universal, 29; masculinity of, 210–1, 242, 284; mathematical method, 1, 3, 28, 37, 85, 109, 385; practical applications of, 381–2; required curriculum, 321, 369; responsible for progress, 376, 378, 380. *See also* Algebra; Calculus; Geometry; Systematic model; Wolffianism

Matthiae, Georg, 52

Maupertuis, Pierre Louis Moreau de, 180, 181

Mauskopf, Seymour H., 89

Measurement, 7, 249; controversies over accuracy of, 188–90, 194–5; increase in accuracy of, 6, 156–7, 170–1, 181–2; rhetoric of accuracy, 195–8, 199. *See also* Instrumentalism; Instruments

Measurement of arcs of meridian, 180, 181, 183, 204, 216; establishing baselines, 200–1, 224, 233; Ordnance Survey, 179, 184–5; Paris-Greenwich triangulation, 186–95, 216–7; peasants resisted, 232

Mécanique philosophique (Prony), 104

Méchain, Pierre François, 193, 226–7, 231, 232

Mechanical communication, 20, 125–31, 139

Mechanical energy: industries dependent on, 292–3

Mechanics: analytic, 18, 105–6; mathematization of, 88

Medical nomenclature. *See* Nosology

Medical nosology (Sauvages), 98

Medicine: effect of systematic model on, 88, 96–100; public, 392. *See also* Nosology

Medullar layers, 274

Mémoires (Koelreuter), 287

Mercantilism, 8, 351–2, 355, 358, 361

Metallurgy, 247, 251. *See also* Mineralogy; Mining

Metaphors, 70–1, 394n, 394; language as mechanical, 134; of minerals and plants, 88–9, 284; of seeds and salt crystals, 275; of the statistical body, 346n, 346

Metaphysics (Baumgarten), 38

Meteorological instruments: barometers, 6, 148–9, 157, 169; development and precision of, 156–7, 170–1, 176; lack of, 148–9; magnetic compass, 149, 157, 158, 166

Meteorological networks: 9–10, 146–7, 158–73; the "weather watch," 147, 149, 150, 151

Meteorological parallels, 150, 151

Meteorology, 9–10, 143–5; failures of, 174–5; hygrometry, 104, 157, 159, 175; perception of space and time in, 151–2. *See also* Climatology

Meter: definitive, 231–2, 234, 237; provisional, 227, 229, 231, 234, 237

Methode de nomenclature chimique, 94

Metric system: acceptance and spread of, 211, 227, 238–42; conflict between familiar terms and system, 215, 237; field measurements pertaining to, 226–33; national and international standardization of, 222, 233, 234–5; politics of, 208–9, 212–5, 238–9; proposals for reform of weights and measures, 209–10, 219–24; resistance to, 235–8. *See also* Measurement of arcs of meridian

Meyer, Friedrich Johann Lorez, 128

Microscope, 53

Military: fostered standardization, 17; institutional character of, 310, 312; Napoleon extended the Carte de Cassini, 204; used and encouraged cartography, 201–5.

Mineralogy: approaches to classification of minerals, 89–92, 93; blowpipes, 255–6; and emergence of analytic quantification, 253, 255; existence of mineral species denied, 90; influenced by systematic botany, 88–92; interaction with chemistry, 254–6. *See also* Metallurgy

Mining, 89, 247, 250–1, 295–300

Modest proposal (Swift), 343

Molecular physics, 4, 5, 23

Molécule intégrante (Haüy), 92

Monceau, Duhamel, 155

Monge, Gaspard, 102–3

Montesquieu, Charles Louis de Secondat, 144

Moon, 163, 171, 213–4

Morphogenesis, 268

Morphology, 69

Morveau, Guyton de, 94

Moser, Wilhelm Gottfried von, 320

Moser, Heinrich Christoph, 340

Muscle power, 292, 306–7; attempts to measure output of, 307–8, 309–10

Musée d'Histoire Naturelle, 66

Museum naturalium academiae upsaliensis (Thunberg), 66

Museums, 65, 66

Musschenbroek, Pieter van, 53

Mycology, 69

Napoleon, 14, 204, 214

Napoleonic wars, 20

National Assembly (French), 227, 228, 229, 234

Natural history, 65, 68–9, 70; epistemology of, 52n, 175; diversity and anomaly in, 267–8, 272–3; based on notion of stability, 267

Natural resources, 3

Nature, 234, 235, 379; Book of, 28, 46; irrational side of, 267; plenitude of being, 53, 54–61, 62–3; reproductive intentions of, 273, 282; shift in metaphors concerning, 292; uniformity of, 333–4, 340–1, 382; void universe, 62

Nature mathematized (Shea), 386

Naudé, Gabriel, 52

Networks. *See* Meteorological networks

New instruments for the Royal Prussian forest-geometers and forest-assessors (Hartig), 331

Newtonian theory, 183, 192, 246, 256, 261

Newton, Isaac, 19, 210, 377

Nomenclature: *Grand nomenclateur*, 123; *Petit nomenclateur*, 122–3. *See also* Binomial nomenclature

Non-verbal technology, 305

Norberg, Johan Eric, 309, 310, 312

Normalbaum (standard tree), 15, 329, 332, 340

Northmore, Thomas, 130

Nosographie philosophique: ou la méthode de l'analyse appliquée à la médicine (Pinel), 99

Nosology, 87, 94, 97–100

Nova plantarum genera (Thunberg), 66

Numeracy-innumeracy, 211, 240, 242, 375

Numeric tables, 103–4, 211, 249, 329; for constructing artificial languages, 122–3, 135; tabular statistics, 9, 13, 345–6, 347, 358; in telegraphic codebooks, 129–30

"Observations botanico-météorologiques" (Monceau), 155

Observation, 376–7

Oettelt, Carl Christoph, 326, 327, 328

Office of Tables (Swedish), 358

Ökonomische Gesellschaft, 161

Olby, Robert, 269

On the science of method (Coleridge), 67n

Optical telegraphs, 126–30

Optics: geometrical, 144; mathematization of, 88

Optics (Newton): 31st Query to, 256

Optimization (hydrostatic), 296

Oratio de telluris habitabilis (Linnæus), 54

Ordinance of 1785, 15–6

Ordnance Survey, 179, 184–5, 203

Ornithology, 70

Oxygen theory, 265

Palatine Society, 161, 168–9, 170

Pallas, Peter Simon, 282, 287

Parameter variation, 296

Parental contributions (theories of), 275, 277, 282–4

Paris Academy of Sciences, 11, 180, 216, 388

Paris Greenwich triangulation, 186–95, 216–7

Paris baseline, 200–1

Paris clinical school, 99

Parish priests, 358

Paris Observatory, 221

Pasigraphy. *See* Universal languages

Patent Law of 1791, 218

Pendulum, 219–22, 224, 225

Pendulum clock, 7

Petit nomenclateur, 122–3

Petty, William, 343, 348–9

Pharmaceutical science, 107–8, 249–50; training of apothecaries, 250

Philosophia botanica (Linnæus), 74, 78

Philosophia practica universalis (Wolff), 32

Philosophia prima sive ontologia (Wolff), 32

Philosophical essay on the theory of probability (Laplace), 378–9

Philosophical transactions, 163, 173, 187

Philosophie chimique (Fourcroy), 106

Philosophy (influence of). *See* Aristotelian tradition; Condillac

Philosophy (mathematization of). *See* Rationalism; Wolffianism

Phlogiston theory, 93, 94, 260, 264–5

Photometry, 144

Physicotheology, 35, 36, 41–2, 54

Physics: defined, 103, 256, 382; experimental, 157–8; physical properties, 105–6; and the systematic model, 103–5. *See also* Chemical revolution

Physics reduced to systematic tables (Barruel), 103

Picard, Jean, 180

Pietism, 31–2

Pinel, Philippe, 99–100

"Pinte," 207

Plant reproduction, 268, 275; animalcule theory, 272–3, 277, 287; Aristotelian doctrine, 277; limited variability theory, 286, 287–8; problem of plant sexuality, 269–71; theory of inconsistency of form, 282

Plants: binomial nomenclature of, 21, 74, 79, 80

Plenitude of being: dilemma of, 53, 63; measurements of, 54–61, 62–3

Pneumatics, 143

Polestar observations, 190

Polhem, Christopher, 56, 294, 295, 296–7, 298; background of, 300

Political arithmetic, 10–11, 14, 349, 350, 359–60. *See also* Statistics

Political economy: cartography, surveys, and government, 201; conservative influence of institutions, 218; democracy and the decimal, 213; internationalism, 222; locus for the quantifying spirit, 22–3, 392. *See also* French Revolution; Political arithmetic; Statistics

Politics: of the metric system, 208–9, 212–5, 219–24, 238–9; of the Paris baseline, 200–1

Politisk arithmetica (Berch), 356

Pollination (of hybrids), 278–9

Polygraphie, 125

Population, 351–2, 354. *See also* Statistics

Porter, Roy, 382

Pott, Johann Heinrich, 249, 257–8

Practicality, 3; applications of statistics, 350, 355–7, 379–80. *See also* Instrumentalism

Practical proof that mathematics serves indispensable services for forestry (Oettelt), 327

Praecepta grammatica atque specimina linguae philosophicae sive universalis, ad omnevitae genus adcommodatae (Kalmár), 120

Preformation theory, 287

Price, Richard, 350

Priestly, Joseph, 64, 258

Prieur de la Côte d'Or, Comte Claude Antoine, 212, 219, 229, 231

Principia mathematica (Newton), 19, 183, 377

Principles of forest-economy (Moser), 320

Probability, 385, 386

Prony, Gaspard, 105–6

Proportions: doctrine of, 356

Prostitutes' rates, 214

Proust, Joseph Louis, 255

Provisional meter, 227, 229, 231, 234, 237

Prytologia oder Kiess-Historie (Henkel), 257

Psychologia empirica (Wolffe), 32

Psychologia rationalis (Wolffe), 32

Public health, 154

Pyrometer, 199

Quantifying, 45, 384; in analysis of population, 354; analytic, 249, 250, 252, 253–4, 259; in chemistry, 265–6; "neutrality" of valued by social science, 343; synthetic, 116–7, 249, 250. *See also* Calculus; Deductive method; Mathematics; Systematic model

"Quantification of power," 382

Ramsden, Jesse, 157, 192, 197–8, 199, 217; achievements of, 196. *See also* Theodolites

Rationalism, 30, 120, 348; as a function of language, 117, 133–4; movement of, 86–8; reaction against in Sweden, 40–1, 42. *See also* Wolffianism

Rationality. *See* Reason

Ratio praelectionum Wolfianarum in mathesin et philosophiam universam (Wolff), 31

Ray, John, 53

Reason: deductive method of, 33–4, 293–7, 377–8; inductive method of, 293, 297, 377; "naturalization" and mathematization of, 28

Réaumur, René-Antoine Ferchault de, 61

Recherches (Deluc), 160

Reciprocal crosses: identity of, 277

Religion, 376

Repeating circle, 193, 194–5, 200, 217; accuracy of, 224–5

Reproduction. *See* Plant reproduction

Resmilor, 302–3

Reynolds, Terry, 293

Rhetoric, 237; of accuracy, 195–8, 199, 234; Enlightenment "buzz words," 209; of numbers, 2, 259–60; revolutionary, 229

Rondelet, Jean-Baptiste, 239

Royal Society of London, 146, 173; collected meteorological reports, 9–10, 161–2; dissension within, 185–6

Royal Swedish Academy of Sciences, 43, 299

Roy, William, 179, 184–6, 187, 191–2; rhetorical resources of, 195–6, 198; rift with Ramsden, 197–8; role in Paris-Greenwich triangulation, 186, 188–90

Runeburg, E.O., 353–5

Sachs, Julius von, 77

Sauvages, François Boissier de, 97–98

Scheele, Carl Wilhelm, 250

Schofield, Robert E., 313

Schröter, Johan Samuel, 58

Scientific academies and societies: role of, 387. *See also* by individual name

Scientific Society of Copenhagen, 183

Semaphore, 130

Seven Years' War, 318

Shea, William R., 386

Signs, 132. *See also* Language

Singer, Charles, 246

Smith, Adam, 343, 382

Smith, Sir James Edward, 65

Societas meteorologica palatina, 161, 168–69, 170

Société royale d'agriculture, 160

Société royale d'médecine (Society of Medicine), 164–73

Space: characterlessness of, 151–2; changing perceptions of, 167, 172, 177

Späth, Johann Leonhard, 340

Species: constancy of from the beginning, 54, 55–6, 62, 272–3; existence of mineral species denied, 90; number of estimated, 53, 54–6; transformation of, 283, 284

Species planatarum (Linnæus), 74

Sprengel, Christian, 287

Staatenkunde, 343–7

Stahl, Ernst, 256–7

Stahl, J.F., 319

Standard Model (of molecular physics), 4, 5, 23

Standard tree, 15, 329, 332, 340

Stars: aberration of, 7

States: comparison of, 344

Statistics: application of, 350, 355–7, 379–80; collection of vital statistics, 349–50, 358; demographic, 11–4, 23, 392; derivation and emergence of (*Staatenkunde*), 3, 344–5, 344n, 346–7, 349, 357; descriptive, 357–60; emigration measured, 359; as political arithmetic, 10–11, 14, 349, 350, 358–60; tabular, 9, 13, 345–6; Utopian, 348, 353–5. *See also* Census; Population

Stearn, W.T., 76–7

Steiglehner, Coelestin, 172

Stereometry, 322, 329

Sufficient reason (principle of), 33

Surveys. *See* Census; Land Surveys, Statistics

Svendenstierna, Eric Thomas, 295

Sweden, 352

Swedish Ironmasters' Association (*Jernkontoret*), 19, 294–5, 299–300

Swift, Jonathan: satire on mathematics, 343, 395–6

Sydenham, Thomas, 97

Synonyms, 122

Syntax, 114

Systema naturae (Linnæus), 46, 55, 64, 74, 77; scheme for mineral "kingdom" included in the, 88–9

Systematic instruction for the assessment of woods (Cotta), 330

Systematic model, 83–5, 86, 108; in anatomy, 107; associated with mathematics, 11, 109; in botany, 52, 70, 73, 108, 110; in chemistry, 93–6; decline of, 110; defined, 45, 46, 73; in medicine and nosology, 88, 98–9, 107–8; philosophical-linguistic basis, 93–4, 99, 109; in physics, 103–5; in study of hybrids, 269

Systematics. *See* Systematic model

Tabellverket, 13

Tables. *See* Numeric tables

Tabular affinitatum animalium (Hermann), 63

Tabular statistics, 9, 13, 345–6, 347, 349, 358–9

Talleyrand, 219, 233–4

Tavern staff, 359

Taxonomy, 63

Technology, 299, 306, 311, 314; copied social organization of science, 312–3; non-verbal, 305; quantification of, 291–2, 308, 310–2

Telegraphische Chiffren-Tabelle, 129

Telegraphs, 20, 125–31, 139

Teleology, 35–6, 38

Telescopes, 6, 370

The Lunar Society of Birmingham (Schofield), 313

Theodolites, 6, 200; accuracy in design of, 192, 194–5, 217

Theologia naturalis (Wolff), 32, 36–7

Theologia polemica, 42

Thermometers, 6, 149, 157, 158, 166
Thiébault, Diedonné, 132–3
Three-dimensional geometry, 102–3
Thümmig, Ludwig Philipp, 37
Thunberg, Karl Peter, 66
Tibell, Gustaf Wilhelm, 204
Timber: consumption and depletion of, 301, 302; methods of stacking, 303–4
Time, 152, 172, 338
Toise, 212, 236, 237, 238
Tooke, John Horne, 113
Townships: divisions into, 16
Toynbee, Arnold, 22
Tracy, Destutt de, 133
Trade, 8, 352, 356
Traité analytique des probabilité (Laplace), 364
Treadwheels, 306, 308
Treatise on the classes of diseases (Sauvages), 97
Trees. *See* Forests; *Normalbaum*
Triangulation. *See* Measurements of arc of meridian; Paris-Greenwich triangulation
Trigonometry, 101–2, 102n, 175, 213
Truth, 4, 199, 210
Turgot, Anne-Robert-Jacques, 133

Uhr, Carl David af, 302
Ullén, Petrus, 41–2
Uncertainty, 378
Universal languages, 19–20, 120–5, 130, 132–3; evaluation of proposals for, 137–9
Untersuching über die Deutlichkeit der Gründsätze der natürlichen Theologie und der Moral (Kant), 43
Utopia, 212

Verbascum, 285–6
Vernünfftige Gedancken von der Absichten der natürlichen Dinge (Wolff), 36
Vernünfftige Gedancken von der Menschen Thun und Lassen: zu Beförderung ihrer Glückseligkeit (Wolff), 35
Vierenklee, Johann, 326, 328
Vital statistics, 249–50, 357
Volta, Alessandro, 8–9
Voltaire, 42
Vorläufige Nachricht (Koelreuter), 274

Wallerius, Erik, Nils, and Johann Gottschalk, 39
Wallerius, Nils, 42
Water mills, 293
Water power, 293, 295
Waterwheels, 293–4
Watt, Fran, 18
Weather: lunar influence on, 163, 171, 213–4; *See also* Climatology; Meteorology
"Weather watch," 147, 149, 150, 151
Weights and measures, 207–8, 211, 212, 298; aune, 207–8. *See also* Metric system
Welt-Weissheit, 34
Werner, Abraham, 89–90
Wisdom of God manifested in the works of creation, The (Ray), 53
Wolff, Christian: character and life of, 29–30, 31–3; role in propagating mathematical method, 1, 28, 30, 37–8; theology of, 35–7
Wolffianism, 29, 33–5, 48; champion of, 37–8, 40, 41–2; debated in Sweden, 39–43; influence of receded, 44; mathematical method of, 33–5, 37, 44

Women, measuring emotions of, 386n, 386; prostitutes' rates, 214; value of, 354

Woodlands. *See* Forests; Timber

Wood. *See* Timber

Xylometry, 322

Zanthier, H.D. van, 319

Zedler, Johann Heinrich, 20, 47

Zeno's paradox, 63

Zimmermann, Eberhard, 58–60